# THE DIGITAL CONTROL OF SYSTEMS

**Coordination of the Work**

| | | |
|---|---|---|
| M. GAUVRIT | C.E.R.T./D.E.R.A.<br>2, avenue Edouard-Belin | 31055 TOULOUSE Cedex |

**Authors**

| | | |
|---|---|---|
| C. CASTEL | C.E.R.T./D.E.R.A.<br>2, avenue Edouard-Belin | 31055 TOULOUSE Cedex |
| J.P. CHRETIEN | C.E.R.T./D.E.R.A.<br>2, avenue Edouard-Belin | 31055 TOULOUSE Cedex |
| G. FAVIER | L.A.S.S.Y.<br>41, boulevard Napoléon III | 06041 NICE Cedex |
| M. FLIESS | LSS, Ecole Supérieure d'Electricité<br>Plateau du Moulon | 91190 GIF-SUR-YVETTE |
| A.J. FOSSARD | C.E.R.T./D.E.R.A.<br>2, avenue Edouard-Belin | 31055 TOULOUSE Cedex |
| M. GAUVRIT | C.E.R.T./D.E.R.A.<br>2, avenue Edouard-Belin | 31055 TOULOUSE Cedex |
| B. GIMONET | C.E.R.T./D.E.R.A.<br>2, avenue Edouard-Belin | 31055 TOULOUSE Cedex |
| M. LABARRERE | C.E.R.T./D.E.R.A.<br>2, avenue Edouard-Belin | 31055 TOULOUSE Cedex |
| A. LIEGEOIS | L.A.M./USTL<br>Place E.-Bataillon | 34060 MONTPELLIER Cedex |
| J.F. MAGNI | C.E.R.T./D.E.R.A.<br>2, avenue Edouard-Belin | 31055 TOULOUSE Cedex |
| A. PIQUEREAU | C.E.R.T./D.E.R.A.<br>2, avenue Edouard-Belin | 31055 TOULOUSE Cedex |
| C. REBOULET | C.E.R.T./D.E.R.A.<br>2, avenue Edouard-Belin | 31055 TOULOUSE Cedex |
| S. STEER | I.N.R.I.A.<br>Domaine de Voluceau<br>Rocquencourt | 78150 LE CHESNAY |
| J.L. TESTUD | ADERSA/GERBIOS<br>7, boulevard Mal. Juin | 91370 VERRIERES-LE-BUISSON |

**Contributors**

| | | |
|---|---|---|
| M. ESPIAU | IRISA<br>Campus de Beaulieu | 35042 RENNES Cedex |
| J.P. FORESTIER | C.E.R.T./D.E.R.A.<br>2, avenue Edouard-Belin | 31055 TOULOUSE Cedex |
| M. LLIBRE | AKR Robotique<br>6, rue Maryse-Bastié | 91031 EVRY Cedex |
| A. FOURNIER | L.A.M./USTL<br>Place Eugène-Bataillon | 34060 MONTPELLIER Cedex |
| B. TONDU | L.A.M./USTL<br>Place Eugène-Bataillon | 34060 MONTPELLIER Cedex |

**Administration**

| | | |
|---|---|---|
| L. ARENDO | C.E.R.T./D.E.R.A.<br>2, avenue Edouard-Belin | 31055 TOULOUSE Cedex |

# THE DIGITAL CONTROL OF SYSTEMS

## APPLICATIONS TO VEHICLES AND ROBOTS

EDITED BY C FARGEON

Authors: C. CASTEL, J.P. CHRETIEN, G. FAVIER,
M. FLIESS, A.J. FOSSARD, M. GAUVRIT, B. GIMONET,
M. LABARRERE, A. LIEGEOIS, J.F. MAGNI, A. PIQUEREAU,
C. REBOULET, S. STEER, J.L. TESTUD

VNR VAN NOSTRAND REINHOLD
New York

First published in French under the title *Command Numérique des Systèmes* 

Library of Congress Catalog Card Number 89-5797
ISBN 0-442-23942-4



Printed in Great Britain

First published in Great Britain
by North Oxford Academic Publishers Ltd
a subsidiary of Kogan Page Ltd
120 Pentonville Road, London N1 9JN

First published in the USA by
Van Nostrand Reinhold
115 Fifth Avenue
New York, New York 10003

Macmillan of Canada
Division of Canada Publishing Corporation
164 Commander Boulevard
Agincourt, Ontario M1S 3C7, Canada

16 15 14 13 12 11 10 9 8 7 6 5 4 3 2 1

*Library of Congress Cataloging-in-Publication Data*
Fargeon, C.
Digital control systems.
Bibliography: p.
Includes index.
1. Digital control systems 2. Robotics.
3. Aeronautical instruments. 4. Astronautical instruments. I. Title
TJ223.M53F37 1989 629.8 89-5797
ISBN 0-422-23942-4

# CONTENTS

## PART TWO: APPLICATIONS OF DIGITAL CONTROL TO VEHICLES AND ROBOTS

# PREFACE

During the first quarter of 1983, following numerous comments from industry and universities concerning the limited distribution of the results of research contracts financed by the DRET, the idea of publishing these results first took root. Using the fourteen chosen authors mentioned on the cover page, all of whom work under contract for the DRET, the research results have been brought together and elaborated on in this book.

A second idea, reflected in the structure of this work, was the desire to make certain control techniques, which had hitherto been restricted to specialist researchers, available to a larger number of engineers and students. Thus the second part of this book, concerned with practical applications, illustrates the use of algorithms whose theory is described in the first part. The first part is the result of a large amount of work which brings together the many different strands of current work. The applications to machines and robots enhance the book by providing non-academic illustrations.

Each application may form the subject of a case study in an engineering course. Readers engaged in teaching may contact the authors if they require more detailed information.

The first version of the book was compiled at the beginning of 1984. It then became clear that, despite attempts to create a uniform text, the manuscript remained somewhat inaccessible because of the large number of authors. At this point, one of the authors, Michel Gauvrit, Doctor of Automation, took on the task of remodelling the whole of the book, and in particular of rewriting the pivotal chapters. The eventual appearance of this work is largely due to his work, and DRET wishes to take this opportunity to express its thanks to him.

The theoretical part of this work, consisting of seven chapters, is concerned with the synthesis of numerical controllers. The unity of this part will be immediately apparent from the chapter titles.

Following an introduction which provides a clear description of the set of problems encountered in closed loop control, Chapter 1 deals with the various continuous and discrete representations of signals and systems. The choice of the form of modelling is, in fact, a prerequisite for any subsequent synthesis, and an appropriate choice is often the decisive factor in the success of control techniques.

Before dealing with digital control as such, the authors describe, in the second chapter, the links between continuous and discrete control. The time control techniques applicable to multivariable systems (modal methods, linear quadratic and linear quadratic Gaussian methods, and systems with time delays) are analysed in great detail. The inclusion of techniques of digitisation of continuous laws in this chapter provides an indispensable aid to the implementation of these laws by means of a digital computer. The original feature of this second part is its presentation of a very relevant critical analysis of the various digitisation methods, an analysis that is often ignored in current literature.

The third chapter is essentially concerned with the digital control of deterministic linear systems. We wish to emphasise here the original and contemporary nature of the method adopted by the authors. The digital systhesis of controllers minimising either the response time or criteria of the quadratic type, or those placing the poles of the closed loop system at the required points, is achieved by a polynomial algebraic approach. The simplicity and power of this method (it can easily be generalised to the control of multivariable systems) make it highly attractive, and this impression will be amply borne out by the contents of the fifth chapter.

Chapter 4 describes a robust and powerful method of control by internal model which is well suited to multivariable and non-linear processes. The author clearly describes the change from a conventional concept of regulation to control using an internal model for the compensation of a measured difference between the desired behaviour and the observed behaviour of the process. He then establishes the basic principles of this form of control, emphasising the robustness of the technique.

The fifth chapter deals with the discrete control of stochastic systems. Two essential tools for the synthesis of digital controllers are used: the polynomial matrix method and the state vector method. The authors' very important and instructive work not only covers the whole set of problems (open loop and closed loop control minimising various criteria, and generalisation to multi-variable systems), but also illustrates the theoretical results with numerous examples. The chapter ends with the comparison of these two techniques of synthesis, which forms one of the more novel features of this theoretical work.

Self-adaptive control, and more particularly self-tuning controllers, form the subject of the sixth chapter. The first part outlines the basic concepts in order to clarify the frequently confused notions of dual self-adaptive control, open loop and closed loop control, and feedback control, and provides a very useful classification of the various self-adaptive laws. The second part is a simple description of the adaptive algorithms (of the self-tuning regulator type) which have already been used in many industrial applications.

We considered it appropriate to complete the theoretical part of the book with an introduction to the algebraic and geometrical methods used in non-linear automation. Chapter 7 provides concise information on these new techniques which will not fail to stimulate the reader interested in non-linear control and identification.

Chapter 8, the first in the part of the book concerned with applications, deals with attitude control in satellites, and demonstrates the usefulness of the techniques of digitisation and digital methods of synthesis of the controllers described in Chapters 2 and 3.

The following chapter, on the piloting (movement about the centre of gravity) of a launch vehicle, is concerned with the application of the linear quadratic Gaussian technique explained in Chapter 5 (in the discrete form) and Chapter 2 (in the continuous form).

Chapter 10 provides a solution to the problems of turbulence on board passenger aircraft, making it possible to reduce the accelerations induced at different points of the aircraft, and thus to improve passenger comfort. The development of the method forms an example of the stage of digitisation described in Chapter 2 (DCAé Direction des Constructions Aéronautiques contract).

The automatic pilot for a ship, proposed in the next chapter, is a very successful application of control by internal model as formulated in Chapter 4. A study concerned with the development and automatic updating of the ship model follows this stage (under a DRET contract with ADERSA GERBIOS and the Compagnie Générale d'Automatismes) and will make it possible to develop a self-tuning automatic pilot on the basis of rudimentary information about the ship.

Although not referring to the precise contents of Chapter 4 on control by internal model, Chapter 12 describes a method based on the reference model for increasing the stability of a helicopter and providing good manoeuvrability. The resulting control law has the merit of being robust over the whole flight domain (contract between Aerospatiale/DH and CERT).

Chapter 13 deals with a direct application of the self-adaptive control techniques analysed in Chapter 6. The application in question is the reduction of the vibrations induced by the rotor of a helicopter (DRET contract with Aerospatiale/DH and CERT). The balancing of the rotors by means of passive devices, such as hanging dampers, does not make it possible to meet the requirements of the civil and military authorities as regards the vibration level (± 0.02G). Since the blade dynamics vary with the flight conditions, the multivariable self-adaptive regulator developed here provides an active solution which is greatly superior to the techniques used up to the present.

Chapter 14, on the minimisation of the operating cost of an aircraft flight, is somewhat atypical in this book, since it is not related to any technique described in the theoretical part. However, it does have relevance to the use of the method of singular perturbations for the optimisation of trajectories. It would have been undesirable to deprive the reader of this useful study of the management of flight profiles, which forms an excellent introduction to the problems of optimisation which are frequently connected with those of control.

Chapter 15, on the guidance of a pursuit missile which has to intercept a target in minimum time in given terminal orientation conditions, is an obvious application of the maximum principle which is well-known in the field of optimal control (DCN contract with CERT).

Chapter 16, concerned with the topical subject of robots, provides a careful in-depth study and description of the control methods that are not widely described in the current literature principally concerned with robot modelling. The reader will find it easy to see why the geometric, kinematic and dynamic models are all used at different times in robot control. The argument is clearly set out, as regards both the space of the position variables of each joint of the robot and the space of the coordinates of the final element (DRET contract with LAM, AMD-BA and Dassault Systèmes).

We are sure that the combination of theoretical descriptions and applications in one book which has led, or will lead, to industrial implementation will be very stimulating for the reader (student, teacher, or engineer) interested in the digital control of systems.

Professor Lallemand
*Scientific Director of DRET*

# PART ONE

# DIGITAL CONTROL OF MULTIVARIABLE LINEAR SYSTEMS

## *INTRODUCTION TO AUTOMATIC CONTROL*

Before defining the basic structures characterising a control system, we consider it important to locate the closed loop control of processes within the general framework of the "systems" approach – an expression which has become very familiar to automatic control engineers. In fact, the term "system" refers to a combination of elements which act together to achieve one or more objectives. It may consist of anything from an electrical circuit, comprising a set of resistances, inductances and capacitances, to a complex industrial plant. Each element of the system is commonly represented by a "box" (see Fig. 1) which receives the input signals: these give rise to output signals representing the measurable effects of the inputs on the system. A dynamic system may thus be envisaged as a dynamic transformation of inputs into outputs. Such a transformation is described as dynamic (as opposed to static, or without memory) when the previous and present values of the input u(t) have an effect on the output y(t).

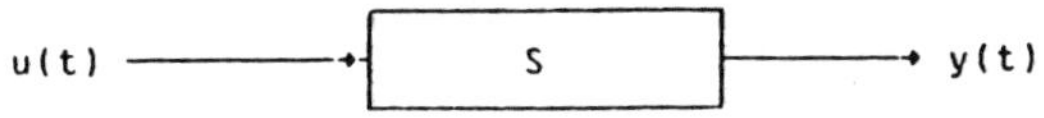


Fig. 1

The relationship between the inputs and the outputs may be found either by mathematical analysis of the physical phenomena characterising the behaviour of the element (transparent box) or, in the absence of precise knowledge, by imposed modelling (black box). For example, a voltage applied to a resistance R causes the circulation of a current I for which the input–output relationship, or causal relationship, is i = V/R. These concepts of input–output relationship and black box are fundamental in the modelling of systems, which is the stage which must precede the synthesis of any control system.

### *I – SYSTEMS MODELLING*

In order to achieve automatic control of an aircraft, stabilisation of a satellite, control of an industrial production system, etc., it is first necessary to establish a mathematical input–output model which best approximates the physical reality with which we are concerned. There are two immediately apparent levels of classification for the description of such modelling:

a) The nature of the mathematical equations obtained enables an initial classification to be established. Fig. 2 shows the main branch leading to the invariant linear systems which are the subject of many of the chapters in this book.

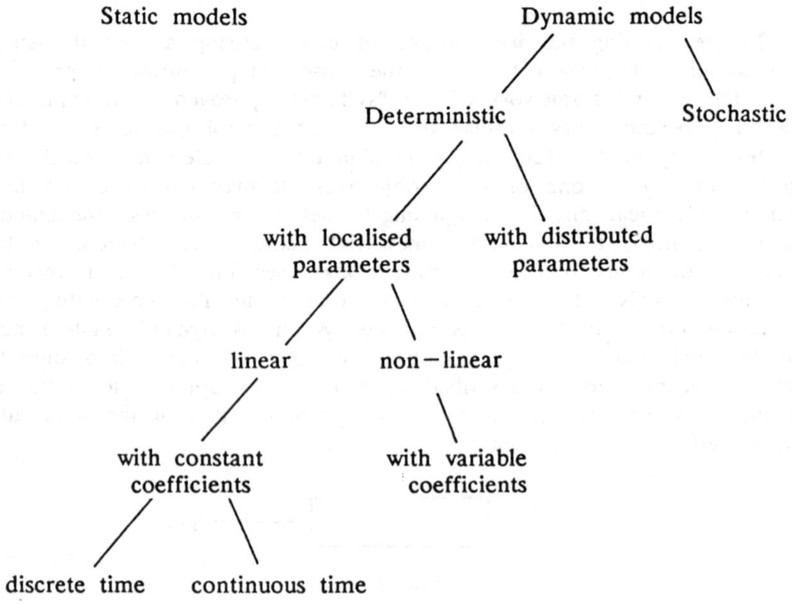


Fig. 2

Thus a sequence of important dichotomies appears:

- dynamic or static models
- stochastic or deterministic models, depending on whether or not the random aspects of disturbances or certain parameters are modelled
- models with distributed parameters, defined by equations with partial derivatives, or models with localised parameters governed by differential or recurrent equations
- linear or non-linear models, with constant or variable coefficients
- continuous or discrete models.

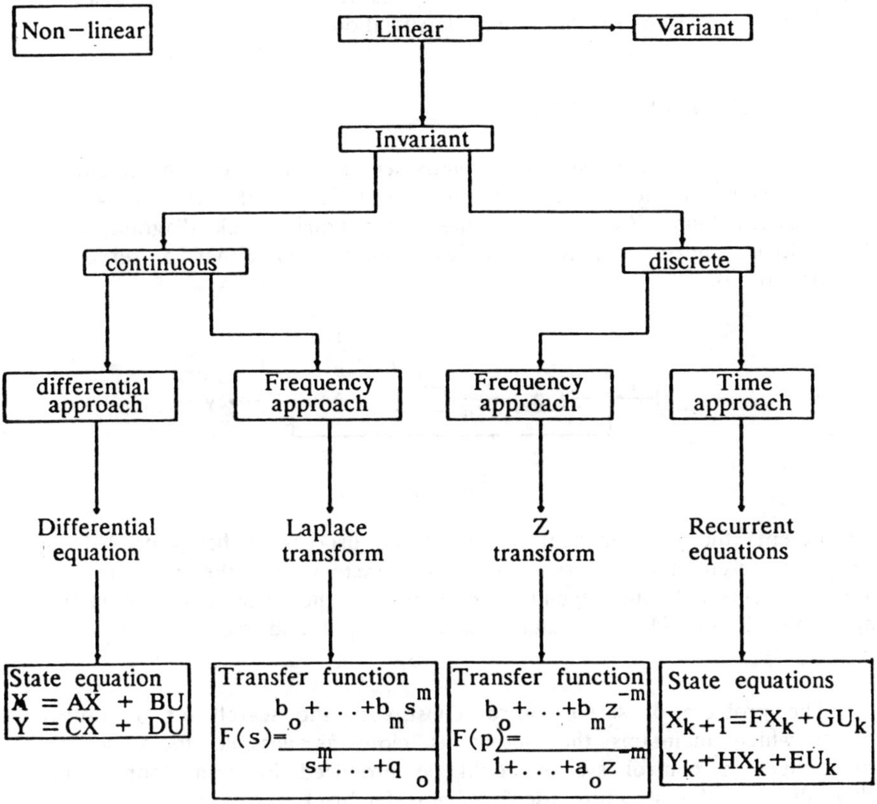


Representation of invariant linear systems

Fig. 3

b) After the intrinsic nature of the model has been defined in this way, the problem of mathematical representation is posed. The richness of the form of representation chosen inevitably affects the future synthesis of the control law. Since this problem will be dealt with in Chapter 1, the standard representations are shown here in summary form only: for invariant deterministic linear systems, they include s or z transfer functions, systems of differential or recurrent equations, and continuous or discrete state vector equations. This classification, shown in detail in Fig. 3, is a sub-level of the initial classification.

## *II – THE FEEDBACK LOOP*

The characteristic and fundamental feature of an automatic control system is the concept of the closed loop, otherwise known as the feedback loop. Fig. 4 shows the conventional block diagram of a closed loop control system which can operate as either a servo or regulator (e=0).

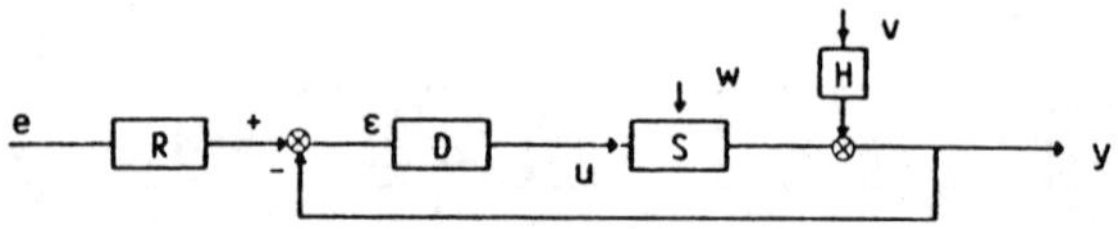


Fig. 4

S represents the mathematical model of the process to be controlled, w and v represent the disturbances of the system and the measurement noise, D represents the regulator, $\epsilon$ represents the error signal, e is the input, and R and H are models related to input and measurement noise respectively.

The problem of servo control consists in the search for a control system which maintains the output as close as possible to a desired value. Such a control system could be produced in open loop form. Why, then, is this structure used? Why is feedback necessary?

This question may appear to be obsolete after several decades of automation. However, in view of the current revival of theories of robust control, it appears that the answers to the question have been forgotten to some extent. Is it not the case that some of the modern theories of closed loop control are actually hypersensitive to imprecisions in the model and therefore contradict some of the properties of the closed loop?

Two basic properties provide sufficient justification for preferring closed loop to open loop control.

## II.1 BETTER ROBUSTNESS TO EXTERNAL DISTURBANCES

The very simple example shown in Fig. 5 will illustrate this point better than any lengthy discussion. A static open loop control system and the corresponding closed loop control system are shown:

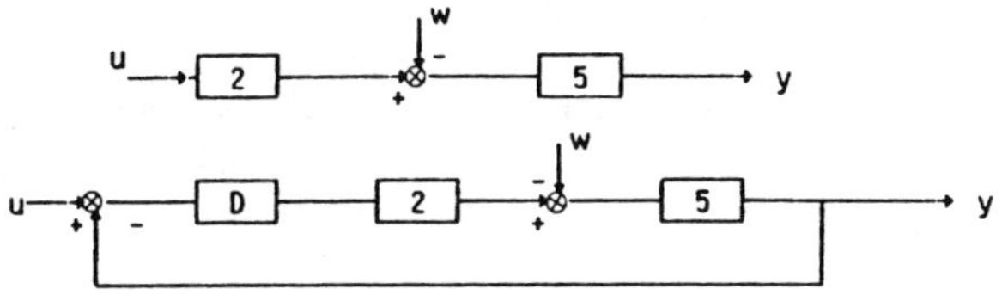


Fig. 5

In the open loop, y = 10u–5w.

The input–output relationship for the closed loop may be written:

$$y = \frac{10D}{1 + 10DK} u - \frac{5w}{1 + 10DK}$$

By suitable choice of parameters D and K, we can ensure that the static gains of closed loop and of open loop are the same, and can also specify that no more than 10% of the disturbance will appear in the output. It is sufficient to specify the two following conditions:

$$\frac{10D}{1 + 10DK} = 10 \quad \text{and} \quad \left|\frac{5w}{1 + 10DK}\right| \leqslant |0.1\, w|$$

and therefore

$$D = 50 \quad \text{and} \quad K = \frac{49}{500}$$

The highly beneficial effect of the feedback in absorbing the external disturbance is immediately apparent. Clearly, however, an open loop control system is justified when the disturbances can be measured. Moreover, the problems of stability inherent in the closed loop must be emphasised.

## II.2 BETTER ROBUSTNESS TO VARIATIONS OF THE SYSTEM PARAMETERS

The important feature of a regulator or servo system is that it allows a small deterioration in the performance of the system when the parameters of the process are partly known and/or vary with time. It could be shown in a similar way that any parametric variation of the gain of the preceding system causes a much greater deterioration of the output in an open loop system than in a closed loop system.

Other properties can be quoted in addition to the above, in particular certain effects of linearisation caused by the closed loop in the control of non-linear systems.

## *III - BASIC PROBLEMS OF CONTROL*

Having detailed the essential properties of the closed loop, we can now formulate, and classify in order of increasing complexity, the principal problems encountered in process control.

*** Deterministic optimal or non-optimal control**

Given a deterministic mathematical model of a system S, with known parameters, the objective is to find the control law relating the measurements y to the input u in such a way as to provide (Fig. 6):

a) overall stability of the closed loop system

b) required constraints on performance (precision, response time, damping, pass band) or minimisation of general criteria in time domain (linear quadratic criterion) or frequency domain (positioning of poles, mode control). Chapters 2, 3 and 4 describe this search procedure, which is essentially concerned with the synthesis of digital controllers (Chapters 3 and 4).

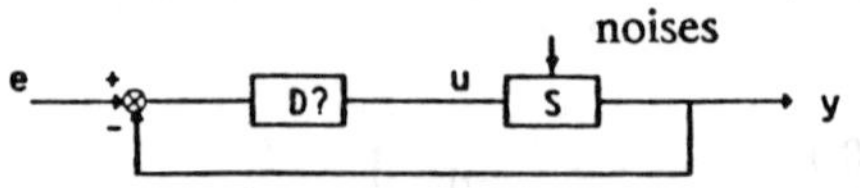


Fig. 6

*** Optimal stochastic control**

Given the stochastic mathematic model, with known parameters, of a system S, the digital controller D which minimises a linear quadratic criterion is sought. Chapter 5 deals with this problem in detail, using the polynomial approach and the state variable method. The essential difference from the previous control system consists in the allowance made for the random nature of the noise.

*** Identification and estimation in real time**

Beginning with a mathematical model of an imperfectly known system S, the objective is to use the available inputs and measurements to obtain estimates of the state vector x and the unknown parameters, $s_i$, of the system (Fig. 7).

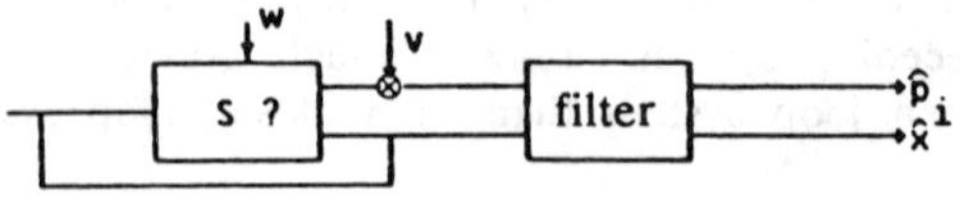


Fig. 7

This technique of filtering and/or identification is used in the stochastic optimal control and self-adaptive control systems discussed in Chapter 6.

* **Adaptive control**

Adaptive control is taken to refer to adaptive regulators which minimise a performance criterion for systems with unknown parameters which are constant or variable with time.

A fairly general structure of this control system which combines the two preceding problems is shown in Fig. 8.

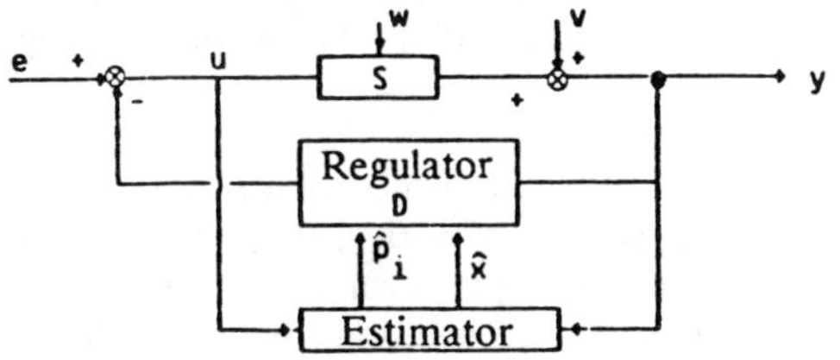


Fig. 8

An estimator generates the state and the identified parameters of the system, which will subsequently be used in the structure of the adaptive regulator.

Chapter 6 is entirely concerned with the synthesis of self-tuning regulators.

## *IV - CONCLUSION*

In the previous discussion of the problems of control, no reference was made to the dichotomy between digital and continuous controllers. Present-day automation is essentially digital, due to recent developments in data-processing and especially microcomputer technology, and therefore this book will be mainly concerned with the synthesis of digital controllers. The problems discussed in Chapter 2 are related to the transformation of continuous controllers to discrete controllers.

Many applications developed under contracts with industrial organisations are described in the second part of this book. They illustrate the theories of control set out in the first seven chapters.

# *CHAPTER 1*

# *THE REPRESENTATION OF SIGNALS AND SYSTEMS*

## *1.1 - INTRODUCTION*

Modelling is the initial phase in the search for digital controllers to control any given process. The complexity of this digital synthesis often arises from the nature of the initial choice of the representation to be used.

In this chapter, the different mathematical representations of continuous and discrete signals (Fourier, Laplace, and Z transforms) are presented initially. Their principal properties are stated (usually without any proof), together with the connections between them.

The second part is concerned with representations of continuous and discrete deterministic systems. The different forms of modelling mentioned in the general introduction are analysed here. The relationships between the different representations obtained, and particularly the transformation from continuous to discrete, form the introduction to Chapter 2.

The second part deals with the introduction of random signals into linear systems; the procedure given here is used as the framework for the optimal stochastic control of discrete systems described in chapter 5.

## *1.2 - THE REPRESENTATION OF DETERMINISTIC SIGNALS*

Three types of signal are considered in this book:

a) Continuous signals are defined by $R \rightarrow R^s$, where the variable assigned to R is denoted as t, designating time. It should be noted that the word "continuous" signifies continuous in steps in the mathematical sense.

b) Discrete signals are defined by $Z \rightarrow R^s$, where the variable assigned to Z is denoted n, and $n\Delta t$ designates instants of time (Fig. 1).

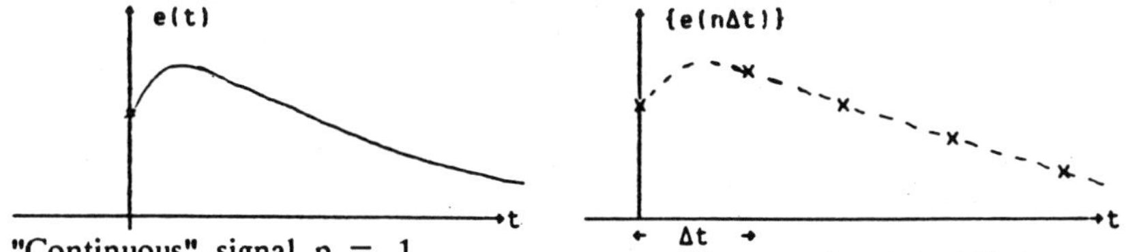


"Continuous" signal p = 1

Discrete signal (p=1)

Fig. 1

These two types of signal have an obvious physical nature. There is, however, a third type of signal: the sampled signal. Such signals have no physical nature, but act as an intermediate calculation unit enabling discrete signals to be processed with the procedures used for continuous signals. They are defined as follows:

c) Sampled signals are defined by discrete signals in a one-to-one way as follows. A sampled signal e*(t) is related to a discrete signal represented by the sequence e(nΔt) by the following mapping:

$$\left[ \left| \begin{matrix} e_1(n\Delta t) \\ \\ e_s(n\Delta t) \end{matrix} \right| , \; n \in Z \right] \longrightarrow e^*(t) = \sum_{n=-\infty}^{+\infty} \left| \begin{matrix} e_1(n\Delta t) \\ \\ e_s(n\Delta t) \end{matrix} \right| \delta(t-n\Delta t)$$

Fig. 2 shows the form of a signal of this type, consisting of a sequence of Dirac pulses:

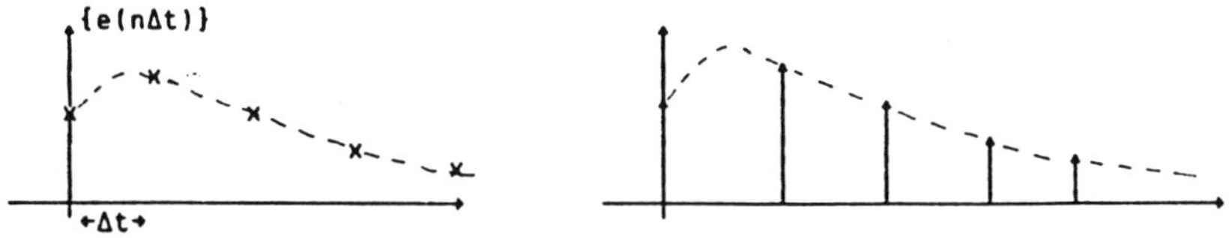


Discrete signal

Sampled signal

Fig. 2

Note that if the signal e(t) is not continuous in t=nΔt, it is assumed that it is continuous to the right of this instant, and e(nΔt) is defined as the limit:

$$\lim_{\epsilon \to 0^+} \{e(n\Delta t+\epsilon)\}$$

A signal of any of the three types described above is "causal" if it is zero for any instant preceding its cause. The signals used subsequently will be causal, and this chapter will be concerned only with scalar signals (s=1).

## 1.2.1 - NOTES ON CONTINUOUS SIGNALS

### 1.2.1.1 - Fourier series and transform

Let e(t) be a continuous periodic signal for the period T such that:

$$e(t) = e(t+T)$$
$$\int_{-T/2}^{+T/2} | e(t) | \, dt < \infty$$

Such a signal can be expanded into a **Fourier series**:

$$e(t) = \sum_{n=-\infty}^{n=+\infty} C_n \, e^{jn\frac{2\pi t}{T}} \qquad (1)$$

where

$$C_n = \frac{1}{T} \int_{-T/2}^{+T/2} e(t) \, e^{-jn\frac{2\pi t}{T}} \qquad (2)$$

Note that the values $C_n$ represent the amplitudes of the different components of the pulse $n2\pi/T$ forming the signal e(t). Now consider a continuous signal with a summable modulus; the Fourier transform of e(t) is then defined by the expression:

$$E(\omega) = \int_{-\infty}^{+\infty} e(t) \, e^{-j\omega t} dt \qquad (3)$$

It then follows that:

$$e(t) = \frac{1}{2\pi} \int_{-\infty}^{+\infty} E(\omega) \, e^{j\omega t} \, d\omega \qquad (4)$$

Note that in the expression above, $E(\omega)\,d\omega$ has the same significance as $C_N$ in expression (1). In addition, if e(t) is a continuous signal except at certain points, $t_0$ for example, then:

$$\frac{e(t_o^+) + e(t_o^-)}{2} = \frac{1}{2\pi} \int_{-\infty}^{+\infty} E(\omega) \; e^{j\omega t_o} \, d\omega \qquad (5)$$

The following notations are equivalent: $E(\omega) = F(e(t))$.

1.2.1.2 - **Laplace transform**

$E(s) = L(e(t))$ designates the Laplace transform (or one-sided Laplace transform) of a continuous signal $e(t)$ defined by:

$$E(s) = \int_{o}^{+\infty} e(t) \; e^{-st} \, dt \qquad \text{for } R_e(s) > a \qquad (6)$$

where a represents the abscissa of convergence of the integral.

The inverse expression providing the time signal e(t) is written as follows:

$$e(t) = \frac{1}{2\pi j} \int_{C-j\infty}^{C+j\infty} E(s) \; e^{st} \, ds \qquad \text{for } C \geqslant a \qquad (7)$$

The calculation of e(t) by the inverse formula (7) requires the use of the residues theorem and the Jordan lemma which will now be introduced.

Residues theorem

Let E(s) be a complex function which is holomorphic on a contour and within this contour except at a countable number of poles. It follows that:

$$\int_{\Gamma} E(s) \, ds = 2\pi j \; \Sigma \text{ residues for the poles contained within the contour } \Gamma$$

If $s_o$ is a pole of order n, the residue E(s) in $s_o$ is given by the formula:

$$\frac{1}{(n-1)!} \lim_{s \to p_o} \left[ \frac{\partial^{n-1}}{\partial_s^{n-1}} (s-p_o)^n \; E(s) \right]$$

Jordan lemma

Consider the two contours $\gamma_1$ and $\gamma_2$ indicated in Fig. 3:

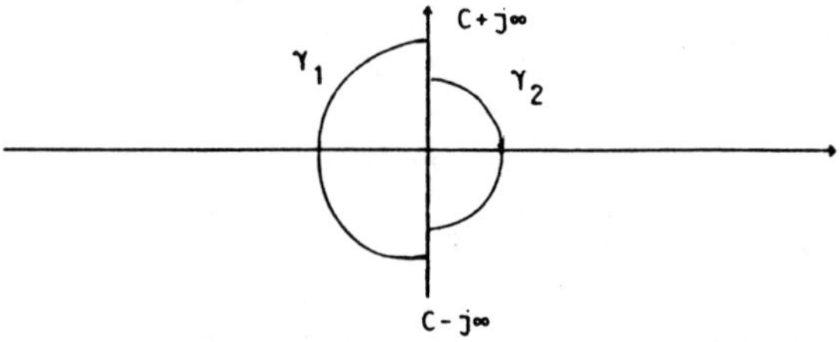


Fig. 3

If $E(s) \to 0$ when $|s| \to \infty$
then:

$$\lim_{\text{radius } \gamma_2 \to \infty} \int_{\gamma_2} e^{st} E(s)\, ds = 0 \qquad \text{for } t < 0$$

$$\lim_{\text{radius } \gamma_1 \to \infty} \int_{\gamma_1} e^{st} E(s)\, ds = 0 \qquad \text{for } t > 0$$

If $sE(s) \to 0$ for $|s| \to \infty$
then:

$$\lim_{\text{radius } \gamma \to \infty} \int_{\gamma} E(s)\, ds = 0 \qquad (\gamma = \gamma_1 \text{ or } \gamma_2)$$

## 1.2.2 - NOTES ON DISCRETE SIGNALS

### 1.2.2.1 - Definition of the Z transform

The Z transform is one of the basic tools for the analysis of signals and digital systems. Starting with the sequence $\{e(n\Delta t)\}$ of a discrete signal, the Z transform of this sequence is the limit when it exists of the series:

$$E(z) \triangleq \sum_{n=0}^{\infty} z^{-n} e(n\Delta t) \qquad \text{for } |z| > a \tag{8}$$

where a is the radius of the convergence.

For example, if $e(n\Delta t) = e^{-\alpha n\Delta t}$
then:

$$E(z) = \sum_{n=0}^{\infty} z^{-n} e^{-\alpha n\Delta t} = \frac{1}{1 - e^{-\alpha\Delta t} z^{-1}}$$

with $|e^{-\alpha\Delta t} z^{-1}| < 1$, i.e. a radius of convergence equal to $e^{-\alpha\Delta t}$.
The following notations will be commonly used in the rest of this text:

$$E(z) = Z(e(t)) = Z(E(s)) = Z\{e(n\Delta t)\}$$

1.2.2.2 - **Properties of the Z transform**

Certain basic properties will now be stated. The proofs which are often provided have been omitted. The interested reader can consult the following works: Boudarel, Delmas, Guichet (1969); Jury (1964).

- The initial and final value theorem:

$$e(o) = \lim_{z\to\infty} E(z)$$

$$\lim_{n\to\infty} e(n\Delta t) = \lim_{z\to 1} (1-z^{-1}) E(z)$$

- Derivation theorem:

$$Z\{n\Delta t\, f(n\Delta t)\} = -\Delta t\, z \frac{dF(z)}{dz}$$

- Parseval relation

$$\sum_{n=0}^{\infty} e^2(n\Delta t) = \frac{1}{2\pi j} \int_\Gamma \frac{E(z)\, E(z^{-1})\, dz}{z} \qquad (9)$$

where $\Gamma$ must belong to the common convergence domain of E(z) and $E(z^{-1})$.

This crucial relationship in the synthesis of digital controllers establishes a link between the time and frequency domains. It enables a quadratic criterion to be expressed on an infinite horizon as a function of the Z transformation of the discrete signal.

- Inverse transformation by the complex integral.

The objective is to obtain the sequence $\{e(n\Delta t)\}$ on the basis of the knowledge of the transform E(z).
According to the Cauchy theorem,

$$\int_\Gamma z^m \, dz = 0 \qquad \text{if } m \neq -1$$
$$\qquad\qquad = 2\pi j \qquad \text{if } m = -1$$

for any contour $\Gamma$ surrounding the origin.
From formula (8), it follows that

$$\int E(z)\, z^m \, dz = 2\pi j \; e((m+1)\,\Delta t)$$

which naturally leads to the inverse transform formula:

$$e(n\Delta t) = \frac{1}{2\pi j} \int_\Gamma E(z) \; z^{n-1} dz \qquad (10)$$

where $\Gamma$ belongs to the convergence domain of E(z)

This result can be used, for example, to demonstrate the Parseval relation shown above.

$$\sum_{n=0}^{\infty} e^2(n\Delta t) = \frac{1}{2\pi j} \sum_{n=0}^{\infty} e(n\Delta t) \int_\Gamma z^{n-1} \, E(z)\, dz$$
$$= \frac{1}{2\pi j} \int_\Gamma \Big( \sum_{n=0}^{\infty} e(n\Delta t) z^{n-1} \Big) E(z)\, dz$$
$$= \frac{1}{2\pi j} \int_\Gamma \frac{E(z^{-1})\; E(z)}{z} \, dz$$

– Inverse transformation by expansion into whole series of $z^{-1}$

Using definition (8), the sequence $\{e(n\Delta t)\}$ may be obtained by dividing E(z) with a progressive increase in the power of $z^{-1}$

if $E(z) = \dfrac{1}{1-e^{-\alpha\Delta t} z^{-1}}$

$$\begin{array}{l|l} 1 & 1 - e^{-\alpha\Delta t} z^{-1} \\ \hline 0 + e^{-\alpha\Delta t} z^{-1} & 1 + e^{-\alpha\Delta t} z^{-1} + e^{-2\alpha\Delta t} z^{-2} + \ldots \\ \quad 0 + e^{-2\alpha\Delta t} \; z^{-2} & \end{array}$$

It is also possible to analyse E(z) into simple elements whose expansion is assumed to be known.

1.2.2.3 – **Relationships between the Fourier–Laplace transforms and the Z transform**

These relationships are crucial, since they form the basis of all digitisation. The following should be noted: e*(t), defined by the expression:

$$e^*(t) = \sum_{n=0}^{\infty} \delta(t-n\Delta t)\ e(t) = \sum_{n=0}^{\infty} \delta(t-n\Delta t)\quad e(n\Delta t)$$

and E*(s), the Laplace transform of e*(t), which is called the **sampled Laplace transform** of e(t).

Let us express E*(s) as a function of E(z). We may write:

$$E^*(s) = \int_o^{\infty} e^*(t)\ e^{-st} dt = \int_o^{+\infty} \sum_{n=0}^{\infty} \delta(t-n\Delta t)\ e(t)\ e^{-st} dt$$

i.e.

$$E^*(s) = \sum_{n=0}^{\infty} \int_o^{\infty} \delta(t-n\Delta t)\ e(t)\ e^{-st} dt$$

$$E^*(s) = \sum_{n=0}^{\infty} e(n\Delta t)\ e^{-sn\Delta t}$$

We finally obtain the relationship:

$$\boxed{E^*(s) = |E(z)|_{z=e^{s\Delta t}}} \qquad (11)$$

Let E*(ω) represent the Fourier transform of e*(t); then, if it exists, the following must be true:

$$E^*\ (\omega) = |E(z)|_{z=e^{j\omega\Delta t}} \qquad (12)$$

These relationships demonstrate the importance of the imaginary sampled signals: discrete signals transformed in an obvious way into sampled signals can be treated by the same procedures (Laplace and Fourier transforms) as continuous signals.

The relationships connecting the E*(s) or E(z) transforms to the Laplace transform E(s) will now be found. To do this, the E(z) transform is calculated from the sampled signal e(t):

$$e^*(t) = e(t)\ P(t)$$

where P(t) is the train of unit pulses:

$$P(t) = \sum_{n=0}^{\infty} \delta(t-n\Delta t)$$

Clearly, the Laplace transform of P(t) is:

$$P(s) = \sum_{n=0}^{\infty} e^{-n\Delta t s} = \frac{1}{1-e^{-\Delta t s}} \quad \text{where } R_e(s) > 0$$

E*(s) is then deduced from E(s) by the "frequency" convolution

$$E^*(s) = \frac{1}{2\pi j} \int_{C-j\infty}^{C+j\infty} E(u)\ P(s-u)\ du$$

where E(u) is the Laplace transform of e(t).

The conditions for the existence of this integral are that C should be not only less than the real part of s, so that the integration takes place in the definition domain of R, but also greater than the abscissa of convergence of E(s) denoted by a.

E*(s) may be written in the following form:

$$E^*(s) = \frac{1}{2\pi j} \int_{C-j\infty}^{C+j\infty} E(u) \frac{1}{1 - e^{-\Delta t s} e^{\Delta t u}}\ du \qquad (13a)$$

Note that this formula is valid only if e(t) has no discontinuities at the moments of sampling. If, for example, e(t) is discontinuous for t = 0, a correcting term must be used, and the transform is written:

$$E^*(s) = \frac{1}{2\pi j} \int_{C-j\infty}^{C+j\infty} E(u) \frac{1}{1 - e^{-\Delta t s} e^{\Delta t u}}\ du + \frac{1}{2}\ e(0^+) \qquad (13b)$$

To calculate this integral, the residues theorem will be used, and in order to do this the set of poles of the integrating term must be determined.

We have, on one hand, a set of poles of E(u), all situated to the left of the abscissa a, since a is the abscissa of convergence of E, and on the other hand the zeros of $1-e^{-\Delta ts}\ e^{e\Delta tu}$, in other words an infinite number of zeros such that:

$$u_k = s + \frac{2k\pi j}{\Delta t} \qquad \text{where k varies from } -\infty \text{ to } +\infty$$

To calculate the integral 13a, the integration line may be completed either by a contour $\gamma_1$ to the left or by a contour $\gamma_2$ to the right (see Fig. 4):

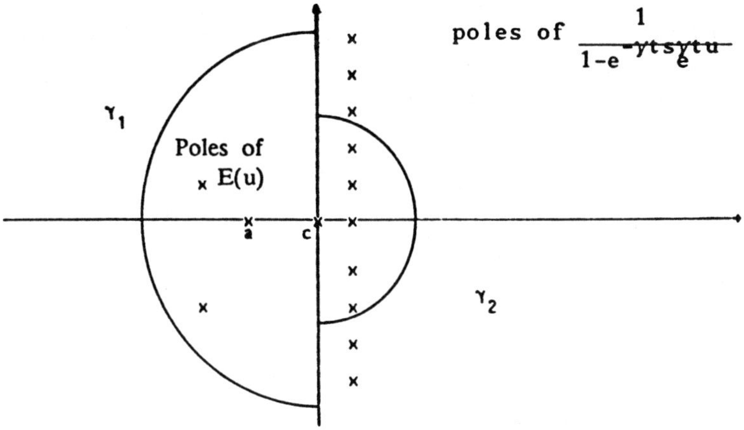


Fig. 4

a) **Integration along the contour $\Gamma=\gamma_1 \cup [C-j\infty, C+j\infty]$**

If the expression $uE(u) \to 0$ tends towards zero when $|u| \to \infty$ on $\gamma_1$, then the integral of E(u) on the semicircle $\gamma_1$ tends towards zero, while the radius of $\gamma_1$ increases indefinitely. By integrating along the contour $\Gamma$, we obtain:

$$E^*(s) = \sum_{\text{poles of } E} \text{residues of } \frac{E(u)}{1 + e^{\Delta t(u-s)}}$$

If uE(u) does not tend towards zero (in other words $e(0^+) \neq 0$), formula 13b must be used.

It is thus shown that under certain conditions

$$\int_{\gamma_1} F(u)\, du \to \frac{1}{2}\, e(0^+)$$

and the previous relationship is then found.

The above results may therefore be summarised as follows:

If $uE(u) \to 0$ when $|u| \to \infty$ on $\gamma_1$ or if E(u) is a rational function, the degree of whose numerator is strictly less than that of its denominator, then

$$E^*(s) = \sum_{\text{poles of } E} \text{residues of } \frac{E(u)}{1-z^{-1} e^{\Delta t u}} \qquad (14)$$

b) **Integration along the contour $\Gamma=\gamma_2 \cup [C-j\infty, C+j\infty]$**

If the expression uE(u) tends towards zero when $|u| \to \infty$ on $\gamma_2$, then the integral of E(u) on $\gamma_2$ tends towards zero, while the radius of $\gamma_2$ tends towards infinity. Taking certain precautions necessitated by the infinite number of poles, we may write:

$$E^*(s) = \sum_{u_k = s + \frac{2k\pi j}{\Delta t}} \text{residues} \frac{E(u)}{1 - e^{\Delta t(u-s)}}$$

and therefore:

$$E^*(s) = \frac{1}{\Delta t} \sum_{-\infty}^{+\infty} E\left(s + \frac{2k\pi j}{\Delta t}\right) \qquad (15a)$$

A more general and rigorous proof is given by D.A. Pierre and R.C. Kolb (1964); the result of this is as follows:
If $e(0^+)$ exists and is the only discontinuity for $t \geqslant 0$, then

$$E^*(s) = \frac{1}{2} e(0^+) + \frac{1}{\Delta t} \sum_{-\infty}^{+\infty} E\left(s + \frac{2k\pi j}{\Delta t}\right) \qquad (15b)$$

Thus two expressions, (14) and (15), which are radically different from the Z transform, are obtained. The first of these enables the Z transform of the discrete signal to be calculated easily if the Laplace transform of the continuous signal is known. The second demonstrates the periodic nature of the spectrum of the sampled signal with all its consequences for the reconstruction of the original continuous spectrum. We shall return to this point in the next section.

### 1.2.2.4 - Generation of continuous signals from discrete signals

The control of a process is generally continuous in nature. the discrete nature of digital controllers therefore makes it necessary to reconstruct a continuous signal from a discrete signal. Three standard types of extrapolator, also called sampler and hold, perform this function.

a) **Pulse extrapolator**

This is found in practice in an approximate form in the control of the attitude of a satellite by means of gas jets. It consists in the combination of the sequence {e(nΔt)} with the signal e*(t) such that

$$\{e(n\Delta t)\} \to e^*(t) = \sum_{n=0}^{\infty} e(n\Delta t)\delta(t-n\Delta t) \qquad (16)$$

The pulsed nature of the command requires the use of such an extrapolator, which is rarely used in practice.

b) **Zero order hold**

This is by far the most commonly used type, and consists of the combination of the sequence $\{e(n\Delta t)\}$ with a signal $e_b(t)$ defined thus:

$$e_b(t) = e(n\Delta t) \qquad \forall t \in [n\Delta t, (n+1)\Delta t] \tag{17}$$

The transformation may be written thus:

$$\{e(n\Delta t)\} \rightarrow e_b(t) = \sum_{n=0}^{\infty} u(t-n\Delta t)\, e(n\Delta t)$$

where

$$\begin{cases} u(t) = 1 & \text{if } 0 \leqslant t \leqslant \Delta t \\ u(t) = 0 & \text{if } t > \Delta t \end{cases}$$

and therefore

$$e_b(t) = \sum_{n=0}^{\infty} \int_0^{\infty} u(\tau)\, e(n\Delta t)\, \delta(t-\tau-n\Delta t)\, d\tau$$

where

$$e_b(t) = \int_0^{\infty} u(\tau)\, e^*(t-\tau)\, d\tau$$

Moving to the Laplace transforms, we immediately find that

$$L(e_b(t)) = L(u(t))\, L(e^*(t))$$

where

$$L(u(t)) = \frac{1-e^{-\Delta t s}}{s}$$

Consequently, the zero-order hold can be represented as follows (Fig. 5):

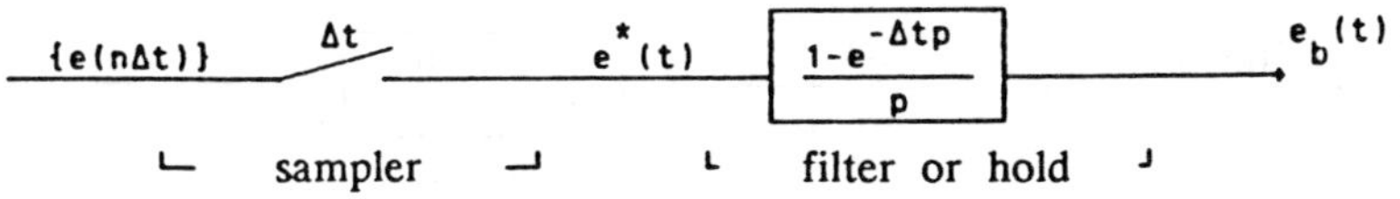


Fig. 5

More generally, a linear extrapolator has the general form:

$$e_b(t) = \sum_{i=-\infty}^{i=+\infty} e(i\Delta t)\, u(t-i\Delta t)$$

As a consequence of this definition, the value u(t) represents the pulse response. The system is illustrated in Fig. 6 where the Laplace transform of the zero order hold is shown.

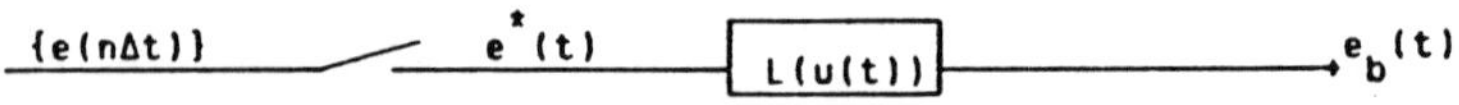


Fig. 6

For information, it is worth noting the first order hold, defined by its response to the discrete pulse $\{\delta(n)\}$ designated by $u_1(t)$ (Fig. 7):

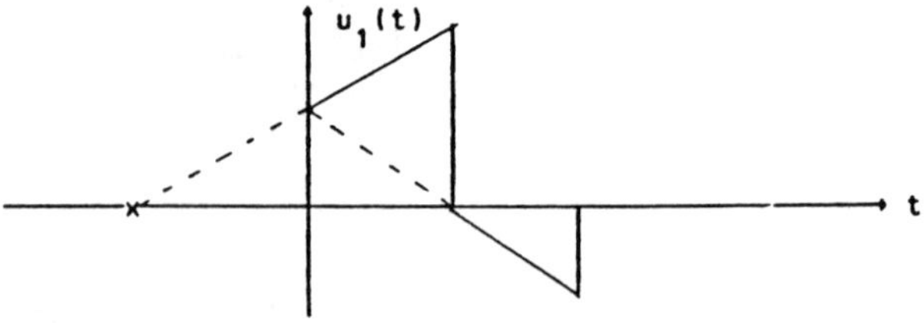


Fig. 7

Another particularly interesting example which will now be analysed is the Shannon extrapolator.

### c) The Shannon extrapolator

This is an extrapolator which reconstructs the signal e(t) exactly from the sequence $\{e(n\Delta t)\}$ under certain conditions.

The previous extrapolators were defined initially in the time domain and then translated into frequency terms, i.e. transformed into a sampler followed by a continuous filter. We shall now proceed in the reverse direction.

Consider the following diagram:

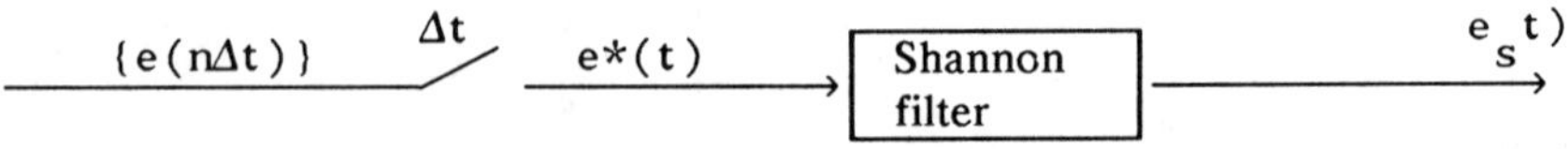


For the two signals e(t) and $e_s(t)$ to be equal at almost every point, they must have the same Fourier transform.

If $B_s(\omega)$ is used to designate the unknown Fourier transform of the Shannon filter, we may write:

$$E_s(\omega) = B_s(\omega)\ E^*(\omega)$$

with $$E^*(\omega) = F\left(e(t) \sum_{-\infty}^{+\infty} \delta(t - nT)\right)$$

where

$$E^*(\omega) = \frac{1}{2\pi} \int_{+\infty}^{+\infty} E(u)\ P(\omega - u)\ du$$

In the last expression, $P(\omega)$ is the Fourier transform of the pulse train.

According to the theory of distribution:

$$P(\omega) = \frac{2\pi}{\Delta t} \sum_{-\infty}^{+\infty} \delta\left(\omega - n \frac{2\pi}{\Delta t}\right)$$

Given the last relationship, it is easily found that

$$\boxed{E^*(\omega) = \frac{1}{\Delta t} \sum_{-\infty}^{+\infty} E\left(\omega - n \frac{2\pi}{\Delta t}\right)} \qquad (18)$$

This expression should be compared with relationship (15) for the Laplace transforms.

If the frequency spectrum $E(\omega)$ is limited to the band

$$\left[ \frac{-\pi}{\Delta t} , \frac{\pi}{\Delta t} \right]$$

we see (Fig. 8) that $E(\omega)$ can be extracted from $E^*(\omega)$ by using an ideal rectangular filter $B_s(\omega)$ defined by

$$B_s(\omega) = \Delta t \qquad \text{for } |\omega| < \frac{\pi}{\Delta t} \qquad (19)$$

$$B_s(\omega) = 0 \qquad \text{for } |\omega| > \frac{\pi}{\Delta t} \qquad (20)$$

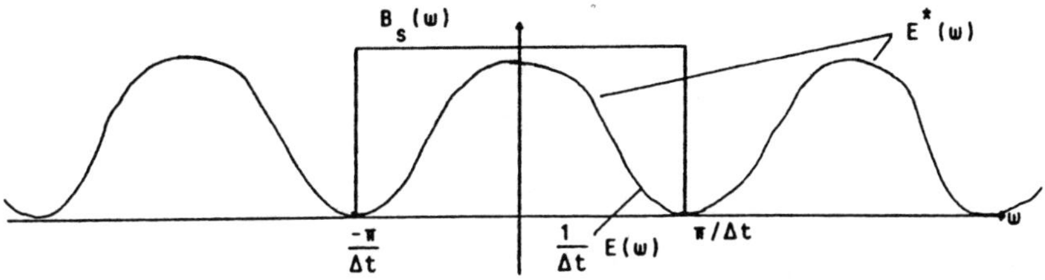


Fig. 8

The inverse of the Fourier transform will now be used to generate the time signal having the rectangular form $B_s(\omega)$ as its Fourier transform. The result is:

$$\frac{1}{2\pi}\int_{-\infty}^{+\infty} B_s(\omega)\, e^{j\omega t} d\omega = \frac{1}{2\pi}\int_{-\pi/\Delta t}^{\pi/\Delta t} B_s(\omega)\, e^{j\omega t} d\omega = \frac{\sin(\pi t/\Delta t)}{\pi t/\Delta t}$$

The output from the Shannon filter then has the following mathematical form:

$$e\ (t) = \int_{-\infty}^{+\infty} \frac{\sin(\pi t/\Delta t)}{\pi t/\Delta t} \sum_{-\infty}^{+\infty} e(n\Delta t)\,\delta(t-\tau-n\Delta t)\ d\tau$$

and therefore

$$\boxed{e_s(t) = \sum_{-\infty}^{+\infty} \frac{\sin[\pi(t-n\Delta t)/\Delta t]}{[\pi(t-n\Delta t)/\Delta t]}\ e(n\Delta t)} \qquad (21)$$

where $e_s(t) = e(t)$ at almost all points.

Formula (21) represents a linear extrapolator with

$$u(t) = \frac{\sin(\pi t/\Delta t)}{\pi t/\Delta t}$$

This Shannon extrapolator cannot be produced in physical form since it requires an infinite number of points $e(n\Delta t)$ to reconstruct $e(t)$. Furthermore, it cannot be used in real time, since its response to a discrete pulse is not causal. In fact,

$$e_s(t) = \sum_{i=-\infty}^{+\infty} e(i\Delta t)\ u(t-i\Delta t)$$

where $e_s(t)$ depends on the values $e(i\Delta t)$ for $i\Delta t > t$.

We shall now check that a zero-order hold approximates the Shannon filter when $\Delta t \to 0$.

Consider the system in Fig. 9 where the input is a periodic input:

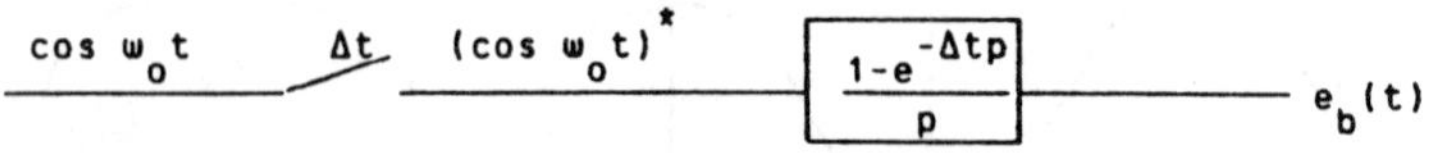


Fig. 9

The Fourier transform of $\cos \omega_0 t$ is written:

$$F(\cos \omega_0 t) = \pi\ [\delta(\omega-\omega_0) + \delta(\omega+\omega_0)]$$

According to relationship (18),

$$F(\cos^* \omega_o t) = \frac{\pi}{\Delta t} \sum_{-\infty}^{+\infty} [\delta(\omega-\omega_o-n\frac{2\pi}{\Delta t}) + \delta(\omega+\omega_o - n\frac{2\pi}{\Delta t})]$$

The frequency spectrum of the zero-order hold is:

$$\left|\frac{1-e^{-\Delta t s}}{s}\right|_{s=j\omega} = 2\left|\frac{\sin(\omega \Delta t/2)}{\omega}\right|$$

Fig. 10 shows $F(\cos^*\omega_o t)$ and the spectrum of the hold.

Note that when $\Delta t \to 0$ the Fourier transform of the input signal $\cos \omega_o t$ is present.

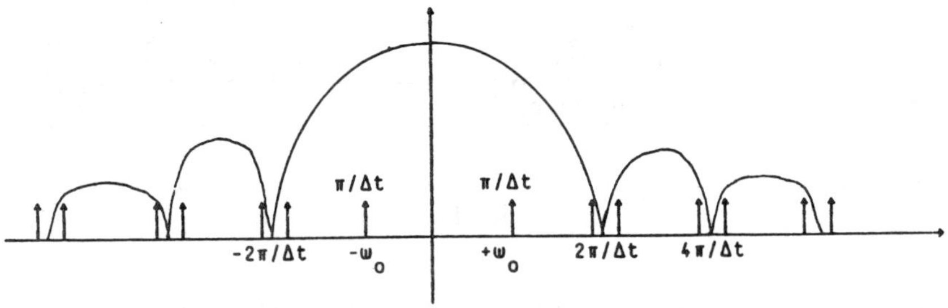


Fig. 10

## *1.3 - REPRESENTATION OF DETERMINISTIC LINEAR SYSTEMS*

The modelling of many processes naturally leads to a set of differential equations which are the mathematical representation of the physical and/or chemical laws governing the observed behaviour of the system. These relationships lead to representations in continuous form; the two best known types of these are state space representation and modelling by s transfer function matrices.

In addition, the digital control of processes, which is continuous in nature, requires the discrete representation of the systems. The two aspects of modelling referred to above are found here, namely the representation of the discrete state and the matrix representation of the Z transfer function.

The essential characteristics of these four modelling procedures are described in this section. A more detailed description will be given of the relationships connecting them and shown as 1, 2, and 3 in Fig. 11; these relationships are:

- the transformation from continuous state form to the discrete state form;
- the transformation from the s transfer function matrix to the Z transform;
- the relationships between state representation and transfer function matrix representation.

The concept of the vector signal will be used in the following text. Such signals, like scalar signals, may be continuous, discrete, or sampled. Their Fourier, Laplace, Z, and sampled Laplace transforms are vectors whose components are the transforms of the corresponding elements of the vector signal.

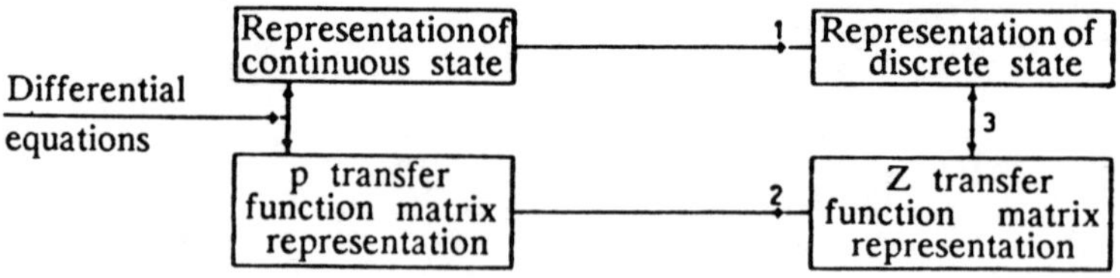


Fig. 11

## 1.3.1 - REPRESENTATION OF SYSTEMS IN CONTINUOUS FORM

### 1.3.1.1 - Form of state vectors

The usual form of representation of the state of linear systems is as follows:

$$\begin{aligned} \dot{x} &= A\,x + B\,u \\ y &= C\,x + D\,u \end{aligned} \qquad (22)$$

where

$$x \in R^n, \; u \in R^q, \; y \in R^s$$

u is the system command, in other words the signal which the designer can vary, y is the measured output, in other words the information available to the designer for the definition of u, and x is the state of the system which represents the internal behaviour.

This form of representation has given rise to many very important concepts (the notions of controllability, observability, invariants, and minimal representation). Our purpose here is not to describe all these concepts in detail; there is abundant literature on this subject. We will simply state the form of the solution of the state vector equation (22).

Using $x_o$ to designate the initial condition of the state at the instant $t = t_0$, then for $t \geqslant t_0$

$$x(t) = e^{A(t-t_0)} x_o + \int_{t_0}^{t} e^{A(t-\tau)} Bu(\tau)\, d\tau \qquad (23)$$

$$y(t) = C\, x(t) + D\, u(t)$$

1.3.1.2 - **The form of the transfer function matrix**

Differential equations generally lead directly to a representation in the form of a transfer function matrix without using equation (22). However, this form is used to simplify the introduction of the concept of the transfer function matrix.

Let X(s), Y(s) and U(s) be the Laplace transforms of x(t), y(t) and u(t). We may write:

$$sX(s) = A\, X(s) + B\, U(s) + x(o)$$

$$Y(s) = C\, X(s) + D\, U(s)$$

i.e.

$$Y(s) = (C(sI-A)^{-1}\, B + D)\, U(s) + C(sI-A)^{-1}\, x(o) \qquad (24)$$

The transfer function matrix S(s) represents the system when the initial conditions are zero

$$S(s) = C(sI-A)^{-1}\, B + D \qquad (25)$$

and

$$Y(s) = S(s)\, U(s)$$

If D = 0, the preceding equation may be reformulated in the time domain as follows:

$$y(t) = S(t) * u(t)$$

where the operator * designates the convolution product, or alternatively

$$y(t) = \int_{t_0}^{t} S(t-\tau)\, u(\tau)\, d\tau$$

Using inputs of the form $e_i\, \delta(t)$ ($e_i$ = i-th vector of the canonical base of $R^q$), it can immediately be seen that the columns of the matrix S(t) represent the pulse responses of the system.

### 1.3.1.3 – Stability of continuous systems

The state representation (22) and its general solution (23) can be used to define the stability.

Definition

A linear system is stable (internally) if and only if the effect of the initial conditions disappears when time is increased indefinitely.

Using (23), a necessary and sufficient condition for asymptotic stability can be formulated thus:

- a necessary and sufficient condition for internal stability is that all the eigenvalues of the matrix A are negative in their real parts.

By putting matrix A into the Jordan form by means of a suitable transformation, it can be shown that the poles of S(s) are eigenvalues of matrix A. The above condition for internal stability may therefore be transposed to the representation by a transfer function matrix. This leads to the following condition:

> A system described for a matrix S(s) can be called stable in the input–output sense if the real parts of the poles of S(s) are all negative.

These two characteristic properties of stability are not equivalent, since the number of eigenvalues of the matrix A can be greater than the number of poles of S(s) if the system is uncontrollable or unobservable (Fossard, 1972).

## 1.3.2 – REPRESENTATION OF DISCRETE SYSTEMS

In the continuous form, the representation is based on the system in isolation. For a discrete system, however, it is not possible to dissociate the system from the filter of the extrapolator, and the whole filter–system combination is modelled (Figs. 12 and 13).

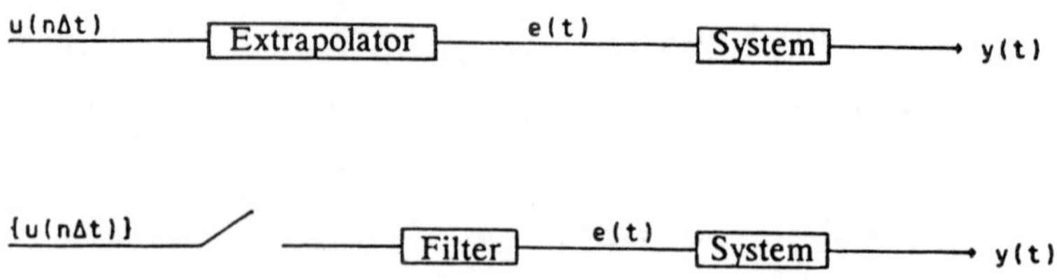


Figs. 12 and 13

### 1.3.2.1 – Transfer from the continuous state form to the discrete state form

A discrete system is controlled by a discrete signal extrapolated between the sampling instants, for example by a zero-order hold. More

generally, if e(t) is used to designate the extrapolated form of the sequence {u(nT)}, then according to equation (23) we may write

$$x(n\Delta t) = e^{A\Delta t} x((n-1)\Delta t) + \int_{(n-1)\Delta t}^{n\Delta t} e^{A(n\Delta t - \tau)} B\, e(\tau)\, d\tau$$

If the extrapolator is a zero-order hold, then

$$e(t) = u(n\Delta t) \qquad \forall t \in [n\Delta t,\ (n+1)\Delta t]$$

and therefore

$$x(n\Delta t) = e^{A\Delta t} x((n-1)\Delta t) + u((n-1)\Delta t) \int_{0}^{\Delta t} e^{A(\Delta t - \tau)}\, B\, d\tau$$

Finally, the digitised form of the state equation is obtained:

$$\begin{aligned} x(n\Delta t) &= F\ x((n-1)\Delta t) + G\ u((n-1)\Delta t) \\ y(n\Delta t) &= H\ x(n\Delta t) + E\ u(n\Delta t) \end{aligned} \qquad (27)$$

The system (27) will form an important basis for modelling in the synthesis of digital controllers which will be discussed in later chapters. These equations of the recurrent type are already directly usable for digital simulation of the process. The state at the instant n+1 can also be expressed as a function of the initial state and of the commands at different instants, thus:

$$x((n+1)\Delta t) = F^{n+1}\ x(o) + \sum_{j=0}^{n} F^{n-j}\ G\ u(j\Delta t) \qquad (28)$$

### 1.3.2.2 – Transfer from the s transfer matrix to the z transfer matrix

A relationship between the sequences {u(nΔt)} and {y(nΔt)} will now be sought.

Returning to Fig. 13, the transfer function matrix of the extrapolating filter and of the system has the folowing form:

$$S_b(s) = S(s)\ B(s) \qquad (29)$$

where B(s) is the transfer function matrix of the extrapolating filter.

We may therefore write

$$y(t) = s_b(t) * u^*(t)$$

i.e.

$$y(t) = \sum_{i=o}^{\infty} \int_{o}^{\infty} s_b(t-\tau)\ u(i\Delta t)\delta(\tau-i\Delta t)\ d\tau$$

$$y(t) = \sum_{i=o}^{\infty} s_b(t-i\Delta t)\ u(i\Delta t)$$

and therefore

$$y(n\Delta t) = \sum_{i=o}^{\infty} s_b((n-i)\Delta t)\ u(i\Delta t) \tag{30}$$

Now, the columns of the matrix $s_b(t)$ (which is the inverse of $S_b(s)$) represent pulsed responses. Thus equation (30) provides a discrete convolution between the input and the pulsed responses. By analogy with continuous systems, it is then possible to develop a representation in the form of a transfer function matrix.

Let $S_b(z)$ designate the Z transform of the sequence $\{s_b(n\Delta t)\}$ and let $U(z)$ designate the Z transform of the sequence $\{u(n\Delta t)\}$;

then,

$$S(z) = \sum_{i=0}^{\infty} s_b(i\Delta t)\ z^{-i}$$

$$U(z) = \sum_{i=0}^{\infty} u(i\Delta t)\ z^{-i}$$

It is then evident from equation (30) that:

$$\boxed{Y(z) = S_b(z)\ U(z)} \tag{31}$$

As in the case of continuous systems, it has now been shown that discrete systems are described by a digitised pulse response or by a Z transfer function. The most important difference is that the extrapolator preceding the system cannot be dissociated from it.

– Special cases of discrete systems

* Consider the single–input single–output system whose transfer function is $z^{-1}$:

$$Y(z) = z^{-1} U(z)$$

$$\{y(n\Delta t)\} = \{u(n-1)\Delta t\}$$

This is a case of a simple delay between the input and the output.

* Now consider a single–input single–output system whose z transfer function is a rational function of the form:

$$Y(z) = \frac{a_o + a_1 z + \ldots\ldots + a_m z^m}{b_o + b_1 z + \ldots\ldots + z^n} U(z)$$

The realisability condition requires that the degree n be greater than or equal to m; otherwise the output y would depend on inputs u applied subsequently.

This transfer function can be rewritten in the form

$$Y(z) = \frac{a_o z^{-n} + a_1 z^{-n+1} + \ldots\ldots + a_m z^{-n+m}}{b_o z^{-n} + b_1 z^{-n+1} + \ldots\ldots + 1} U(z)$$

where

$$Y(z) = -(b_o z^{-n} + \ldots + b_{n-1} z^{-1})Y(z) + (a_o z^{-n} + \ldots + a_m z^{-n+m})U(z)$$

According to the above extrapolation of the simple delay $z^{-1}$, the last equation is the translation of the following recurrent time equation:

$$y(k\Delta t) = -\sum_{i=0}^{n-1} b_i\{y((k-n+i)\Delta t)\} + \sum_{i=0}^{m} a_i\{u((k-n+i)\Delta t)\} \quad (32)$$

Rational functions in z can therefore be immediately interpreted in terms of recurrent equations.

1.3.2.3 – **Relationships between state representations and z transfer function representations**

Consider the following system of discrete recurrent equations:

$$x(n\Delta t) = F\ x((n-1)\Delta t) + G\ u((n-1)\Delta t)$$

$$y(n\Delta t) = H\ x(n\Delta t) + E\ u(n\Delta t)$$

According to the definition of the z transform,

$$Z\{x((n+1)\Delta t)\} = z\ Z\{x(n\Delta t)\} - z\ x(o)$$

We may then write:

$$Z\ \{x(n\Delta t)\} = \sum_{i=0}^{\infty} x(i\Delta t)\ z^{-i}$$

$$Z\{x((n+1)\Delta t)\} = \sum_{i=0}^{\infty} x((i+1)\Delta t)\ z^{-i} = z\left[\sum_{i=0}^{\infty} x(i\Delta t) z^{-i}\right] - zx(o)$$

From this it may be deduced by taking the Z transforms of the sequences appearing in the initial system that

$$zX(z) = F\ X(z) + G\ U(z) + z\ x(o)$$

$$Y(z) = H\ X(z) + E\ U(z)$$

i.e. $$Y(z) = (H(zI-F)^{-1}\ G + E)\ U(z) + H\ (zI-F)^{-1}\ x(o) \qquad (33)$$

The Z transfer function of the system is defined as for continuous systems:

$$S_b(z) = H(zI-F)^{-1}\ G + E \qquad (34)$$

1.3.2.4 – **Stability of discrete systems**

As with continuous systems, the internal stability is defined from the state representation. Referring to equation (28), the following stability condition is obtained:

A necessary and sufficient condition for the internal stability of the linear discrete system is that each of the eigenvalues of the development matrix F has a modulus of less than unity.
By putting the matrix F into the Jordan form by means of a transformation to the appropriate state vector, the stability condition

applicable to the transfer function matrix is obtained.

A system described by a transfer function matrix $S_b(z)$ will be called stable in the input–output sense if and only if each of the poles of $S_b(z)$ has a modulus of less than unity.

As for the continuous case, these two definitions are not equivalent if the system is uncontrollable and/or unobservable.

### 1.3.3 – ANALYSIS AND MANIPULATION OF BLOCK DIAGRAMS

By analogy with the sampled Laplace transform of a signal, the sampled s transfer function is the sampled Laplace transform of the pulsed response which will be designated by $S_b^*(s)$.

In the same way as for continuous signals,

$$S_b^*(s) = |S_b(z)|_{z=e^{\Delta ts}} \tag{35}$$

According to equation (31),

$$Y^*(s) = S_b^*(s)\ U^*(s)$$

Since there is a sampler present between the command u and the system to be controlled,

$$Y(s) = S_b(s)\ U^*(s)$$

We therefore deduce that

$$Y^*(s) = (S_b(s)\ U^*(s))^*$$

and this leads to the following basic relationship:

$$(S_b(s)\ U^*(s))^* = S_b^*(s)\ U^*(s) \tag{36}$$

Conversely, if there is no sampler between the control u and the system,

$$Y(s) = S_b(s)\ U(s)$$

and

$$(S_b(s)\ U(s))^* \neq S_b^*(s)\ U^*(s) \quad \text{in general} \tag{37}$$

The essential difference between the manipulation of continuous and sampled block diagrams lies in this inequality which is almost always true. To demonstrate some of the difficulties of representation, the different possible manipulations will be presented in an example.

* Example of application

Consider the standard diagram of a control circuit shown in Fig. 14.

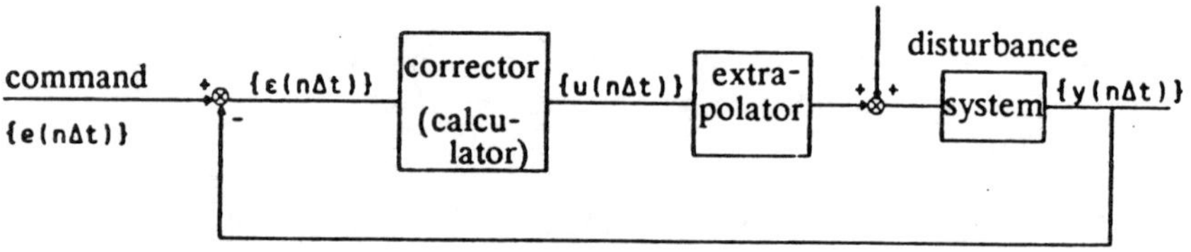


Fig. 14

A digital controller is to be found by using the z transform procedure.

a) **Analysis of the circuit**

The above diagram is simple enough to be put directly into the form of a "z block diagram" by using Z transfer functions. To introduce the analysis of more complex diagrams, in other words those containing discrete and continuous signals, the procedure for sampled signals will be used.

The calculator may be considered as an imaginary transfer function D(s) which performs calculations on a sampled signal $\epsilon^*(t)$ rather than on the physical signal $\{\epsilon(n\Delta t)\}$. The output of this circuit is a continuous signal u(t) such that the digitisation is exactly the sequence $\{u(n\Delta t)\}$ shown in Fig. 14.

Figs. 15 and 16 show the blocks in which the sampled and continuous signals appear:

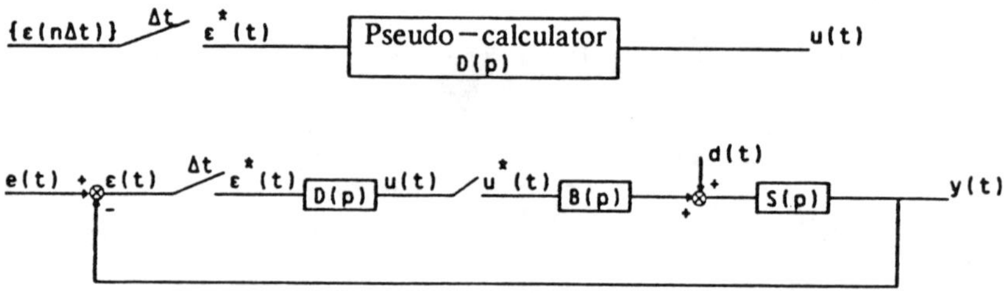


Figs. 15 and 16

The following four equations can be written for the latter diagram:

$Y(s) = S(s)\ d(s) + S(s)\ B(s)\ U^*(s)$

$Y^*(s) = (S(s)\ d(s))^* + (S(s)\ B(s))^* U^*(s)$ according to

$\epsilon(t) = e(t) - y(t)$ equation (36)

$\epsilon^*(t) = e^*(t) - y^*(t)$

By replacing the sampled Laplace transforms with their Z transforms, the following equivalent diagram is obtained (Fig. 17):

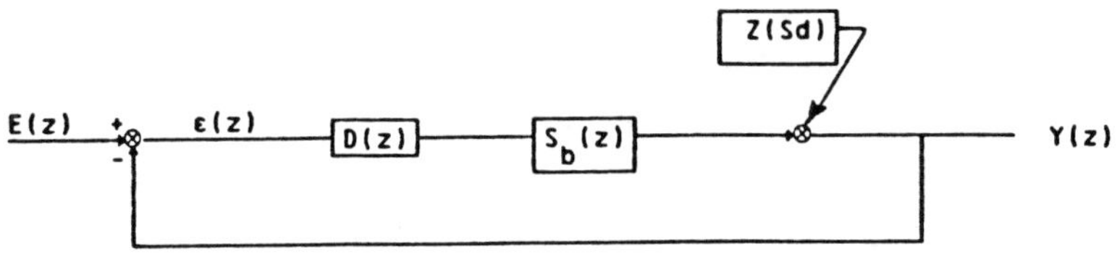


Fig. 17

b) **Synthesis and application of the command**

The problem of the synthesis of D(z) forms the principal subject of this book. Once the structure of the controller is known, all that is required for application is to replace D(z) with a recurrent equation obtained by realisation in the state vector form. The programming of this equation on the computer forms the last stage of the real-time digital control procedure.

## *1.4 - INTRODUCTION OF ALEATORY SIGNALS INTO LINEAR SYSTEMS*

The representation of discrete linear systems receiving deterministic signals has now been described in detail. A number of disturbances affecting the system are random in nature (state noises and measurement noises). The digital synthesis of the controllers controlling stochastic linear systems is dealt with in Chapter 5. It is first necessary to describe the random signals and to define mathematically the transformations that they undergo when they are introduced into linear systems. This is the subject of this section.

### 1.4.1 - CONTINUOUS RANDOM SIGNALS

Let e(t) be a stationary stochastic process (or stochastic function). At any instant, this stochastic variable e(t) can be characterised by its moments, thus:

- its mean, defined by $m_e(t) = E(e(t)) = m_e$

- its autocorrelation function $\varphi_{ee}(\tau) = E(e(t)e^T(t+\tau))$

The autocorrelation spectrum, which is the bilateral Laplace transform of the autocorrelation function, is used as a practical means of characterising these signals after their introduction into a linear system.

$$\Phi_{ee}(s) = \int_{-\infty}^{+\infty} \varphi_{ee}(\tau)\, e^{-s\tau}\, d\tau \tag{38}$$

The inversion formula is written:

$$\varphi_{ee}(\tau) = \frac{1}{2\pi j} \int_{-j\infty}^{+j\infty} \Phi_{ee}(s)\, e^{s\tau} ds \tag{39}$$

Let y(t) designate the signal obtained by introducing the signal e(t) into the transfer function matrix system S(s).

$$Y(s) = S(s)\, E(s)$$

Using the causality hypothesis,

$$y(t) = \int_{0}^{\infty} S(\tau)\, e(t-\tau)\, d\tau = \int_{-\infty}^{+\infty} S(\tau)\, e(t-\tau)\, d\tau$$

The static properties of the output signal will now be defined.

a) **Mean value of the output signal**

The mean can be defined by:

$$E(y(t)) = \int_{0}^{\infty} S(\tau)\, e(t-\tau)\, d\tau$$

If the system is stable, the two integrals above are convergent and can be permutated.

$$E(y(t)) = \int_{-\infty}^{+\infty} S(\tau)\, E(e(t-\tau))d\tau$$

From this it is deduced that

$$E(y(t)) = K\, E(e(t)) \tag{40}$$

where K designates the static gain of the system

$$K = \int_{-\infty}^{+\infty} S(\tau)\, d\tau$$

**b) Autocorrelation spectrum and function of the output signal**

To define the autocorrelation spectrum, we can write:

$$\Phi_{yy}(s) = \int_{-\infty}^{+\infty} \underbrace{E\left[\left[\int_{-\infty}^{+\infty} s(\mu)\ e(t-\mu)\ d\mu \int_{-\infty}^{+\infty} e^{T}(t+\tau-\nu)\ S^{T}(\nu)\ d\nu\right]\right]}_{\varphi_{yy}(\tau)} e^{-\tau s} d\tau$$

By permutating the integrals and regrouping the terms, we obtain:

$$\boxed{\Phi_{yy}(s) = S(-s)\Phi_{ee}(s)\ S^{T}(s)} \qquad (41)$$

Using $\varphi_{ye}(\tau)$ to designate the autocorrelation function and $\Phi_{ye}(s)$ to designate the corresponding spectrum,

$$\varphi_{ye}(\tau) = E(y(t)\ e^{T}(t+\tau))$$

$$\Phi_{ye}(s) = \int_{-\infty}^{+\infty} \varphi_{ye}(\tau)\ e^{-s\tau}\ d\tau$$

The following results can be obtained in a similar way:

$$\boxed{\begin{aligned} \Phi_{ye}(s) &= S(-s)\ \Phi_{ee}(s) \\ \Phi_{ey}(s) &= \Phi_{ee}(s)\ S^{T}(s) \end{aligned}} \qquad (42)$$

Note

Before demonstrating the above results in an example, two very useful notes on the bilateral Laplace transform will be given.

The bilateral Laplace transform, denoted $L_{II}$, of a function $\varphi(t)$, is written

$$L_{II}(\varphi(t)) = \int_{-\infty}^{+\infty} \varphi(\tau)\ e^{-s\tau}\ d\tau$$

$$= \int_{0}^{+\infty} \varphi(\tau)\ e^{-s\tau} d\tau + \int_{0}^{+\infty} \varphi(-\tau)\ e^{s\tau}\ d\tau$$

$$= L_{I}\ (\varphi(\tau)) + [L_{I}\ \ (\Phi(-T))] \quad \text{s changed into -s}$$

where $L_I$ designates the monolateral Laplace transform. This result shows that $L_{II}$ can easily be expressed as a function of $L_I$.

In order to be able to use the following inversion formula for the bilateral transform, it is important to know its definition domain:

$$f(t) = \frac{1}{2\pi j} \int_{\alpha-j\infty}^{\alpha+j\infty} F(s)\, e^{st} ds$$

where the abscissa line $\alpha$ is within the definition domain.

**c) Example of application**

Given a white noise, e(t) with a mean of zero, passing through a linear system with a transfer function S(s), the objective is to find the standard deviation of the output signal y.

It is known that

$$E(e(t)) = 0$$

$$\varphi_{ee}(\tau) = \delta(\tau)$$

and

$$S(s) = \frac{\sqrt{10}\; s + 4}{(s+1)(s+2)}$$

the mean of the output signal is zero according to equation (40) where $K = \sqrt{10}$.

Using equation (41), the autocorrelation spectrum of the output is obtained:

$$\Phi_{yy}(s) = \frac{(-\sqrt{10}s+4)\;(\sqrt{10}s+4)}{(-s+1)(-s+2)(s+1)(s+2)} = \frac{-10\; s^2 + 16}{(1-s^2)\;(4-s^2)}$$

Factorisation of $\Phi_{yy}(s)$ gives:

$$\Phi_{yy}(s) = \left(\frac{1}{s+1} + \frac{2}{s+2}\right) + \left(\frac{1}{1-s} + \frac{2}{2-s}\right)$$

It is then easy to deduce $\varphi_{yy}(\tau)$ and in particular the variance:

$$\sigma_y^2 = \varphi_{yy}(o) - m_y^2$$

According to the note above, we can write

$$\Phi_{yy}(s) = \int_0^\infty \varphi_{yy}(\tau)\, e^{-\tau s}\, d\tau + \int_0^\infty \varphi_{yy}(-\tau)\, e^{\tau s}\, d\tau$$

$$= F(s) + F(-s)$$

Consequently,

$$F(s) = \frac{1}{s+1} + \frac{2}{s+2}$$

By applying the initial theorem,

$$\varphi_{yy}(o) = \lim_{s \to \infty} sF(s) = 3$$

and therefore

$$\sigma_y = \sqrt{3}$$

### 1.4.2 - DISCRETE RANDOM SIGNALS

$\{e(n\Delta t)\}$ is an input signal of a discrete linear system with a transfer function matrix S(z) and $\{y(n\Delta t)\}$ designates the corresponding output signal.

With the notation introduced for the continuous case, the bilateral z transform, denoted $Z_{II}$, of the sequence $\{\varphi_{ee}(k\Delta t)\}$ is defined by the relationship

$$\Phi_{ee}(z) = \sum_{k=-\infty}^{k=+\infty} \varphi_{ee}(k\Delta t)\, z^{-k} \tag{43}$$

As in the deterministic case (relationship (10)) it is demonstrated that the inverse transformation relationship is written thus:

$$\varphi_{ee}(k\Delta t) = \frac{1}{2\pi j} \int_{\Gamma} \Phi_{ee}(z)\, z^{k-1}\, dz \tag{44}$$

where $\Gamma$ represents a closed contour within the definition domain of $\Phi_{ee}(z)$.

The standard transformation formulae are given below:

$$\boxed{\begin{aligned} \Phi_{yy}(z) &= S(z^{-1})\, \Phi_{ee}(z)\, S^T(z) \\ \Phi_{ye}(z) &= \Phi_{ee}(z)\, s^T(Z) \\ \Phi_{ey}(z) &= S(z^{-1})\, \Phi_{ee}(z) \end{aligned}} \tag{45}$$

### 1.4.3 – APPROXIMATION OF A CONTINUOUS RANDOM SIGNAL BY A DISCRETE RANDOM SIGNAL

The white noise e(t) is defined by

$$E(e(t)) = 0$$

$$\varphi_{ee}(\tau) = M\ \delta(\tau)$$

and therefore $\Phi_{ee}(s) = M$

A pseudo-white noise $\{d(n\Delta t)\}$ is defined by the following relationships:

$$E(d(n\Delta t)) = 0$$

$$\varphi_{dd}(n\Delta t) = N\ \delta(n\Delta t) \rightarrow \Phi_{dd}(z) = N$$

#### 1.4.3.1 – Approximation by the transfer matrix method

Given a white noise e(t) passing through a linear system H(s), let y(t) designate the corresponding continuous output noise having the following second order characteristic:

$$\Phi_{yy}(s) = H(-s)\ M\ H^T(s)$$

The continuous random signal y is to be approximated by a discrete random signal r (Fig. 18):

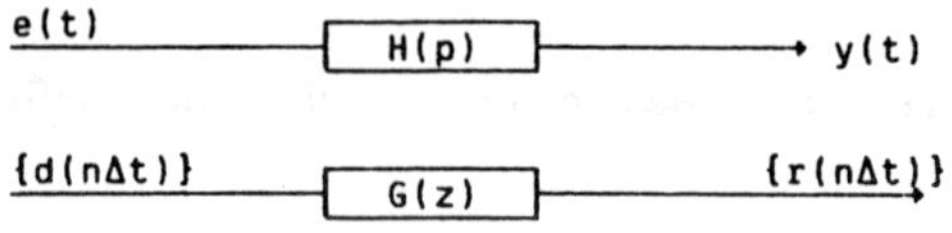


Fig. 18

The sequence r is said to approximate the signal y if the second order statistic of both signals is the same at the instants of sampling. This condition is written thus:

$$G(z^{-1})\ N\ G^T(z) = Z_{II}[H(-s)\ M\ H^T(s)] \qquad (46)$$

Application

Assume that

$$\Phi_{yy}(s) = \frac{\omega_o^2}{\omega_o^2 - s^2}$$

and the sequence $\{r(n\Delta t)\}$ such that $\Phi_{rr}(z) = Z_{II}(\Phi_{yy}(s))$ is to be found. It is known that

$$\Phi_{rr}(z) = Z_{II}(\Phi_{yy}(s)) = \frac{0.5\ (1 - e^{-2\omega_o \Delta t})}{(1 - e^{-\omega_o \Delta t} z^{-1})\ (1 - e^{-\omega_o \Delta t} z)}$$

$$= G(z)\ G(z^{-1})$$

where

$$G(z) = \frac{\sqrt{0.5(1-e^{-2\omega_o \Delta t})}}{(1 - e^{-\omega_o \Delta t} z^{-1})}$$

Therefore the sequence $\{r(n\Delta t)\}$ is given by the following recursion:

$$r_n = e^{-\omega_o \Delta t}\ r_{n-1} + \sqrt{0.5\ \omega_o (1-e^{2\omega_o \Delta t})}\ d(n\Delta t)$$

### 1.4.3.3 – Approximation by the state vector method

The continuous linear system is assumed to be subjected to a white noise $e(t)$ and described by the following state equations:

$$\dot{x}(t) = Ax(t) + e(t)$$
$$y(t) = Cx(t) \tag{47}$$

with

$$\varphi_{ee}(\tau) = M\ \delta(\tau)$$

The aim is to find a sequence $\{r(n\Delta t)\}$ in the discrete state vector form subjected to a pseudo-white noise $\{d(n\Delta t)\}$:

$$x((n+1)\Delta t) = Fx(n\Delta t) + d(n\Delta t)$$
$$r(n\Delta t) = H\ x(n\Delta t) \tag{48}$$

where

$$\varphi_{dd}(\tau) = N\ \delta(\tau)$$

Equations (47) can be used to calculate $y(n+1)\Delta t)$ as a function of $y(n\Delta t)$ by using relationship (23):

$$x(t+\Delta t) = e^{A\Delta t}\, x(t) + \int_0^{\Delta t} e^{A(\Delta t-\tau)}\, e(\tau)\, d(\tau)$$

$$y((n+1)\Delta t) = C\, e^{A\Delta t}\, x(n\Delta t) + C \int_0^{\Delta t} e^{A(\Delta t-\tau)} e(\tau)\, d(\tau)$$

Identification with relationships (48) gives:

$$\begin{aligned} F &= e^{A\Delta t} \\ H &= C \\ d &= \int_0^{\Delta t} e^{A(\Delta t-\tau)}\, e(\tau)\, d(\tau) \end{aligned} \tag{49}$$

The value of N can now be calculated:

$$N = E(d_n^2) = E \int_0^{\Delta t} \int_0^{\Delta t} e^{A(\Delta t-\tau)} e(\tau)\, e(\tau')\, e^{A^T(\Delta t-\tau')} d\tau\, d\tau'$$

and therefore

$$N = \int_0^{\Delta t} e^{A\tau}\, M\, e^{A^T\tau}\, d\tau \tag{50}$$

It is now necessary to calculate N by algebra. It is known that

$$AN = \int_0^{\Delta t} A\, e^{A\tau}\, M\, e^{A^T\tau}\, d\tau$$

$$= \left[e^{A\tau}\, M\, e^{A^T\tau}\right]_0^{\Delta t} - \int_0^{\Delta t} e^{A\tau}\, M\, e^{A^T\tau}\, A^T\, d\tau$$

therefore

$$AN = FMF^T - M - NA^T$$

N is then found by solving the following Lyapunov equation:

$$AN + NA^T = FMF^T - M \tag{51}$$

Finally, the discrete noise is provided by the state equation (48), where

$$\left[\begin{array}{l} F = e^{A\Delta t} \\ H = C \\ N \text{ solution of (51)} \end{array}\right.$$

## *1.5 - CONCLUSION*

The principal definitions and basic relationships required for the study of discrete signals and systems have been given in this introductory chapter. The concepts introduced in the first two sections will be used intensively in the rest of the book. The calculations will frequently make use of the Z transform without reference to the real structures with computers and extrapolators. The random aspect introduced in the last section provides the basis for the stochastic interpretation of the criteria which will be minimised in Chapter 5.

### *REFERENCES*

R. Boudarel, J. Delmas, and P. Guichet, Commande optimal des processus, Editions Dunod, 1969.

A. Fossard, Commande des systèmes multidimensionnels, Editions Dunod, 1972.

E.I. Jury, Theory and applications of the z-transform method, John Wiley and Sons, 1964.

M. Labarrère, J.P. Krief, and B. Gimonet, Le filtrage et ses applications, Editions CEPADUES, 1978.

Pierre, Kolb, Concerning the Laplace transform of sampled signals, IEEE on automatic control, April 1964.

Y. Sevely, Systèmes et asservissements linéaires échantillonnés, Editions Dunod, 1973

## *CHAPTER 2*

# *FROM CONTINUOUS CONTROL TO DISCRETE CONTROL*

## *2.1 - SYNTHESIS OF CONTINUOUS LAWS*

As indicated in the introduction, recent developments in data processing naturally lead to the digitisation of control laws. However, many systems may be synthesised in the continuous domain, the digitisation of the continuous laws being the final stage. It therefore appeared important to us to provide a chapter on this type of method in a book orientated towards digital control.

In the first part of the chapter, the most up-to-date methods of continuous synthesis are presented, while the techniques of digitisation (transformation from continuous to discrete) are described in the second part.

No attempt is made here to produce an exhaustive survey of all the continuous methods. We have restricted ourselves to a description, within the general framework of multivariable systems, of certain techniques which are particularly important from a practical viewpoint, without detailing the numerous theoretical developments of these techniques.

The classification of the range of methods proposed in the literature remains highly problematic. There is a conventional distinction between frequency-based and time-based methods; only the latter will be discussed in this book on digital control.

For these methods, the following classification has been adopted:

* **Modal methods**: this term is used to signify the set of methods dependent on an a priori choice of eigenvalues, the synthesis being carried out by assigning the corresponding eigenvectors in order to attain one or more objectives. An algebraic study (zero calculation) can facilitate the a priori choice of these eigenvalues.

* **Optimal methods**: this term designates methods dependent on an a priori choice of one or more criteria to be minimised. Most optimal methods make use of nonlinear programming. Some details of linear quadratic and Gaussian linear quadratic commands, obtained by algebraic optimisation (without nonlinear programming) will be given.

However, this classification remains ambiguous. In modal commands, the choice of the eigenvalues may be governed by the minimisation of a time criterion; moreover, the time formulation adopted here may often be translated into a frequency approach. In fact, there are links between the time and frequency analyses; the Parseval relation, for instance, is the transformation of time criteria into frequency terms.

## 2.1.1 - MODAL METHODS

### 2.1.1.1 - General

Consider a deterministic linear system described by its continuous state vector equation:

$$\dot{x} = A\,x + B\,u \qquad x \in R^n$$
$$y \in R^p \qquad (1)$$
$$y = C\,x \qquad u \in R^q$$

It is required that the set of eigenvalues of the **closed loop** system should comprise the self-conjugate* set of complex numbers specified in advance
$\{\lambda_1, \lambda_2, \ldots, \lambda_n\}$. In other words, the matrix for the feedback of the output K (u = K y) is to be found, such that

$$\sigma(A + BKC) = \{\lambda_1, \lambda_2, \ldots, \lambda_n\}^{**} \qquad (2)$$

The importance of such a condition is evident. To ensure stability and certain performance levels required by the schedule of specifications (precision, response time, damping, etc.) it is necessary, but not sufficient, to assign the eigenvalues in a closed loop system.

Before a possible solution of this problem is given, some results will be stated. In a single-variable case (within the set of single-input systems) the use of state feedback (u = Kx) makes it particularly simple to solve the problem of the assignment of the eigenvalues (2). The interested reader can refer to the extensive literature on this subject, e.g. Fossard (1972). In this case, the controllability of the system gives rise to a unique

---

* $\Delta$ is described as self-conjugate if $\lambda \in \Lambda \rightarrow \bar{\lambda} \in \Lambda$.

** $\sigma(A)$ is the set of eigenvalues of matrix A.

solution. For a system with multiple inputs, the same problem has an infinite number of solutions, and the objective is to define one solution which is particularly useful. In this section, a method generating an output feedback satisfying the relationship (2) will be described, while the infinite number of possible solutions will be demonstrated. The choice of a solution to achieve certain other control objectives will not be discussed. It should also be noted that the assignment of poles by output feedback (incomplete state feedback) for a single-variable system has no general solution; if a solution exists in a particular case, it is unique (provided that the system is controllable and observable). However, such systems can be dealt with by the method shown below, by making them artificially multivariable by adding a dynamic compensator (see relationships (4) and (5) below).

It will be assumed from now on that the number of inputs and outputs is greater than the number of states, i.e.

$$n < q + p \tag{3}$$

If this relationship is not true, the order of the system is increased by using a dynamic compensator of order r in the following form:

$$\begin{bmatrix} \dot{x} \\ \dot{x}_c \end{bmatrix} = \begin{bmatrix} A & 0 \\ 0 & 0 \end{bmatrix} \begin{bmatrix} x \\ x_c \end{bmatrix} + \begin{bmatrix} B & 0 \\ 0 & I_r \end{bmatrix} \begin{bmatrix} u \\ u_c \end{bmatrix}$$
$$\begin{bmatrix} y \\ y_c \end{bmatrix} = \begin{bmatrix} C & 0 \\ 0 & I_r \end{bmatrix} \begin{bmatrix} x \\ x_c \end{bmatrix} \tag{4}$$

In these conditions, the dimensions n, p and q become n+r, p+r and q+r. It is then only necessary to select the dimension r of the compensator such that

$$r > n - q - p \tag{5}$$

It is therefore not restrictive to assume that the inequality (3) is satisfied; if this is not the case, a dynamic compensator of a fairly high order r can be used to satisfy it. Note that the partitioning of the matrix of the gain K according to equation (2) shows that the system (A, B, C) is then regulated simultaneously by a fixed gain $K_{11}$ and by a dynamic compensator of order r in which the three gains $K_{21}$, $K_{22}$ and $K_{12}$ appear (see Fig. 1).

In fact,

$$\begin{bmatrix} u \\ u_c \end{bmatrix} = \begin{bmatrix} K_{11} & K_{12} \\ K_{12} & K_{22} \end{bmatrix} \begin{bmatrix} y \\ y_c \end{bmatrix}$$

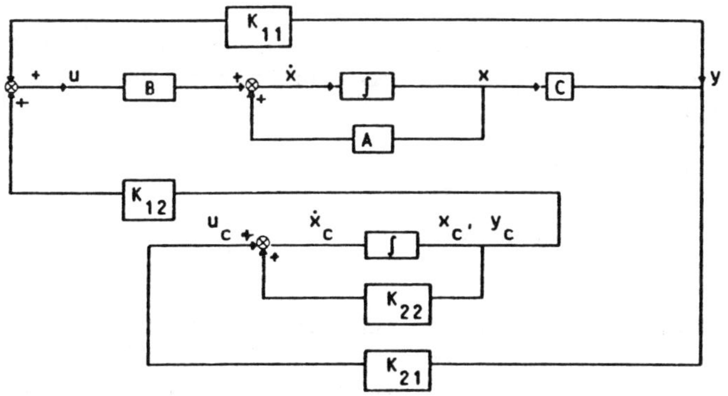


Fig. 1

2.1.1.2 - **Assignment of the eigenvalues**

Given the system (1) satisfying the inequality (3), the aim is to find a matrix for the output feedback K satisfying condition (2). For this purpose an algorithm whose details can be found in Fletcher (1981) is used.

Instead of an exact description of this algorithm, the principal concepts used in it will be given, but, in order to make it easier to understand, the different stages will be shown in detail with reference to an example.

Let the system be of the closed loop type, governed by the following system:

$$\begin{aligned} \dot{x} &= (A + B\ K_o\ C)\ x + B\ u \\ y &= C\ x \end{aligned} \qquad (6)$$

The choice of the matrix $K_o$ has already enabled $q_1 + q_2$ self-conjugate eigenvalues to be positioned in such a way that

a) - $\Lambda_1$ is a set of $q_1$ self-conjugate eigenvalues corresponding to the eigenvectors to the right $[v_1,..., v_{q1}]$ denoted $V_1$;

b) - $\Lambda_2$ is a set of $q_2$ self-conjugate eigenvalues corresponding to the eigenvectors to the left $[u_1,..., u_{q1}]$ denoted $U_2^T$.

Note that $x_1$ and $x_2$ are two diagonal matrices whose eigenvalues are the elements of $\Lambda_1$ and $\Lambda_2$ respectively. According to the definition of $V_1$ and $U_2$,

$$(A + B K_o C) V_1 = V_1 X_1 \tag{7}$$

$$U_2^T (A + B K_o C) = X_2 U_2^T \tag{8}$$

The method of the algorithm will now be described; it consists in assigning a $(q_1 + q_2 + 1)$th real eigenvalue $\lambda$ while maintaining the eigenvalues $\Lambda_1 \cup \Lambda_2$ together with their associated eigenvectors (to the right or left). This method will be justified subsequently by the proposition which will be stated below.

Description of the method

1/ – The first stage is the calculation of the matrices $D_1$ and $D_2$ such that

$$D_1 C V_1 = 0 \quad \text{where } D_1 \in \mathbf{R}^{(p-q_1)xp} \quad \text{and } D_1 \text{ is of maximum rank} \tag{9}$$

$$U_2^T B D_2 = 0 \text{ where } D_2 \in \mathbf{R}^{qx(q-q_2)} \quad \text{with } D_2 \text{ of maximum rank} \tag{10}$$

2/ – The vectors v and w are then chosen in such a way that the following relationship is satisfied:

$$(A + B K_o C - \lambda I, BD_2) \begin{bmatrix} v \\ w \end{bmatrix} = 0 \tag{11}$$

– A matrix G is calculated in such a way that

$$G D_1 C v = w \quad \text{where } G \in \mathbf{R}^{(q-q_2)x(p-q_1)} \tag{12}$$

– finally, the required gain $K_1$ can be found:

$$K_1 = K_o + D_2 G D_1$$

3/ – Alternatively, vectors u and s are chosen in such a way that the following relationship is satisfied:

$$[u^T \; s^T] \begin{bmatrix} A+BK_oC-\lambda I \\ D_1C \end{bmatrix} = 0 \tag{13}$$

- A matrix G' is then calculated in such a way that

$$u^T B D_2 G' = s^T \text{ where } G' \in \mathbf{R}^{(q-q_2)\times(p-q_1)}$$

- finally, the required gain $K_1'$ is found from the equation

$$K_1' = K_o + D_2 G' D_1 \qquad (14)$$

The following basic proposition can now be stated:

<u>Proposition</u>

If the vector v is calculated from equations (11) and (12), the preceding procedure is such that:

1/ – $|V_1, v|$ represent the eigenvectors to the right of the matrix $A+BK_1C$, associated with the eigenvalues $\Lambda_1 \cup \lambda$;

2/ – $|U_2|$ are the eigenvectors to the left of the matrix $A+BK_1C$, associated with the eigenvalues $\Lambda_2$.

The second part of this proposition may be formulated thus: if the vector u is calculated from equations (13) and (14), the following procedure is such that:

1/ – $|V_1|$ represent the vectors to the right of the matrix $A+BK_1'C$, which correspond to the eigenvalues $\Lambda_1$;

2/ – $|U_2, u|$ are the eigenvectors to the left of the matrix $A+BK_1'C$.

<u>Proof</u>

We may write

$$(A+BK_1C)\, V_1 = (A+BK_oC)\, V_1 + B\, G_2\, G\, D_1\, C\, V_1$$

Using equations (7) and (9), we obtain

$$(A+BK_1C)\, V_1 = V_1\, X_1$$

This relationship shows that the gain $K_1$ ensures that the right-hand eigenvalues $V_1$ are retained.

Furthermore,

$$U_2^T (A+BK_1C) = U_2^T (A+BK_oC) + U_2^T BD_2GD_1C$$

Applying equations (8) and (10),

$$U_2^T\ (A+BK_1C) = X_2\ U_2^T$$

Consequently, the left-hand eigenvectors $U_2$ are conserved. According to equations (11) and (12), we may then write

$$(A+BK_1C)\ v = (A+BKoC)\ v + BD_2GD_1Cv$$

$$(A+BK_1C)\ v = (A+BKoC)\ v + BD_2w$$

$$(A+BK_1C)\ v = \lambda\ v$$

The last relationship shows that, for this chosen gain $K_1$, the vector v represents a right-hand eigenvector associated with the eigenvalue $\lambda$. This result therefore completes the proof of the proposition.

The second part of the proposition is proved in a similar way.

Note

In order to assign a complex eigenvalue $\lambda$, the conjugate eigenvalue $\bar{\lambda}$ must be assigned simultaneously by replacing relationships (12) and (14) with relationships (12') and (14') below, which determine the matrices G and G'.

Matrix G is obtained from the relationship

$$G\ D_1\ C\ (v\ \bar{v}) = (w\ \bar{w})\ ; \quad G \in \mathbf{R}^{(q-q_2)\times(p-q_1)} \tag{12'}$$

or alternatively the following relationship defines matrix G':

$$\begin{bmatrix} u^T \\ \bar{u}^T \end{bmatrix} BD_2G' = \begin{bmatrix} s^T \\ \bar{s}^T \end{bmatrix} \quad ; \quad G' \in \mathbf{R}^{(q-q_2)\times(p-q_1)} \tag{14'}$$

Description of the algorithm

The algorithm for the assignment of the eigenvalues by an output feedback enables them to be assigned in succession while preserving those which have already been assigned.

The algorithm is initialised by specifying

$$K_o = 0$$

$$D_1 = D_2 = I$$

According to the definition of matrices $D_1$ and $D_2$, the following conditions must be true at each stage:

$$p > q_1 \quad \text{and} \quad q > q_2$$

If a complex eigenvalue, associated with two right-hand eigenvectors v and $\overline{v}$, is to be assigned, it is preferable for the matrix $D_1C\,|v\ \overline{v}|$ to be of a rank higher than or equal to 2. This condition requires that

$$\text{rank } D_1 \geqslant 2 \longrightarrow p > q_1 + 1$$

Otherwise the relationship (12') would not provide a solution for the matrix G. Evidently, this condition is neither necessary nor sufficient from a strictly theoretical viewpoint. The details of this will not be pursued as this would be too time-consuming (see Fletcher (1979)).

These considerations confirm relationship (3). It will be found that:

a) If the last real eigenvalue is to be assigned, then n−1 eigenvalues must already have been assigned. Therefore

$$q_1 + q_2 = n-1$$

Since the above discussion showed that $p>q_1$ and $q>q_2$, then

$$q_1 + q_2 < p + q - 1$$

and therefore

$$n < p + q$$

b) If the last two complex values are to be assigned, this implies that n−2 eigenvalues have already been positioned. Therefore

$$q_1 + q_2 = n-2$$

Now, according to the discussion above,

$$p > q_1 + 1 \text{ or } q > q_2 + 1$$

The last two inequalities, when combined with the preceding inequality, lead to inequality (3):

$$n < p + q$$

Note that a rigorous proof of this algorithm will be highly complex, since it is necessary to ascertain, in particular, the existence of a vector v satisfying (11) such that:

$$C\,v \neq 0$$

This vector must also be such that this choice is still possible in the final stages.

### 2.1.1.3 – Example of application

To illustrate the above theoretical results, consider the system defined by the following matrices $A_o$, $B_o$ and $C_o$:

$$A_o = \begin{bmatrix} 0 & 0 \\ 0 & -1 \end{bmatrix} \qquad B_o = \begin{bmatrix} 1 \\ 1 \end{bmatrix} \qquad C_o = [1,\ 1]$$

$$n_o = 2 \qquad q_o = p_o = 1$$

This system does not comply with relationship (3). It is therefore necessary to use a dynamic compensator of order r complying with relationship (5) and having the dimension r = 1.

The expanded system with the compensator is written thus:

$$A = \begin{bmatrix} 0 & 0 & 0 \\ 0 & -1 & 0 \\ 0 & 0 & 0 \end{bmatrix} \qquad B = \begin{bmatrix} 1 & 0 \\ 1 & 0 \\ 0 & 1 \end{bmatrix} \qquad C = \begin{bmatrix} 1 & 1 & 0 \\ 0 & 0 & 1 \end{bmatrix}$$

The aim is to find an output feedback such that the three eigenvalues of the loop system are $\lambda_1 = -1$, $\lambda_2 = -2$ and $\lambda_3 = -3$.

Stage One

To assign the eigenvalue $\lambda_1 = -1$, a vector $v_1$ is chosen to satisfy equation (11) where $K_o = 0$, since no eigenvalue has yet been assigned. The initial choice $D_1 = D_2 = I$ therefore results in the following relationship:

$$(A+I \ \ B) \begin{bmatrix} v_1 \\ w_1 \end{bmatrix} = 0$$

An example of a solution of this equation is:

$$v_1 = \begin{bmatrix} 0 \\ -1 \\ -1 \end{bmatrix} \qquad w_1 = \begin{bmatrix} 0 \\ 1 \end{bmatrix}$$

To obtain the matrix $G_1$, the following equation is solved

$$G_1 \ C \ v_1 = w_1$$

with the possible solution:

$$G_1 = \begin{bmatrix} 0 & 0 \\ 0 & -1 \end{bmatrix}$$

Finally, relationship (12) is used to obtain the feedback gain matrix:

$$K_1 = G_1 = \begin{bmatrix} 0 & 0 \\ 0 & -1 \end{bmatrix}$$

Stage Two

The second eigenvalue $\lambda_2 = -2$ is to be assigned while retaining the preceding eigenvalue and eigenvector.

According to the previous stage:

$$A + BK_1C = \begin{bmatrix} 0 & 0 & 0 \\ 0 & -1 & 0 \\ 0 & 0 & -1 \end{bmatrix}$$

We shall specify a unilinear matrix $D_1$ such that

$$D_1 \ C \ v_1 = 0 \qquad \text{i.e. } D_1 = (1 -1)$$

The vectors $u_2$ and $s_2$ must satisfy the following condition:

$$(u_2^T \ \ s_2^T) \begin{bmatrix} (A+BK_1C)^T+2I \\ D_1 \ C \end{bmatrix} = 0$$

We may, for example, obtain the following solution

$$u_2 = \begin{bmatrix} -1 \\ -2 \\ -2 \end{bmatrix} \qquad \text{and} \qquad s_2 = (2)$$

Finally, relationships (14) can be used to obtain $G_2'$ and the gain matrix $K_2$:

$$u_2^T \ B \ G_2' = s_2^T \longrightarrow G_2' = \begin{bmatrix} 0 \\ 1 \end{bmatrix}$$

$$K_2 = K_1 + G_2' \ D_1 = \begin{bmatrix} 0 & 0 \\ 1 & -2 \end{bmatrix}$$

Stage Three

On completion of the second stage, the closed loop matrix is written thus:

$$A + BK_2C = \begin{bmatrix} 0 & 0 & 0 \\ 0 & -1 & 0 \\ 1 & 1 & -2 \end{bmatrix}$$

The vector $D_1$ conserves the eigenvector $v_1$ and the eigenvalue $\lambda_1 = -1$. To retain the vector $u_2$ and the eigenvalue $\lambda_2 = -2$, a vector $D_2$ must be found such that

$$u_2^T \; B \; D_2 = 0$$

and therefore

$$D_2 = \begin{bmatrix} 2 \\ 3 \end{bmatrix}$$

The vectors $v_3$ and $w_3$ must comply with the following relationship:

$$(A + B\,K_2\,C,\; B_2 D) \begin{bmatrix} v_3 \\ w_3 \end{bmatrix} = 0$$

and therefore

$$v_3 = \begin{bmatrix} -2 \\ -3 \\ -4 \end{bmatrix} \quad \text{and} \quad w_3 = [3]$$

Equations (12) are used to determine the final gain matrix $K_3$:

$$G_3\;D\;C\;v_3 = w_3 \longrightarrow G_3 = -3$$

$$K_3 = K_2 + D_2\,G_3\,D_1 = \begin{bmatrix} -6 & 6 \\ -8 & 7 \end{bmatrix}$$

It is then possible to check that the output feedback $K_3$ which has been found allows the closed loop system to have the three eigenvalues −1, −2, and −3. These three values are eigenvalues of the following matrix:

$$A + BK_3C = \begin{bmatrix} -6 & -6 & +6 \\ -6 & -7 & +6 \\ -8 & -8 & 7 \end{bmatrix}$$

In this example it may be seen that the number of possible choices of the vector $v_1$ is infinite.

### 2.1.1.4 – **Assignment of eigenvectors**

In the algorithm above, the eigenvectors are chosen at random without taking the objectives of the control system into account. As emphasised before, the assignment of the eigenvectors for multivariable systems is often necessary if the system is to have good properties of stability, damping, and response time, but it is not sufficient. To achieve good performance, the degrees of freedom appearing in the choice of the eigenvectors relative to the chosen eigenvalues must be used. The description of methods of translating control objectives into eigenvalue terms would take too long; the reader should refer to the literature cited below. Moore (1976) is the first study of the assignment of eigenvalues, and deals with output modelling by state feedback. Sebakhy and Sorial (1977) and Moore, Wierzbicki, and Klein (1979) analyse the problems of insensitivity and the minimisation of a quadratic cost by state feedback respectively. The output feedback approach to these problems is the subject of Magni (1987).

2.1.1.5 - **Special cases**

When the state is entirely available (p = n) the above algorithm is simpler to use. The inequality (3) is satisfied and it is possible to proceed by only placing eigenvectors to the right.

When a numerical compensator of an order between n-p and n can be used, the observer theory found in the standard literature can be applied to avoid using the above algorithm.

## 2.1.2 - LINEAR QUADRATIC OPTIMAL METHODS

The continuous methods described in this section will be developed further in the discrete techniques using state variables which will be discussed in Chapter 5.

2.1.2.1 - **Formulation of the problem**

Consider the linear system (15):

$$\begin{cases} \dot{x} = A\,x + B\,u + w \\ y = C\,x + v \end{cases} \qquad (15)$$

where the initial state $x_0$ is a random variable with a mean of zero and a variance $x_0$, and w and v are stationary centred white noises with no correlation between them (and between them and $x_0$), with intensity M and N.

The following criterion is to be minimised (16):

$$J = \text{Esp} \int_0^T [x^T(t),\ u^T(t)] \begin{bmatrix} Q & S \\ S^T & R \end{bmatrix} \begin{bmatrix} x(t) \\ u(t) \end{bmatrix} dt \qquad (16)$$

2.1.2.2 - **Optimal solutions**

If the following hypotheses are added:

v and w are Gaussian noises
N and R are invertible matrices

then it can be shown by means of fairly lengthy calculations (Davis (1977)) that the problem can be divided into a number of parts. The optimal solution can be found by determining:

a) the optimal observer in the form (17)

$$\hat{x} = A\hat{x} + Bu + K' (y - C\hat{x}) \tag{17}$$

where K' is defined by the Riccati differential equation:

$$AP' + P'A^T + M - P'CNC^TP' = \dot{P}'$$

and the relationship

$$K'(t) = P'(t)\, C^T N^{-1}$$

b) the optimal control of the system x = Ax + Bu+w, assuming that x is completely known. This optimal control is defined by relationship (18)

$$u = -K(t)\, x \tag{18}$$

where

$$K(t) = -R^{-1}\, B^T\, P(t)$$

where P(t) is the solution of the following Riccati differential equation:

$$PA + A^TP + Q - PB^T\, RBP = \dot{P}$$

On the hypothesis that the integration horizon T is infinite (which assumes that A is stable), the optimal solution is obtained by the solution of Riccati algebraic equations (19) and (20):

$$AP' + P'\, A^T + M - P'CNC^TP' = 0 \tag{19}$$

$$PA + AP + Q - PB^TRBP = 0 \tag{20}$$

and the optimal constant gains are given by relationships (21) and (22):

$$K' = P'\, C^TN^{-1} \tag{21}$$

$$K = R^{-1}\, B^T\, P \tag{22}$$

Additionally, if the whole of the state is measured without disturbance, the optimal observer is clearly superfluous. It can be shown by means of frequency considerations that this type of control (linear quadratic) is very robust, which is significantly less true in the general case (Gaussian linear quadratic). To mitigate the lack of robustness in the Gaussian linear quadratic problem, the observer may be simplified by replacing the optimal observer with an observer developed on the basis of the physical meaning and it is possible to attempt to approximate the solution which would have been obtained by the linear quadratic method.

2.1.2.3 – **Sub-optimal solutions**

The algebraic methods described above have a certain number of disadvantages:

if not all the states are exactly measurable, an observer, and consequently a system of complex dynamic compensation, must be brought into use;

the infinite horizon implies that the compensated system is stable, and this constraint may be unnecessary, particularly if the compensation is used to assist a human operator;

in the algebraic context, it is not possible to take into account the constraints on the compensator (limits, linear relationships,...).

We describe below a non-algebraic method of optimum seeking by algorithms of the gradient type which enables these constraints to be eliminated, by considering only the class of static controllers $u = -Ky$ (by extension to the class of dynamic compensators with fixed state dimensions).

**a) Expression of the criterion**

The criterion J on the trajectories (23)

$$x(t) = x_o e^{A_o t} + \int_0^t e^{A_o(t-\alpha)} (w(\alpha)-BKv(\alpha))d\alpha \quad (A_o = A-BKC) \tag{23}$$

of the compensated system may be written thus:

$$J = tr[X_o \int_o^T e^{A_o^T t} Q_1 e^{A_o t} dt] + tr[M \int_o^T \int_o^T e^{A_o^T \alpha} Q_1 e^{A_o \alpha} d\alpha\, dt]$$

$$+ tr[NK^T B^T \int_o^T \int_o^t e^{A_o^T \alpha} Q_1 e^{A_o \alpha} d\alpha\, dt\; BK] - 2tr[NK^T S_1^T \int_o^T e^{A_o t} dt\; BK]$$

$$+ T\, tr\,(NK^T RK)$$

where

$$Q_1 = Q - SKC - C^T K^T S^T + C^T K^T RKC$$

$$S_1^T = RKC - S^T$$

b) **Expression of the gradient of the criterion**

Developing $J(K+\delta K)$ to the first order, and using the following Belement formula:

$$e^{(A_o - B\delta KC)t} = e^{A_o t} - \int_o^t e^{A_o(t-\alpha)} B\delta KC \, e^{A_o \alpha} d\alpha$$

it may then be demonstrated that

$$\frac{1}{2}\frac{\partial J}{\partial K} = S_1^T \left[ \int_o^T e^{A_o \alpha} X_o \, e^{A_o^T \alpha} d\alpha + \int_o^T \int_o^t e^{A_o \alpha} M_1 \, e^{A_o^T \alpha} d\alpha \, dt \right] C^T$$

$$-B^T \left[ \int_o^T \left[ \int_o^{T-\alpha} e^{A_o^T u} Q_1 e^{A_o u} du \right] e^{A_o \alpha} X_o e^{A'_o \alpha} d\alpha \right.$$

$$\left. + \int_o^T \int_o^t \left[ \int_o^{t-\alpha} e^{A_o^T u} Q_1 e^{A_o u} du \right] e^{A_o \alpha} M_1 e^{A_o^T \alpha} d\alpha dt \right] C^T$$

$$+B^T \left[ \int_o^T \int_o^t e^{A_o^T \alpha} d\alpha \, dt \, S_1 + \int_o^T \int_o^t e^{A_o^T \alpha} Q_1 \, e^{A_o \alpha} d\alpha \, dt \, B \right] KN$$

$$+\left[ RN_1 B^T \int_o^T \int_o^t e^{A_o^T \alpha} d\alpha \, dt + B^T \int_o^T \int_o^t \left[ \int_o^{t-\alpha} e^{A_o^T u} du \right] S_1 N_1 \, B^T e^{A_o^T \alpha} d\alpha dt \right] C^T$$

$$+ \, TRKN$$

where $M_1 = M + BKNK^TB^T$ and $N_1 = KNK^T$

c) – **Calculation method**

The above expressions of the cost J and its gradient can be easily calculated, despite their complexity, if the analytic expression of the integrals in the following form can be determined:

$$R_L(T) = \int_o^T \frac{(T-t)^L}{L!} e^{At} Q \, e^{Bt} dt$$

where A, B and Q are any matrices of compatible dimensions.

It may be shown (Steer (1982)) that if $L_L(A,B,Q)$ is the following Lyapunov equation system:

$$\left[ \begin{array}{l} AX_{-L} + X_{-L}B = X_{1-L} \\ \quad\cdots \\ AX_o + X_oB = X_1 - Q \\ AX_1 + X_1B = X_2 \\ \quad\cdots \\ AX_r + X_rB = 0 \end{array} \right.$$

then:

- $L_L(A,B,Q)$ can be solved;

- for any solution $\{X_{-1},\ldots,X_r\}$ of $L_L(A,B,Q)$,

$$R_L(T) = -e^{AT}X_{-L}\,e^{BT} + \sum_{i=-L}^{r} X_i \frac{T^{L+i}}{(L+i)!}$$

The systems of Lyapunov equations can be solved by means of an algorithm based on the one given in Bartels and Stewart (1972).

d) **Extension to the dynamic compensator**

The above procedures are also applicable to the determination of a fixed-order dynamic compensator described by

$$\begin{aligned} \dot{Z}(t) &= FZ(t) + GY(t) \\ U(t) &= HZ(t) + LY(t) \\ Z(0) & \end{aligned}$$

It should be noted that the compensated system can be written in the form

$$\dot{\zeta} = \begin{bmatrix} A & 0 \\ 0 & 0 \end{bmatrix} \zeta + \begin{bmatrix} B & 0 \\ 0 & 1 \end{bmatrix} u + \begin{bmatrix} W \\ 0 \end{bmatrix}$$

$$y = \begin{bmatrix} C & 0 \\ 0 & 1 \end{bmatrix} \zeta$$

$$u = \begin{bmatrix} F & G \\ H & L \end{bmatrix} y$$

where $\zeta$ is the increased state:

$$\zeta = \begin{bmatrix} X \\ Z \end{bmatrix} \qquad \zeta_o = \begin{bmatrix} X_o \\ Z_o \end{bmatrix}$$

This characterises a dynamic system compensated by static negative output feedback.

## 2.1.3 – CONTROL OF SYSTEMS WITH TIME DELAYS

### 2.1.3.1 – Introduction

This section is concerned with the control of continuous systems having time delays, which are also called dead times. The specificity of such systems requires an appropriate method of synthesis. The time delays may arise from:

- the system being controlled
- the actuators
- the sensors.

The last two cases correspond to the delayed control and delayed measurement cases respectively.

The modelling of many industrial processes includes time delays. Without being exhaustive, we may cite the processes of paper and steel production, thermal processes (drying kilns), distillation processes, and in general those systems including the transport of materials or chemical reactions. Certain biological, economic and political systems may be added to this list.

The basic block diagram of the system to be controlled is given in Fig. 2a:

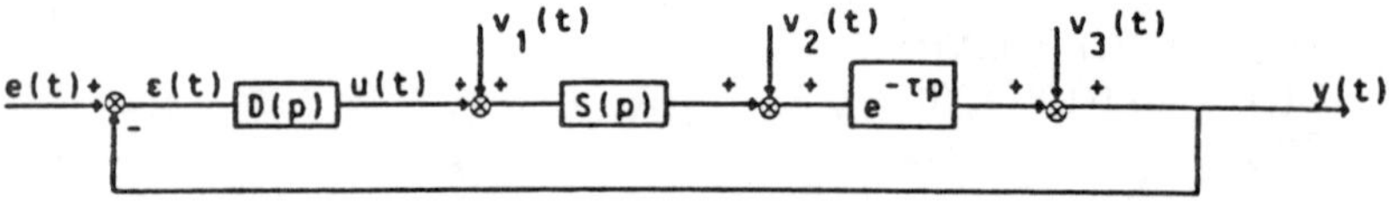


Block diagram with continuous corrector

Fig. 2a

* e(t), $\epsilon$(t), u(t) and y(t) represent the reference, the error, the control signal and the output of the process.

* D(s), S(s) and $e^{-\tau s}$ denote the transfer functions of the continuous controller, the system to be controlled, and the time delay respectively ($\tau$ being expressed in seconds).

* $v_1(t)$, $v_2(t)$ and $v_3(t)$ represent disturbances acting at the level of the control, the system to be controlled, and the measurement respectively. They may be deterministic (such as a constant bias) or random in nature (such as white or coloured noise). A more detailed study of the effect of a disturbance $v_1$ of the impulsive type, corresponding to the introduction of initial conditions in the system, and a deterministic or random disturbance $v_3$ will be made subsequently and in Chapter 3.

It should be noted that Figure 2a specifies the position of the time delay. If the delay occurs at the control level, it is simply necessary to interchange blocks S(s) and $e^{-\tau s}$.

The synthesis of a continuous controller based on the method of the Smith predictor will be presented first in the noise-free case. Subsequently the effects of modelling errors and disturbances will be examined.

### 2.1.3.2 - The Smith's predictor method

The Smith's method is based on the following principles:

1/ - The time delay is located outside the loop (see Fig. 2b) and the control methods (modal methods, for example) are applied to determine the form of the continuous controller $D_c(s)$ in such a way as to comply with the specifications (for degrees of stability, gain and phase margins, overshoot, response time at 5%, etc.) for the delay-free system.

2/ - The controller obtained in this way is unrealisable, since it uses an output of the system which is not accessible. To find the controller D(s), the closed loop transfer functions of the systems shown in Figs. 2a and 2b are equalised. This gives

$$\frac{D\ S\ e^{-\tau s}}{1+D\ S\ e^{-\tau s}} = \frac{D_c\ S\ e^{-\tau s}}{1 + D_c\ S} \tag{25}$$

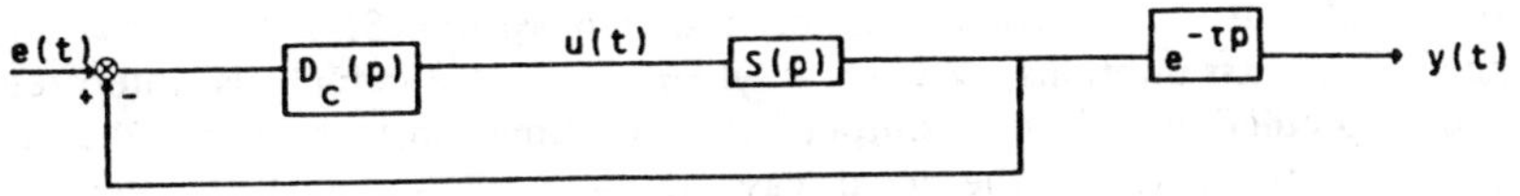


Block diagram with time delay outside the loop

Fig. 2b

Equation (25) is used to find the controller D:

$$D = \frac{D_c}{1 + D_c\ S(1-e^{-\tau s})} \tag{26}$$

The block diagram of the system with the controller D may then be represented as in Fig. 3:

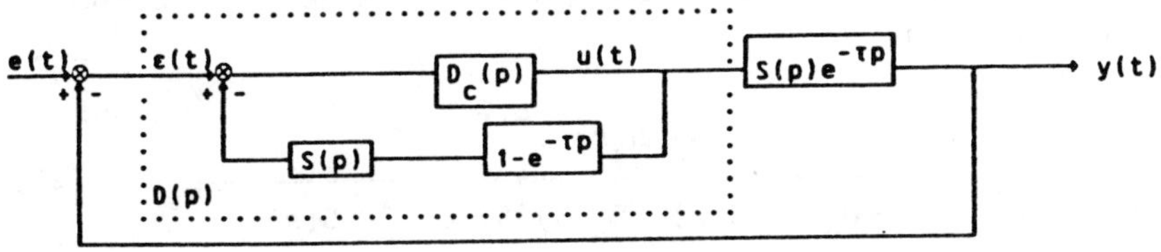


Smith controller

Fig. 3

A particularly interesting block diagram, equivalent to that of Fig. 3 when there are no modelling errors, is shown in Fig. 4; $(S, \bar{\tau}) = (\bar{S}, \tau)$ where $(\bar{S}, \bar{\tau})$ represents the approximate model while $(S, \tau)$ denotes the "exact model" of the system to be controlled.

The effect of a modelling error, $(\bar{S}, \bar{\tau}) \neq (S, \tau)$, will be analysed in section 2.1.3.4. An interpretation of the Smith's method in terms of a predictor will now be examined.

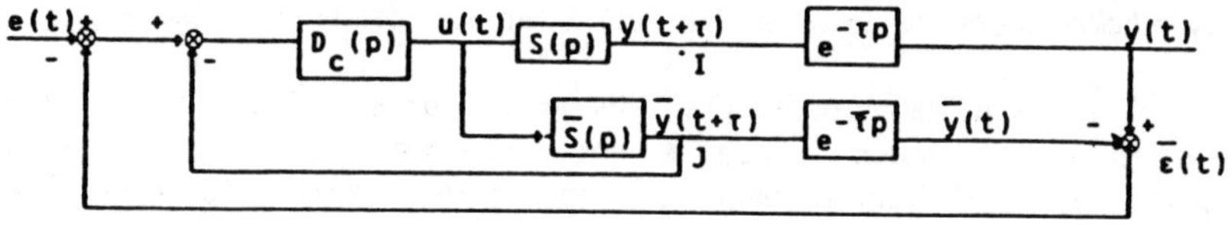


Smith controller – control by internal model
Fig. 4

2.1.3.3 – **Interpretation in terms of a predictor**

The signal $\bar{\epsilon}(t)$ of the outer loop will be zero if the modelling errors and the initial conditions of the sub-system S(s) are also zero. Under these assumptions, $\bar{y}(t+\tau) = y(t+\tau)$, a value which is interpreted as the prediction of the output at the time instant $t+\tau$. We can therefore state that, in the ideal case $(\bar{S} = S, \bar{\tau} = \tau)$, the outer loop can be removed and the control system can be interpreted in terms of an output predictor with transfer function $\bar{S}(s) = S(s)$.

Of course, although the signals appearing at points I and J are completely identical in the ideal case, only point J is accessible. In practice, modelling errors necessitate the presence of an outer loop which has the function of reducing the system's sensitivity to these errors. The control schema shown in Fig. 4 is clearly similar to the internal model control method described in Chapter 4.

2.1.3.4 – **Effects of modelling errors**

The transfer function of the closed loop system can be deduced directly from Fig. 4:

$$\frac{Y(s)}{E(s)} = \frac{D_c(s)\, S(s)\, e^{-\tau s}}{1+D_c(s)\, \bar{S}(s) + D_c(s)[Ss\, e^{-\tau s} - \bar{S}(s)\, e^{-\bar{\tau} s}]} \qquad (27)$$

By comparing equations (25) and (27) it may be concluded that the modelling errors introduce a supplementary term (mismatch term) into the denominator of the closed loop transfer function, and the characteristic equation becomes

$$1 + D_c \bar{S} + D_c (S e^{-\tau s} - \bar{S} e^{-\bar{\tau} s}) = 0 \qquad (28)$$

Thus there is a modification of the positioning of the poles, which becomes smaller as the mismatch term is decreased. In practice, the effect of modelling errors can be reduced by using adaptive control methods (see Chapter 6).

### 2.1.3.5 - Effects of external disturbances

In this section, the effects due to deterministic disturbances $v_1$ and $v_3$, appearing at the level of control u(t) and observation y(t), will be examined.

If these disturbances are measurable, feedforward actions are introduced; these are represented by the transfer functions $K_1(s)$ and $K_3(s)$ in Fig. 5. The transfer function $K_2(s)$ represents the conventional external feedback loop. The new block diagram is shown in Fig. 5.

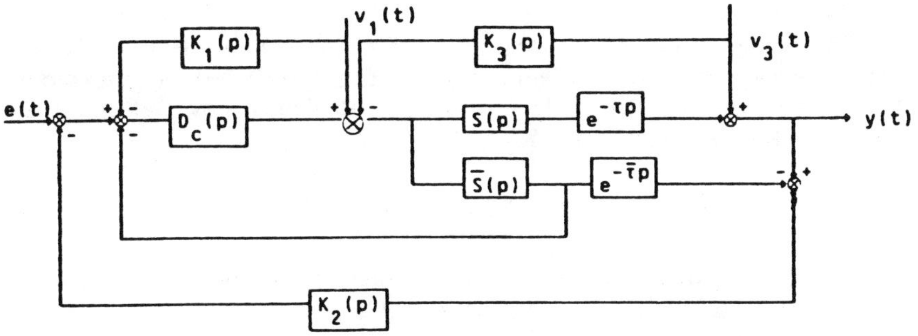


Fig. 5: Controller (case with disturbance)

The output of the closed loop system is then written as:

$$Y(s) = \frac{D_c S e^{-\tau s}}{\Delta} E(s) + \frac{(1-K_1 D_c) S e^{-\tau s}}{\Delta} V_1(s) +$$

$$\frac{(1+D_c \bar{S} e(1-K_2 e^{-\bar{\tau} s}) - K_3 S e^{-\tau s})}{\Delta} V_3(s) \qquad (29)$$

where

$$\Delta = 1 + D_c \bar{S} + K_2 D_c (S e^{-\tau s} - \bar{S} e^{-\bar{\tau} s}) \qquad (30)$$

* *Effects of the disturbance $v_1$*

The location of the point of action of the disturbance $v_1$ before the time delay causes a delay at the output equal to the time delay of the system. If $v_1$ is measurable, this effect can be eliminated by selecting the feedforward action such that

$$K_1(s) = \frac{1}{D_c(s)}$$

Such a selection presupposes that the inverse of the transfer function $D_c(s)$ can be realised. This requirement is met, for example, by pure gains, phase-lag-lead networks, and certain PID controllers. In other cases, an approximate realisation of $D_C^{-1}$ may be used.

Now consider the most frequent case of a non-measurable disturbance $v_1$. The effects of the disturbance $v_1$ in the form of impulses or steps will disappear from the output with the same dynamics as for the reference e(t) (the modes of the dynamics being defined by the characteristic equation $\Delta = 0$). Chapter 3 describes a control system which makes it possible to specify the dynamics relative to e(t) and $v_1(t)$ independently.

* *Effects of the disturbance $v_3$*

Since the point of application is after the time delay, this acts instantaneously on the system output. The effect of a measurable disturbance $v_3$ on the output can be eliminated by selecting the controller $K_3$ such that (with $K_2 = 1$)

$$1 + D_c \bar{S} (1 - e^{-\bar{\tau} s}) - K_3 S e^{-\tau s} = 0 \qquad (31)$$

This equation provides an exact causal solution in the case of a small delay ($\tau \# 0$, $\bar{\tau} \# 0$). It is then found that

$$K_3(s) = 1/S(s) \qquad (32)$$

and there are the same difficulties in realising this controller as was the case with $K_1$..

A second solution, valid for both a measurable and a non-measurable disturbance $v_3$, consists in the introduction of a transfer $K_2$ in the outer loop, such that

$$1 + D_C \bar{S} - K_2 D_C \bar{S} e^{-\bar{\tau} s} = 0 \qquad (33)$$

with $K_3 = 0$.

An approximate causal solution is obtained by making $\bar{\tau} = 0$ in equation (33), thus:

$$K_2 = \frac{1 + D_c \bar{S}}{D_c \bar{S}} \tag{34}$$

However, this solution will not normally be acceptable unless the delay $\tau$ is small. In other cases, the mismatch term $K_2D_c(Se^{-\tau s} - \bar{S} e^{-\bar{\tau} s})$ arising from the modelling errors may lead to a closed loop instability (see equation (30))

Note

In the ideal case with no modelling error, the transfer from $v_3$ to y is:

$$1 - \frac{K_2 \; D_c S \; e^{-\tau s}}{1 + D_c \; S} \tag{35}$$

Consequently, taking into account the choice of controller $K_2$ given by equation (34), it can be seen that a constant bias disturbance is cancelled after $\tau$ seconds.

## 2.2 - *DIGITISATION OF CONTINUOUS LAWS*

The methods of synthesis presented above lead to continuous control laws which are often satisfactory for the analyst. The problem of the digitisation of these continuous laws is encountered if we wish to avoid losing the knowledge gained in this way and having to go again through the whole synthesis of the control laws in the discrete domain; after all, the continuous control law makes the best use (to the extent that the continuous synthesis is considered satisfactory) of the information gathered by the sensors, and it is worth attempting to transfer the use of this information into the discrete domain. Moreover, supplementary elements (various filters, limiters, etc.) are often combined with analogue control laws, requiring transcription into the discrete domain element by element. This is why we initially describe the simple problem of obtaining a digital equivalent of a continuous linear control law independently of the feedback loop itself: the formulation will be similar to that of digital filtering, but with different objectives. We shall then tackle the problem of closed-loop digital control of stationary continuous systems.

### 2.2.1 - DISCRETE EQUIVALENT OF A CONTINUOUS CONTROLLER

Consider a controller C(s) which connects an error signal e(s) to a signal u(s) (Fig. 6a):

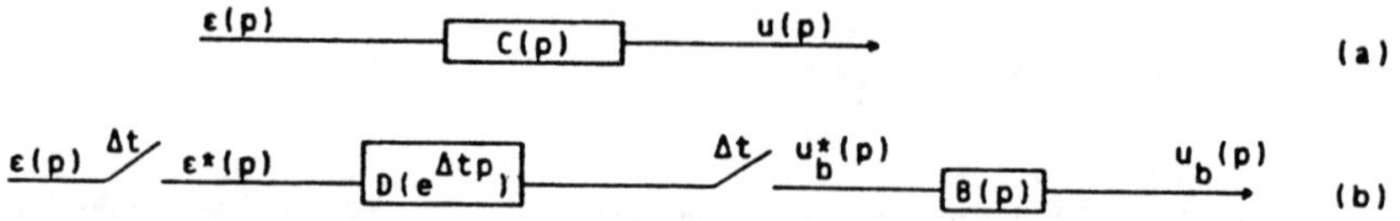


Equivalence between analogue and digital control

Fig. 6

The objective is to replace this continuous law with a system consisting of a digital controller D(z) or its equivalent in a recurrent equation, and a sampler and hold placed after it. One way of digitising the continuous law is to determine D(z) and B(s) in such a way as to preserve the relationship between $\epsilon(s)$ and u(s) as well as possible. In fact, it is impossible to keep this relationship perfectly, since if C(s) represents the transfer functions between $\epsilon(s)$ and u(s) it will no longer be possible to define in diagram 6b a transfer function linking $u_b(s)$ with $\epsilon(s)$. Admittedly, it is possible to calculate the Laplace transform of the signal $u_b(t)$, thus:

$$u_b(s) = D(e^{\Delta t s})\ B(s)\{\frac{1}{2}\ \epsilon(0^+) + \frac{1}{\Delta t}\sum_{n=-\infty}^{+\infty} \epsilon(s + \frac{2\pi jn}{\Delta t})\}$$

Thus it can be seen that it is impossible to define a transfer function independently of the input. The specificity of the problem of digital control appears at this level, since by its nature the error signal $\epsilon(t)$ is unpredictable (a way of making use of the nature of the error signal will be explained later).

It should also be noted that the comparison of the transfer functions C(s) and $D(e^{\Delta t s})$ is inadequate, since the sampled signal u* is not the control signal, as the extrapolation B(s) adds its own dynamic properties to those of the specifically digital element. It is therefore prudent to check that the addition of a zero order hold to the analogue loop does not degrade the performance excessively. This analysis may be performed, for example, by replacing the transcendental function of the zero order hold with a rational equivalent of the form

$$\frac{\Delta t}{1 + s\frac{\Delta t}{2}} \quad \text{or} \quad \frac{\Delta t}{1 + s\frac{\Delta t}{2} + (\frac{s\Delta t}{2})^2} \quad \text{etc.}$$

These equivalents are the Padé approximations.

## 2.2.2 - THE TECHNIQUES OF DIGITISATION

We shall restrict ourselves here to the methods which can be used to obtain, in a simple and direct way, a digital control law whose application will be of a reasonable level of complexity. The methods of digital filtering offer a very wide range of procedures whose power may, however, appear disproportionate to the result required; where filtering is concerned, it is best to comply with patterns specified in the frequency domain. The constraints may appear less powerful in the control problems, where the emphasis is in any case more often placed on the phase properties (in the conventional lead network, for example).

It may also be necessary to digitise unstable systems in open loops, a problem which is outside the scope of filtering methods. We therefore note, simply as a reminder,

- methods based on convolution and deconvolution with the use of windows;

- methods based on identification by non-linear programming, which for our purposes are best applied directly in the closed loop configuration in the digital domain.

Three classes of methods will be distinguished:

a) - **Techniques specifying the nature of the input signal** $\epsilon$

The impossibility of finding a suitable p transfer function for diagram 6b makes it necessary to find a modification of the continuous system such that the discrete equivalent appears naturally. By introducing an imaginary sampler and hold for the sampled signal $\epsilon(t)$ (Fig. 7), it is possible to define a transfer function between the sampled signal $\epsilon^*(t)$ and the sampled signal $u^*(t)$ which depends on the chosen hold.

We then find that

$$D(z) = S(B(s)\ C(s))$$

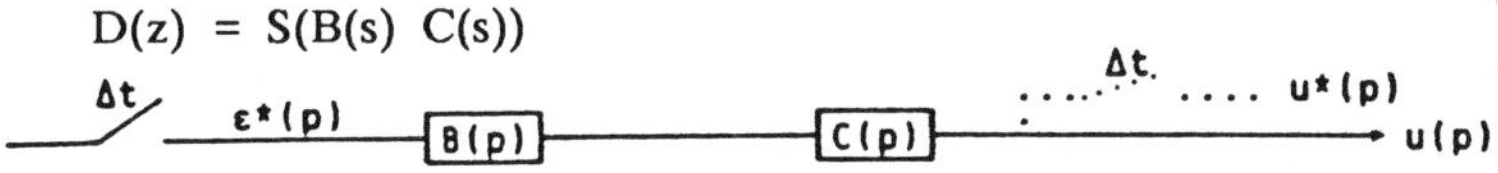


Introduction of an imaginary sampler and hold in diagram 6a

Fig. 7

The designation **impulse invariance** is applied to the transform

$$C(s) \longrightarrow D(z)$$

when $B(s) = 1$.

When $B(s)$ is a zero-order hold, the same transformation is known as index invariance. The use of higher-order holds is not customary, but the procedure of generalisation is simple:

| | |
|---|---|
| $D(z) = Z(C(s))$ | impulse invariance |
| $D(z) = \frac{z-1}{z} Z\left(\frac{C(s)}{s}\right)$ | index invariance |
| $D(z) = \frac{(z-1)^2}{Tz^2} Z\left(\frac{1+Ts}{s^2} C(s)\right)$ | |
| etc. | |

If the modelling of the controller is a representation of state variables:

$$\dot{x} = A x + B \epsilon$$

$$u = C x$$

then clearly the methods of digitisation using this form will give similar results. The digitisation of this state vector equation gives the equation:

$$x_{n+1} = e^{A\Delta t} x_n + \int_{n\Delta t}^{(n+1)\Delta t} e^{A[(n+1)\Delta t - \tau]} B \epsilon(\tau) d\tau$$

$$u_{n+1} = C x_{n+1}$$

The hypothesis as to the nature of the hold implies a precise form of $\epsilon(t)$ between two sampling points, so that the integral can be calculated. For the zero-order hold, we have the equivalent of the calculation of the recurrent equation of a continuous system. The case of the impulse hold is more complex because of the discontinuities of the signal; the instant of application of the control pulse may be chosen arbitrarily within the sampling period. Two limit cases are envisaged here:

1) $\epsilon(\tau) = \delta(n\Delta t)$ implies the following recurrent equation:

$$x_{n+1} = e^{A\Delta t} x_n + e^{A\Delta t} B \epsilon_n = e^{A\Delta t}(x_n + B\epsilon_n)$$

and the transfer function is then written

$$D(z) = C(Iz - e^{A\Delta t})^{-1} e^{A\Delta t} B$$

2) $\epsilon(\tau) = \delta((n+1)\Delta t)$ implies the following digitisation:

$$x_{n+1} = e^{A\Delta t} x_n + B\epsilon_{n-1}$$

The transfer function relating $\epsilon$ to u is then written

$$D(z) = C(Iz - e^{A\Delta t})^{-1} Bz$$

Only this second hypothesis is coherent with the result obtained by the impulse invariance. The Z transform thus requires a definition of the state after the application of a pulsed input (see Fig. 8).

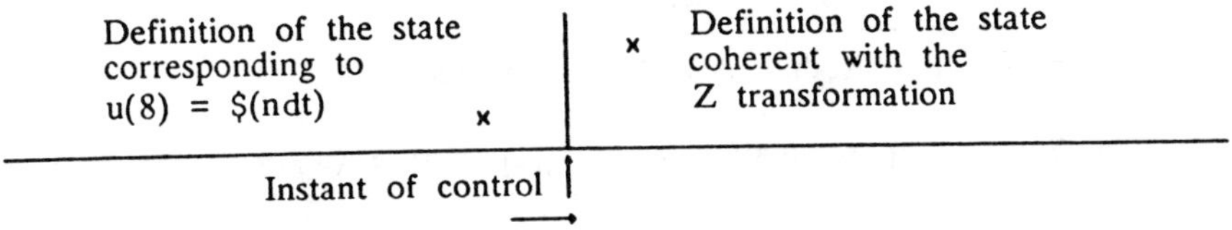


Application of an impulse input and definition of the state

Fig. 8

The modified Z transform can be used to find the result corresponding to $u(t) = \delta(n\Delta t)$ as well as the intermediate results which may be obtained by specifying an arbitrary location of the impulse within each sampling period.

Example

The first-order low-pass filter $1/s+\alpha$ is related to the digital filter $z/z-e^{-\alpha\Delta t}$ by the method of impulse invariance. The modified Z transform is written

$$D(z,m) = \frac{e^{-\alpha m\Delta t}}{z - e^{-\alpha\Delta t}}$$

If $m \to 1$, the Z transform of the signal immediately before the application of the input is obtained, the recurrent equation being written:

$$x_{n+1} = e^{-\alpha\Delta t}\, x_n + e^{-\alpha\Delta t}\, u_n$$

The normal Z transform is obtained by evaluating $zD(z,0)$.

b) **Techniques using quadrature formulae**

Given that s is the inverse of an integration, the different discrete approximations of an integration provide as many ways of replacing the operator s by an "equivalent" in Z. Thus,

- The rectangular approximation $x_n = x_{n-1} + \Delta t\, u_n$ leads to

$$\boxed{\text{replace s by } \frac{1}{\Delta t}\frac{z-1}{z}}$$

This transformation is described as rectangular.

- The trapezoidal approximation $x_n = x_{n-1} + \Delta t/2\, (u_n + u_{n-1})$ leads to

$$\boxed{\text{replace } s \text{ by } \frac{2}{\Delta t}\frac{z-1}{z+1}}$$

The transformation is then described as **bilinear**.

– The Simpson approximation $x_n = x_{n-1} + \Delta t/3\,(u_n + 4u_{n-1} + u_{n-2})$ leads to

$$\boxed{\text{replace } s \text{ by } \frac{3}{\Delta t}\frac{z^2-1}{z^2+4z+1}}$$

It may be noted that these three derivative systems, which are in order of increasing accuracy, are also in order of increasing instability, since the poles of the denominators are respectively 0, –1, and (–3.73; –0.27).

c) – **Heuristic techniques**

Finally, there are other methods which are not based on theoretical arguments. A standard method consists in replacing not only the poles but also the zeros by their sampled equivalents. Thus the poles are replaced by $e^{\Delta t s_i}$ and the zeros by $e^{\Delta t z_i}$, the numerator being completed if necessary by a suitable number of zeros at the point (0, –1) so that the degree of the numerator in Z is smaller by not more than one unit than that of the denominator. This method is known as the "matched Z transform". The gain must normally be adapted subsequently to obtain a suitable digital controller.

d) – **Comments on the techniques**

In general, it may be noted that:

– the impulse invariance and the matched Z–transform are the only transformations which do not conserve the static gain. If the precision of the feedback is fixed by the control specifications, it is important to restore a suitable gain. It should also be noted that the phase may be affected by this correction if the sign of the gain is changed;

– the quadrature formulae and the matched Z–transform have the property of being associative, in other words the transformation of a cascade of transfer functions is the product of the transforms of each function;

– out of the three quadrature formulae, the trapezoidal transformation provides the best compromise between the order of the approximation and the stability of the integration scheme. The frequency distortion of this bilinear transformation may be corrected by the formula:

$$s \longrightarrow \frac{\omega_c}{\operatorname{tg} \dfrac{\omega_c \Delta t}{2}} \; \frac{z - 1}{z + 1}$$

The gain at the continuous frequency is preserved in the discrete frequency response. If the value $\omega_c$ is adapted to each pole or zero, this frequency correction is called "prewarping" in the English-language literature. The operator s+a is transcribed by replacing s with $a(z-1)/(\operatorname{tg} a\,(z+1)\Delta t/2)$. In this way the half straight line of the continuous positive frequencies is made to coincide with the segment $[0, \pi/\Delta t]$ of the discrete frequencies, pole by pole and zero by zero. The effectiveness of this prewarping is found to be low in practice, if the transformed frequencies do not remain close to the frequency $\pi/\Delta t$, a region in which the digitisation of continuous information must anyhow be approached with caution.

e) – **Illustration in an example**

The application of the methods of digitisation will be demonstrated with reference to the control of a system consisting of a double integration (Fig. 9).

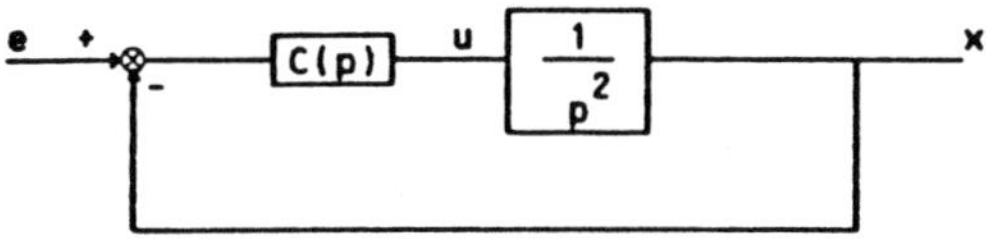


Control of a double integrator

Fig. 9

The continuous controller C(s) is a phase lead system of the form:

$$C(s) = \frac{K(1+as)}{1+a\alpha s}$$

The digitisation of this controller system may be carried out by the various methods described above. Different results will be obtained if the decomposition of the system into simple elements precedes the digitisation. We shall restrict ourselves here to a consideration of the methods of index invariance and rectangular and bilinear transforms and the matched Z transform. The following table shows the results, indicating the gain, pole and zero of the discrete system obtained for each method.

| Transformation | Static gain | Pole | Zero |
|---|---|---|---|
| A/ Index invariance | K | $e^{-\Delta\tau/a\alpha}$ | $1 - \alpha(1 - e^{-\Delta t/a\alpha})$ |
| B/ Rectangular transformation | K | $\frac{a\alpha}{\Delta t+a\alpha}$ | $\frac{a}{\Delta t + a}$ |
| C/ Bilinear transformation | K | $\frac{2a\alpha - \Delta t}{2a\alpha + \Delta t}$ | $\frac{2a - \Delta t}{2a + \Delta t}$ |
| D/ Matched Z-transform | $\frac{K}{\alpha}\,\frac{1-e^{-\Delta t/a}}{1-e^{-\Delta t/a\alpha}}$ | $e^{-\Delta t/a\alpha}$ | $e^{-\Delta t/a}$ |

TABLE

The gain must be corrected in the last case: the result obtained has to be multiplied by

$$\frac{\alpha(1-e^{-\Delta t/a\alpha})}{1-e^{-\Delta t/a}}$$

to restore the initial static gain K of the analogue controller system. In all cases, the numerical controller takes the following form:

$$D(z) = \frac{a_o + a_1 z^{-1}}{1 + b_1 z^{-1}} \qquad \text{with} \quad K = \frac{a_o + a_1}{1 + b_1}$$

### 2.2.3 - ANALYSIS OF THE CLOSED LOOP SYSTEM

#### 2.2.3.1. The classical approach

Simply finding the transcription of the input-output characteristics of the functional block representing the continuous control law appears to be insufficient if the closed loop control system is to be taken into account. In fact, Fig. 6 does not show the whole of the problem, which is much better illustrated in Fig. 10.

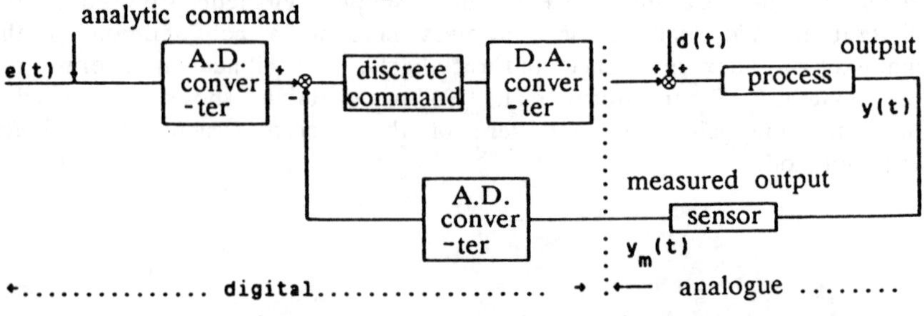


Diagram of computer control of a continuous process

Fig. 10

It will be seen that the effect of the digitisation on the overall stability is important; the standard system which may be deduced from Fig. 10 permits a sampled analysis which will now be examined.

Fig. 10 can either be in the form shown in 11a, if the continuous character of the signals in question is to be retained, or in the form shown in 11b if the procedure is to be entirely in the digital domain.

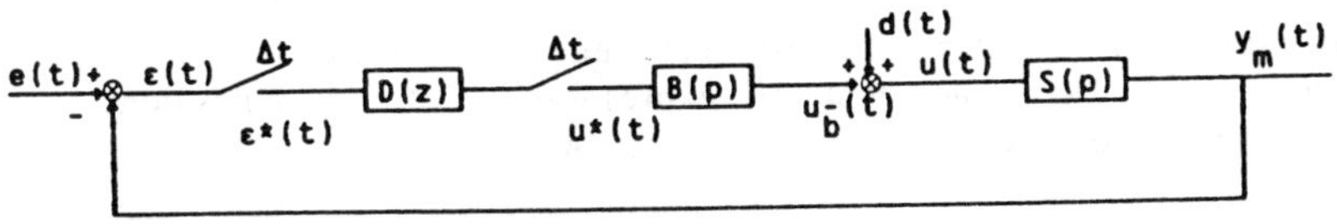


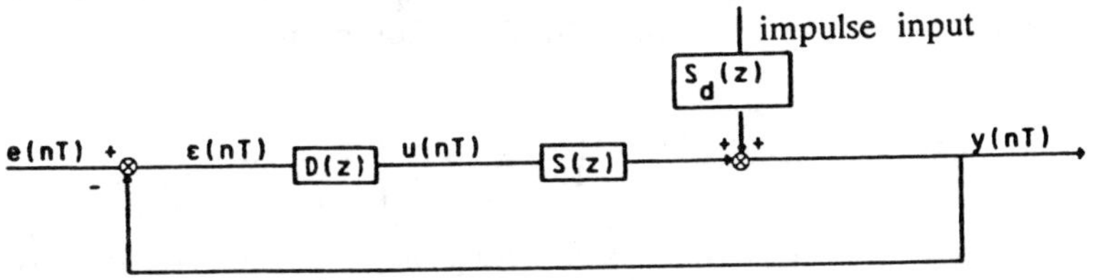


Figs. 11a and 11b

The second diagram can be used to write the relationships at the instant of sampling, thus:

$$\frac{Y(z)}{E(z)} = \frac{D(z)\ S(z)}{1+D(z)\ S(z)} \quad \text{with} \quad \begin{aligned} S(z) &= Z[B(s)\ S(s)] \\ S_d(z) &= Z\,[d(s)\ S(s)] \end{aligned}$$

Conversely, the Z transform of the output can only be calculated for a given disturbance.

If what occurs between the sampling instants is to be investigated, the modified Z transform is a suitable tool for analysing the sampled signal

displaced by a fixed quantity with respect to the nominal instants, for example:

$$Y(z,m) = \frac{D(z)\ S(z,m)}{1+D(z)\ S(z)}\ E(z)$$

where m is the percentage of displacement. The stability analysis of the closed loop is clearly independent of m (except in the case of "hidden oscillations" arising from simplifications between the pole of D(z) and the zero of S(z)). Only the formulation by modified Z transform eliminates these simplifications.

Finally, it is possible to remain in the continuous domain by introducing the star notations for the continuous sampled signals and their Laplace transforms. Thus,

$$Y(s) = \frac{D^*(s)\ B(s)\ S(s)}{1 + S^*(s)\ D^*(s)}\ E^*(s) \quad \text{with} \quad \begin{array}{l} D^* = Z[D(s)] \\ S^*(s) = Z[B(s)\ S(s)] \end{array}$$

expresses the transfer function between the sampled input and the output. It is also known that the expression of Y*(s) is equivalent to the Z representation, when $e^{\Delta ts}$ is replaced by z.

The continuous formulation has the advantage of showing how it is possible to obtain a sampled chain similar to the analogue chain. Indeed, the basic property of the sampling process is to spread the frequencies from zero to infinity, by the following formula:

$$E^*(s) = \frac{1}{2}\, e(0^+) + \frac{1}{\Delta t} \sum_{k=-\infty}^{k=+\infty} E\left[P + \frac{2k\pi}{T}\, j\right]$$

In particular, if e(t) contains a frequency $\omega_o$, e*(t) contains the frequencies $\omega_o + 2k\pi/\Delta t$, $k \in \mathbf{N}$.

Furthermore, a signal containing frequencies beyond $\pi/\Delta t$ can never be reconstructed from its samples, owing to the frequency overlap.

These elements concerning the signals are interpreted in a special way in the case of the closed loop: it is the error signal, and not the set point signal, that is sampled. Even on the hypothesis of a set point signal containing a pure frequency, the sampling followed by transfer through the process will induce all frequencies in the error signal. A suitable method of analysis is provided by the theory of the Z transform, but the overlap due to the closed loop still depends essentially on the system transfer function. In fact, if the system is assumed to be of a perfect low-pass type with respect to the Shannon frequency $\pi/\Delta t$ (for example by connecting

a cardinal filter with a cut-off at $\pi/\Delta t$ in the loop) the transfer function

$$\frac{Y(s)}{E^*(s)} = \frac{D^*(s)\ B(s)\ S(s)}{1 + S^*(s)\ D^*(s)}$$

becomes

$$\frac{Y(s)}{E^*(s)} = \frac{D^*(s)\ B(s)\ S(s)}{1 + D^*(s)\ B(s)\ S(s)}$$

and is again suitable for a standard continuous frequency analysis closed loop on the basis of the open loop transfer function

D*(s) B(s) S(s)

It is therefore important to measure the low–pass properties of the process represented by S(s) with respect to $\varphi$ frequency $\pi/\Delta t$. A method of analysis based on the stability of the characteristic closed loop equation is proposed below.

### 2.2.3.2. Analysis of the low–pass properties of the controller system

Consider the equation

$$1 + D^*(s)\ S^*(s) = 0$$

which, by replacing $e^{\Delta ts}$ with z, becomes

$$1 + D(z)\ S(z) = 0$$

The roots in the plane z are equivalents in the plane s related by the multiform function

$$s = \frac{1}{\Delta t} \log z$$

If the determination such that the absolute value of the imaginary part is less than $\pi/\Delta t$ is denoted the "principal determination", the following results are found:

- a real root $s_1$ for a positive real root $z_1$;

- a pair of complex roots $s_2$ and $\bar{s}_2$ for a pair of complex roots $z_2$ and $\bar{z}_2$;

- two poles having ordinates $\pm\ \pi/\Delta t$ for a negative real root $z_3$ (located on the discontinuity of the multiform functions).

The secondary determinations are obtained by displacements of $2k\pi/\Delta t$ ($k \in Z$) in the imaginary direction (see Fig. 12).

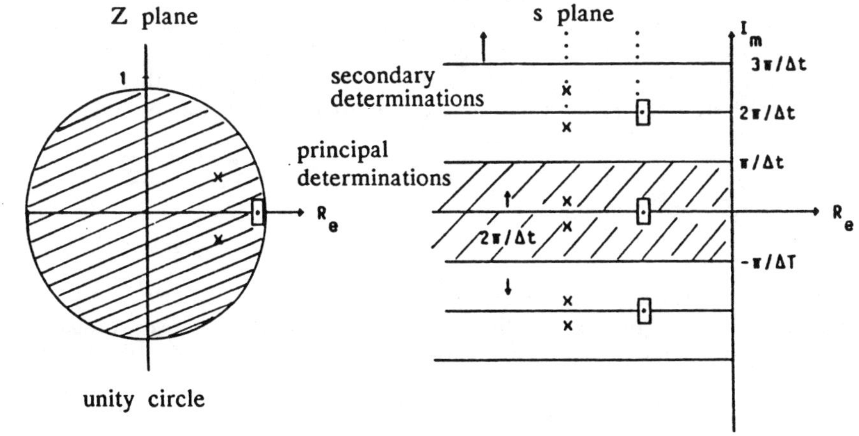


Fig. 12 Transformation s = 1/dt log Z

Consequently, the definition of the stability of a root in s equivalent to a root in Z is more complicated than in the continuous case:

– the absolute stability is the same for all the determinations and is that of the principal determination $|\zeta_0 \omega_0|$;

– the relative stability decreases with k to become very small when k becomes large (Fig. 13).

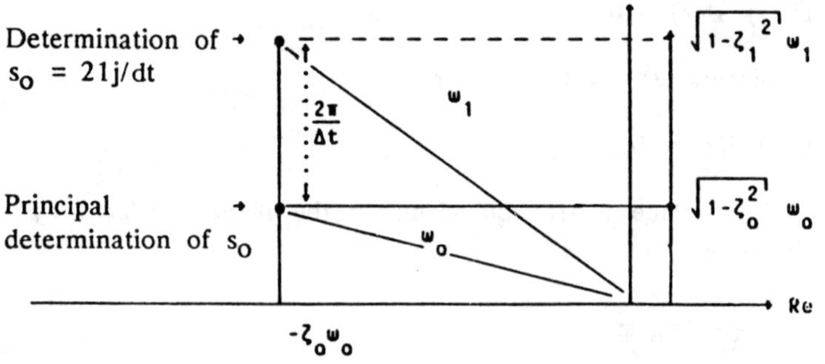


Stability of the roots in s equivalent to the poles in Z

Fig. 13

The frequency contents of the signals in question therefore depend largely on the relative weights of the different determinations of the equivalents in the plane s of the closed loop poles in Z; the residues of the different determinations may be compared. Returning to the transfer function

$$Y(s) = B(s)\ S(s)\ \frac{D^*(s)\ E^*(s)}{1+D^*(s)\ S^*(s)}$$

A residue, $r_i^k$, may be associated with each determination k of each pole $s_i = (1/\Delta t) \log z_i$ of the denominator of Y(s). It is then useful to compare the moduli of these residues for the same pole, in other words to compare the values of $r_i^k$ for different k and for a given i. It appears advisable to take the residue corresponding to the principal determination, i.e. $r_i^o$, as a reference, and to introduce the notations:

$$\rho_k = \frac{|r_i^k|}{|r_i^o|}$$

When the residues are functions of E*(s), the coefficients $\rho_k$ are intrinsic and only depend on S(s). They therefore characterise completely the low-pass properties of the continuous elements of the control loop.

To make this property clear, it is sufficient to perform the following demonstration. Let H* be the characteristic polynomial and assume that all its roots are simple. Rewrite the Laplace transform for the output in the form

$$Y(s) = \frac{N_S(s)}{D_S(s)} \times \frac{P^*(e^{\Delta t s})}{H^*(e^{\Delta t s}) D_E^*(e^{\Delta t s})}$$

where $N_D$ and $D_S$ are the numerator and denominator of the continuous set B(s) S(s), $D_E^*(e^{Dts})$ is the denominator of the sampled set point functions, and $s^*(e^{Dts})$ is the residual numerator (sampled part).

A residue of Y(s) for a root of the expression $H^*(e^{\Delta ts})$ is given by

$$\left. \frac{N_S(s)\ P^*(e^{\Delta t s})}{[D_S(s)\ H^*(e^{\Delta t s}) D_E^*(e^{\Delta t s})]} \right|_s \qquad \text{s being a root of } H^*(e^{\Delta t s})$$

The derivation of the denominator only leaves the following term:

$$[D_S(s)\ \Delta t e^{\Delta t s} \{H^*(e^{\Delta t s})\}'\ D_E^*(e^{\Delta t s})] \qquad \text{s being a root of } H^*(e^{\Delta t s})$$

The value of this residue (for $s = s_{io}$) is:

$$r_i^o = B(s)\ S(s)\ \frac{P^*\ (e^{\Delta t s_o})}{\Delta t\ e^{\Delta t s_o} H_Z'(e^{\Delta t s}) D_E^*(e^{\Delta t s})}$$

and therefore

$$\rho_k = \alpha \left[ \frac{S(s_o + \frac{2k\pi}{\Delta t} j)\ B(s_o + \frac{2k\pi}{\Delta t} j)}{S(s_o)\ B(s_o)} \right]$$

where $\alpha = 2$ if $s_o$ is real
and $\alpha = 1$ if $s_o$ is complex.

In particular, if the digital-analogue converter is a zero-order hold,

$$\rho_k = \alpha \left| \frac{s_o}{s_o + \frac{2k\pi j}{\Delta t}} \right| \left| \frac{S\ (s_o + \frac{2k\pi j}{\Delta t})}{S(s_o)} \right|$$

## *2.3 – CONCLUSION*

As we have seen, the impossibility of obtaining a strict equivalence between an analogue and a digital control law has given rise to a number of methods of approximation, which are rather difficult to classify in order of preference. It may be said, however, that, in the example of a double integration (often encountered in practice) controlled by a phase advance network, the bilinear transform gives the "best" results, in the sense that the range of sampling periods for which the result is satisfactory is the widest; this is, no doubt, not unconnected with the fact that this transformation is the only one making the whole of the segment [–1 +1] correspond to a given pole when $\Delta t$ varies (the other transformations only make the interval [0 +1] correspond). Finally, if the sampling period goes towards zero, in other words when sampling is done sufficiently rapidly, all the methods except that of impulse invariance yield satisfactory results. Conversely, if $\Delta t$ is increased the results of each method deteriorate more or less rapidly.

## *REFERENCES*


Bartels, Steward, A solution of the equation AX+XB = C, COMACM, v. 15, 1972.

J.C. Doyle and G. Stein, Multivariable feedback design, IEEE Automatic Control, v. AC 26, 1981.

L.R. Fletcher, On pole placement for linear multivariable system with direct feedthrough, Int. J. Cont., Part II, VP 33, 1981.

L.R. Fletcher, An algorithm for eigenstructure assignment, Parts I and II, Dept. Mathematics, University of Salford, 1979.

O. Hérail, Commande optimale d'ordre réduit, Doctoral thesis, ENSAE, Toulouse, 1985.

M. Labarrère, J.P. Krief, and B. Gimonet, Le filtrage et ses applications, Editions CEPADUES, 1978.

J.F. Magni, Commande modale des systèmes multivariable, Doctoral Thesis, Toulouse, 1987.

J.E. Marshall, Control of time-delay systems, IEE Control Eng. Series 10, Peter Peregrinus Ltd., 1979.

B.C. Moore, C. Wierzbicki, and G. Klein, Recent developments in eigenvalue and eigenspace assignment, IEEE Conf. on Decision and Control, 1979.

B.C. Moore, On the flexibility offered by state feedback, IEEE AC, v. 21, 1976.

V. Mukhopadhyay, A method for obtaining reduced order control laws for high order systems using optimization techniques, NASA Langley Research Center, 1981.

H.H. Rosenbrock, State space and multivariable theory, Nelson, 1970.

O.A. Sebakhy and N.N. Sorial, Optimization of linear multivariable systems with prespecified closed loop eigenvalues, IEEE AC, V. AC 24, April 1979.

M. Steer, Commande optimale linéaire quadratique par contre réaction de sortie sur un horizon fini, Doctoral Thesis, Ecole des Mines, March 1982.

W.M. Wonham, Linear multivariable control, a geometric approach, Springer Verlag, 1979.

N.H.A. Davis, Linear estimation and stochastic control, Chapman and Hall, 1977.

## *CHAPTER 3*

# *DIGITAL CONTROLLERS OF DETERMINISTIC SYSTEMS THE POLYNOMIAL APPROACH*

## *3.1 – INTRODUCTION*

This chapter will be essentially concerned with the synthesis of digital controllers to optimise discrete closed loop systems with deterministic inputs. The random nature of the disturbances will be introduced at the end of the chapter and dealt with fully in Chapter 5.

Having dealt with the synthesis of continuous laws in the preceding chapter, we shall now use only discrete methods in the rest of this book. These methods are presented here in a unified form and are all based on the algebraic theory of polynomials. For the sake of the reader who is unfamiliar with this theory, Appendix 1 sets out the principal results to facilitate the understanding of the subsequent theoretical developments.

The methods most commonly used in practice may be divided into three main classes:

– those used to obtain a response in finite time in a closed loop system, often with a minimum time criterion which is sometimes combined with a quadratic criterion;

– those based on pole location and designed to specify the dynamics of a closed loop. As in the continuous case, these resemble the modal methods;

– finally, those designed to minimise the effect of noise in the case of stochastic processes (minimum variance method).

Each of these will be discussed at length in this chapter. The discussion of the last class forms an introduction to the discrete optimal stochastic control of multivariable systems, which is dealt with in Chapter 5.

## *3.2 - CONTROLLERS FOR A MINIMUM TIME CRITERION*

Before seeking an optimal minimum time controller, we shall go through a preliminary stage of defining a class of controllers, called stabilising controllers, which provide conditions of stability in the closed loop system and conditions of feasibility which are an essential prerequisite of any proper synthesis.

### 3.2.1 - STABILISING CONTROLLERS

Consider the diagram in Fig. 1, representing a conventional control system in which:

S denotes the process modelled by a $z^{-1}$ transfer function:

$$S = \frac{B}{A}$$

where A and B are prime with respect to each other. The greatest common factor of A and B will henceforth be denoted by (A,B).

D represents the unknown controller of the Y/X transfer function. The G transfer function for the closed loop, relating the discrete input $e_1$ to the discrete output, s, of the system, is written

$$G = \frac{DS}{1+DS} \qquad (1)$$

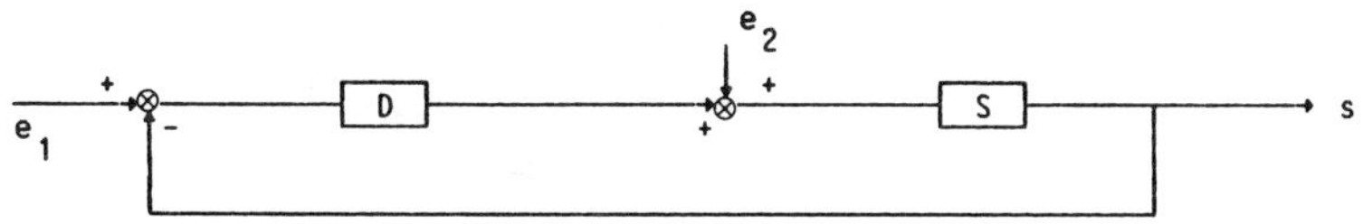


Fig. 1

<u>Notation</u>

The signal notations e and s designate polynomials or rational fractions in $z^{-1}$.

e(t) and s(t) denote the time signals, and $E^+$ denotes the stable domain within the unit circle, i.e.

$\{z^{-1}, |z^{-1}| > 1\}$

$F^+$ represents polynomials in $z^{-1}$ whose roots belong to $E^+$;

$E^-$ represents the unstable zone which is complementary to $E^+$;
$F^-$ is the set of polynomials in $z^{-1}$ whose roots belong to $E^-$.

For example, if $A(z^{-1}) = z^{-1}(1+az^{-1})$ with $|a| < 1$, then

$$A^- = z^{-1}$$

$$A^+ = 1 + a\, z^{-1}$$

Before there can be any actual optimisation, the synthesis of the controllers must meet three very important conditions:

a) The G transfer function for the closed loop must be asymptotically stable.

b) In the process of synthesis, there must be no compensation in the $E^-$ domain.

c) The resulting controller must be feasible (causality condition).

Although the first condition is well known, the second requires some explanation. Suppose that this second condition is not met, in other words consider a controller D = Y/X such that

$$X = B^- X_1$$

$$Y = A^- Y_1$$

The G transfer function is then written

$$G = \frac{BY/AX}{1+BY/AX} \qquad\qquad G = \frac{B^+ Y_1}{A^+X_1 + B^+Y_1}$$

It will be shown that the fact of choosing the polynomials $X_1$ and $Y_1$ as a solution of the equation $A^+ X_1 + B^+ Y_1 = A^*$, where $A^*$ is a stable polynomial, is not sufficient to ensure the asymptotic stability of the closed loop.

Calculate the following transfer functions:

$$\frac{s}{e_2} = \frac{B\ /\ A}{1+BY/AX} = \frac{B\, X_1}{A^-(A^+X_1 + B^+Y_1)}$$

$$\frac{u}{e_1} = \frac{Y\ /\ X}{1+BY/AX} = \frac{A\, Y_1}{B^-(A^+X_1 + B^+Y_1)}$$

It is easily seen that the compensation of an unstable pole of the process by a zero of the controller destabilises the first transfer function, while the compensation of an unstable zero of the process by a pole of the controller destabilises the second transfer function.

The conditions a and b may therefore be condensed by the following condition: X, Y must be the solution of the polynomial equation

$$AX + BY = A^* \text{ in which } A^* \in F^+$$

Finally, condition (c) is met if $X(0) \neq 0$. Let

$$D = \frac{y_o + y_1 z^{-1} + \dots}{x_o + x_1 z^{-1} + \dots}$$

If $x_o = 0$, the controller is not feasible. In fact, this condition is always met to the extent that $A^* \in F^+$.

> To summarise, the set of controllers satisfying the conditions a, b, c is given by
>
> $$D = \frac{Y}{X} \qquad (2)$$
>
> where X, Y are the solution of the equation
>
> $$AX + BY = A^* \quad \text{where} \quad A^* \in F^+$$

We have now found the class of stabilising controllers complying with (2). We now have to find within this class the optimal controller which minimises a finite response time criterion.

## 3.2.2 - SYNTHESIS OF DISCRETE CONTROLLERS MINIMISING THE RESPONSE TIME

### 3.2.2.1 - **Minimal response time without constraints on the** control signal

**a) - Formulation of the problem**

Consider the case of a discrete process for which the recurrent evolution equation is given by

$$As = B\,u$$

where

$$A \text{ and } B \in F \text{ and } (A,B) = 1$$

The system must follow an input sequence e defined by the recurrent equation:

$$Le = R$$

where

$R$ and $L \in F$ and $(R,L) = 1$

To this end, the simplest control structure, defined by the following relationship, is chosen:

$$Xu = Y\,(e-s)$$

where

$$(X,Y) = 1$$

where X and Y represent two polynomials in $z^{-1}$ defining the controller (Fig. 2):

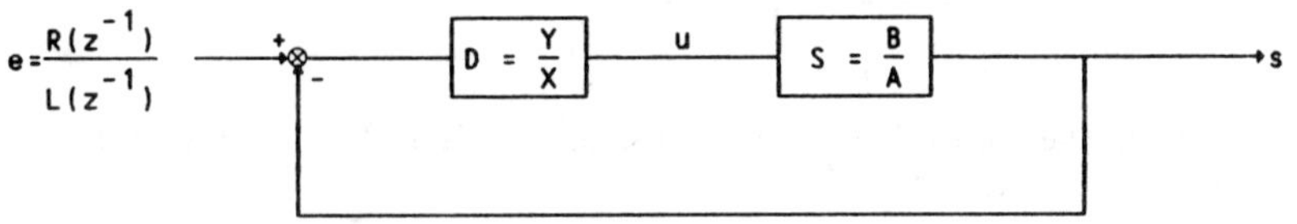


Fig. 2

The objective is to find the controller D such that the error $\epsilon$ is cancelled in minimal time without any constraint on the control signal, u, of the process.

b) – **Finding the optimal controller**

The error sequence $\epsilon$ is given by

$$\epsilon = \frac{AX}{AX + BY}\,\frac{R}{L}$$

Using $A_0$ and $L_0$ to denote the prime polynomials with respect to each other such that

$$\frac{A}{L} = \frac{A_0}{L_0} \quad \text{with } (A_0, L_0) = 1 \tag{3}$$

the sequence $\epsilon$ becomes

$$\epsilon = \frac{A_0 X}{AX + BY}\,\frac{R}{L_0} \tag{4}$$

In accordance with the previous section, this controller will now be sought among the stabilising controllers satisfying the condition

$$AX + BY = A^* \quad \text{where} \quad A^* \in F^+ \qquad \text{(see 2)}$$

The error sequence $\epsilon$ is cancelled after a minimum time if and only if $\epsilon(z^{-1})$ is a polynomial of minimal degree in $z^{-1}$.

Equation (4) then requires that

a) the polynomial X be divisible by $L_o$;
b) the set AX+BY contain and only contain the stable roots in the numerator of expression (4);
c) X is divisible by $B^+$, since the polynomial AX+BY will then also be divisible by $B^+$.

To meet all three conditions, the following relationship must be obtained:

$$AX + BY = B^+ R^+ A_o^+ \qquad (5)$$

by selecting, from all the solutions of polynomial equation (5), the one which gives

$$X = L_o B^+ X_1 \text{ with a minimal degree of } X_1$$

$$Y = A_o^+ Y_1$$

Taking (6) into account, relationship (5) becomes

$$A_o^- L X_1 + B^- Y_1 = R^+$$

> Thus the optimal controller is provided by relationship (7):
>
> $$D = \frac{A_o^+ Y_1}{L_o B^+ X_1} \qquad (7)$$
>
> where $X_1$ and $Y_1$ is the minimum degree solution in $X_1$ such that
>
> $$A_o^- L X_1 + B^- Y_1 = R^+ \qquad (8)$$

This solution exists and is unique if the polynomials $A_o^-$ L and $B^-$ are prime with respect to each other. Since $A_o^-$ and $B^-$ cannot contain common polynomial factors, the solution of minimal degree in $X_1$ exists if

$$(L, B^-) = 1$$

The optimal error sequences $\epsilon(z^{-1})$ and $u(z^{-1})$ are written thus:

$$\epsilon = R^- A_o^- X_1 \qquad (9)$$

$$u = D\,\epsilon = \frac{R^- A_o Y_1}{L_o B^+} \qquad (10)$$

It is thus confirmed that the sequence $\epsilon$ is finite and of minimal degree. On the other hand, the control signal u appears in the form of a rational fraction in $z^{-1}$, and is therefore not cancelled at the end of a finite period.

c) – **The class of stable systems**

To illustrate the above findings, the very important class of stable processes which may be capable of integration will now be considered.

In this case the process S is presented in the form

$$S = \frac{B}{A} = \frac{z^{-1} \prod_{i=1}^{n-1} (1 - z_i z^{-1})}{\prod_{i=1}^{n} (1 - s_i z^{-1})}$$

$\forall\ z_i$ with $z_i \in E^+$
$\forall\ s_i$ with $s_i \in E^+$ (possibly with integration)

We shall consider the response to a ramp input having the following z

$$e(z^{-1}) = \frac{R(z^{-1})}{L(z^{-1})} = \frac{1}{1-z^{-1}}$$

Since polynomials L and A have a common factor, it may be deduced that

$$L_o = 1 \quad \text{and} \quad A_o^- = 1$$

The polynomial equation enabling $X_1$ and $Y_1$ to be found is then written

$$(1 - z^{-1})\, X_1 + z^{-1}\, Y_1 = 1$$

In this particular case, the minimal degree solution in $X_1$ is evident:

$$X_1 = 1 \quad \text{and} \quad Y_1 = 1$$

In accordance with relationship (7), the expression for the optimal controller is as follows:

$$\boxed{D = \frac{A_o^+}{B^+} = \frac{A\ z^{-1}}{B(1-z^{-1})}}$$

The corresponding error sequence has the form $\epsilon(z^{-1}) = 1$, signifying that this error is cancelled from the first instant of sampling ($\epsilon(nT) = 0$ for $n \geqslant 1$).

Similarly, the control sequence becomes

$$u = \frac{A\ z^{-1}}{B(1-z^{-1})}$$

and the closed loop transfer function is written

$$G = z^{-1}$$

since

$$\epsilon = e - s = \frac{1 - G}{1-z^{-1}} = 1$$

d) – **Example of application**

Consider the system shown in Fig. 3, where the sampled signal u enters a zero-order hold $H_0$ to control the second-order process. The sequence

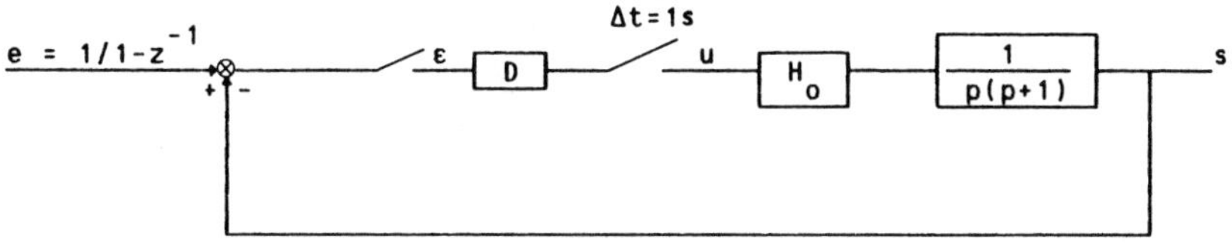


Fig. 3

$\epsilon(z^{-1})$ is to be cancelled in minimal time, in other words the error $\epsilon(t)$ is to be cancelled in minimal time at the moments of sampling.

The z transform of the process is written

$$S = Z\left(H_0\ \frac{1}{s(s+1)}\right) = \frac{B(z^{-1})}{A(z^{-1})}$$

$$= \frac{e^{-1}\ z^{-1} + (1-2e^{-1})\ z^{-2}}{(1-z^{-1})\ (1 - e^{-1}\ z^{-1})}$$

The controller which cancels the error in minimum time is given by the expression

$$D = \frac{A}{B}\ \frac{z^{-1}}{1-z^{-1}}$$

$$= \frac{1 - e^{-1}\ z^{-1}}{e^{-1}+(1-2e^{-1})z^{-2}}$$

The error, input and output sequences then take the following forms:

$$\epsilon = 1$$
$$u = D\,\epsilon$$
$$u = e - \frac{(1-e)^2\, z^{-1}}{1+(e-2)z^{-1}}$$
$$u = e-(1-e)^2 z^{-1}[1+(2-e)z^{-1}+(2-e)^2 z^{-2}+\ldots]$$
$$u = 2.72 - 2.95\, z^{-1} + 2.12\, z^{-2}+\ldots$$
$$s = F\,e = \frac{z^{-1}}{1-z^{-1}}$$

Fig. 4 shows the development of e(t), s(t) and u(t) with time.

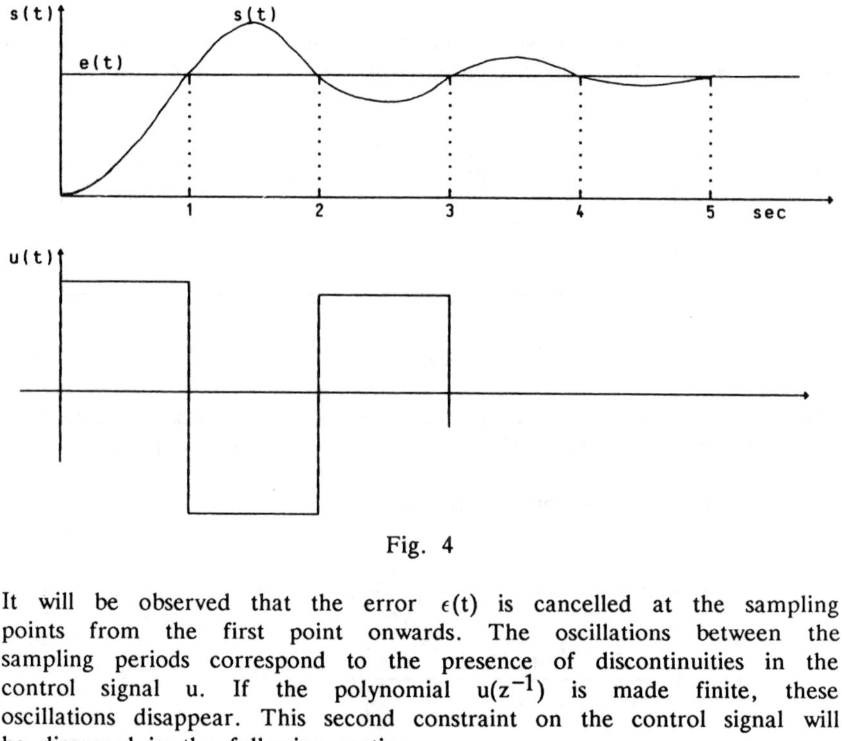


Fig. 4

It will be observed that the error $\epsilon(t)$ is cancelled at the sampling points from the first point onwards. The oscillations between the sampling periods correspond to the presence of discontinuities in the control signal u. If the polynomial $u(z^{-1})$ is made finite, these oscillations disappear. This second constraint on the control signal will be discussed in the following section.

### 3.2.2.2 – Minimal response time with finite control signal

#### a) – Finding the optimal controller

Rewriting the error and control sequences with reference to Fig. 2 gives

$$\epsilon = \frac{A_o X}{AX+BY} \frac{R}{L_o} \tag{11}$$

$$u = \frac{A_o Y}{AX+BY} \frac{R}{L_o} \tag{12}$$

The objective is to find the polynomials X and Y which are prime with respect to each other such that $u(z^{-1})$ and $\epsilon(z^{-1})$ are polynomials in $z^{-1}$ with minimum $\delta(\epsilon)$. The characteristic polynomial AX+BY, by contrast with the preceding case, cannot contain the factor $B^+$ since, if it did, then according to (11) and (12) X and Y would also have to contain it, which is impossible, since (X,Y) = 1. Therefore,

$$AX + BY = R^+ A_o^+$$

Thus the polynomial AX+BY contains the stable poles of the numerators of (11) and (12).

Furthermore, since X and Y cannot be simultaneously divisible by $L_o$, it follows that the problem is insoluble unless $L_o = 1$, in other words unless A is divisible by L.

To summarise, the following expression defines the optimal controller:

$$D = \frac{A_o^+ Y_1}{X} \tag{13}$$

where the polynomials X and $Y_1$ denote the minimal degree solution in X of the polynomial equation (14):

$$A_o^- LX + BY_1 = R^+ \tag{14}$$

**Important note**

The solution of this problem exists and is unique

a) if $L_o = 1$;

b) if B and L are prime with respect to each other. In fact, for relationship (14) to be soluble, it is necessary and sufficient that $A_o^- L$ and B should be prime with respect to each other.

The error and control sequences then become

$$\epsilon = R^- A_0^- X \tag{15}$$

$$u = R^- A_0 Y_1 \tag{16}$$

It may be noted that the duration of the transient is generally greater than in the previous case.

b) – **Application to the class of stable systems**

For stable processes having an integration and subject to a ramp reference input with the following discrete sequence:

$$E(z^{-1}) = \frac{1}{1 - z^{-1}}$$

the polynomial equation (14) becomes, noting that $A_0^- = 1$,

$$(1-z^{-1})\ X + B\ Y_1 = 1$$

The minimal solution in X of this equation is easily obtained:

$$Y_1 = \frac{1}{B(1)} \quad \text{(putting } z^{-1} = 1 \text{ in the preceding equation)}$$

and therefore

$$X = (1 - \frac{B}{B(1)})\ \frac{1}{1-z^{-1}}$$

The optimal controller then becomes

$$\boxed{D(z^{-1}) = \frac{A(z^{-1})}{B(1)-B(z^{-1})}}$$

The optimal error and control sequences are written thus:

$$\epsilon = X$$

$$u = \frac{A\ Y_1}{(1-z^{-1})}$$

It will be noted that the control sequence u is a polynomial in $Z^{-1}$, since according to the hypothesis the process contains an integration. Consequently, A is divisible by $(1-z^{-1})$.

### c) – Example of application

Returning to the example in the preceding section, where the transfer function of the process is written

$$S = \frac{B}{A} = \frac{e^{-1} z^{-1} + (1-2e^{-1})\, z^{-2}}{(1-z^{-1})\,(1 - e^{-1} z^{-1})}$$

the solution of minimal degree in X of the equation is written:

$$Y_1 = \frac{1}{B(1)} = \frac{1}{1-e^{-1}} = 1.58$$

$$X = \left(1 - \frac{e^{-1}z^{-1}+(1-2e^{-1})z}{1 - e^{-1}}\right) \frac{1}{1-z^{-1}}$$

$$X = 1 + x_1\, z^{-1} \quad \text{where } x_1 = \frac{1-2e^{-1}}{1-e^{-1}} = 0.418$$

The rational controller in $z^{-1}$ defining the controller is expressed thus:

$$D = \frac{A_o^{+}\, Y_1}{X} = 1.58\, \frac{(1-e^{-1}z^{-1})}{1+x_1\, z^{-1}}$$

and therefore

$$\epsilon = X$$

$$u = (1-e^{-1}z^{-1})\, Y_1 = 1.58 - 0.582\, z^{-1}$$

$$s(z^{-1}) = (1-x_1)\, z^{-1} + z^{-2} + z^{-3} + \ldots.$$

These results are shown in Fig. 5. The time values $\epsilon(t)$ and u(t) are cancelled after two seconds, and the output strictly follows the input from this point onwards.

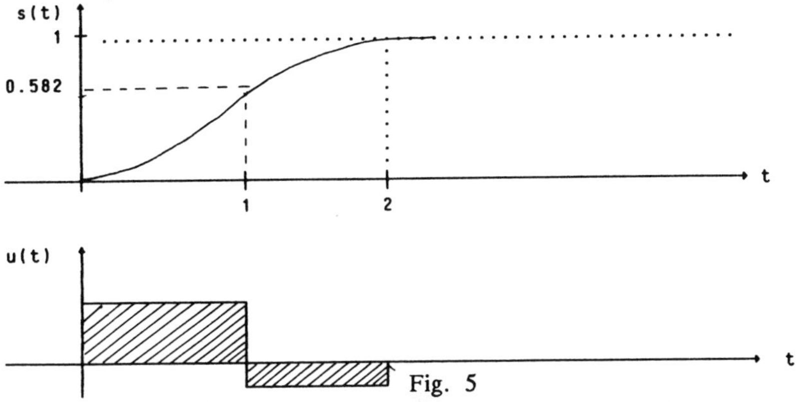


Fig. 5

### 3.2.2.3 – Minimal response time in the initial conditions

This section is concerned with the problem of cancellation in minimal time in the initial conditions which are not zero and are not known.

#### a) – Formulation of the problem

It is known that the effect of the initial conditions on the output of a process S with a transfer function $B(z^{-1})/A(z^{-1})$ may be represented by an additive output $s_2$ (see Fig. 6) having as its $z^{-1}$ transform an expression of the form $C(z^{-1})/A(z^{-1})$ where the structure and the parameters of the polynomial $C(z^{-1})$ depend on the initial conditions.

Given the diagram in Fig. 6:

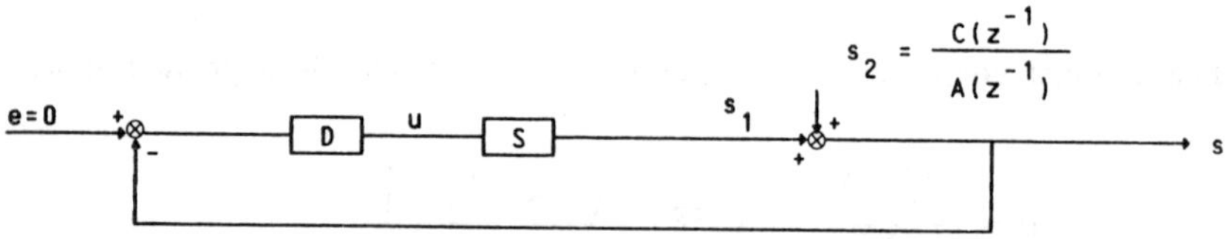


Fig. 6

the objective is to find an optimal controller, D, which cancels the output in minimal time in the following conditions:

* first case: without constraint on the control signal

* second case: by setting a finite control signal.

#### b) – Finding the optimal controller without constraint on the control signal

The z transform of the output is written thus:

$$s = \frac{AX}{AX + BY} \frac{C}{A} = \frac{CX}{AX + BY} \qquad (17)$$

Since the polynomial C is unknown because the initial conditions are unknown,

$$AX + BY = B^+$$

where

$$X = B^+ X_1$$

Thus, by choosing the minimal degree of the polynomial $X_1$ satisfying the polynomial equation, it is ensured that the sequence is polynomial and minimal.

The structure of the optimal controller is given by

$$D = \frac{Y}{B^+ X_1} \tag{18}$$

where $X_1$ and Y are polynomials of minimal degree in $X_1$ forming the solution of the polynomial equation

$$AX_1 + B^- Y = 1 \tag{19}$$

Since A and B are prime with respect to each other, this solution must exist and must be unique.

The output and control sequences take the following forms:

$$s = C X_1 \tag{20}$$

$$u = -\frac{CY}{AX + BY} = -\frac{CY}{B^+} \tag{21}$$

Since the polynomial Y cannot contain the expression $B^+$ unless X and Y are no longer prime with respect to each other, the control signal u cannot be finite.

c) – **Finding the optimal controller with a finite control signal**

Referring to Fig. 6, it will be seen that

$$s = \frac{CX}{AX + BY} \quad \text{and simultaneously} \quad u = -\frac{CY}{AX + BY} \tag{22}$$

and the optimal controller must simultaneously put into polynomial form the sequences $s(z^{-1})$ and $u(z^{-1})$ resulting from relationships (22).

The only possible condition, therefore, is

$$AX + BY = 1 \tag{23}$$

where

$$D = \frac{Y}{X} \tag{24}$$

where the polynomials X and Y represent the solution of minimal degree in X of equation (23).

Consequently,

$$s = CX$$

and

$$u = -CY \tag{25}$$

### 3.2.2.4 – Optimal controller using a fuller control structure

This section will demonstrate the ease with which the polynomial theory can be adapted to different control structures.

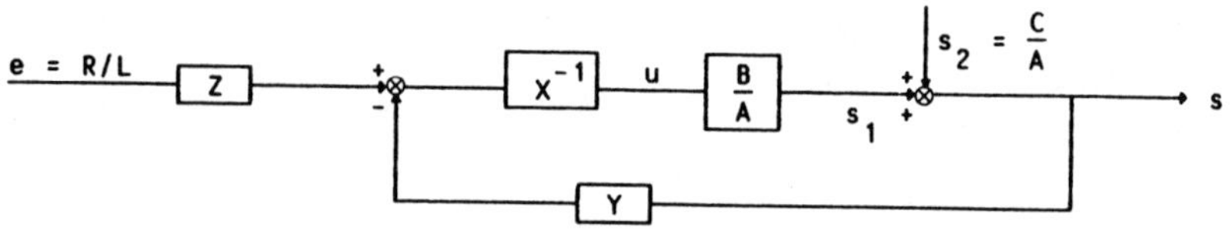


Fig. 7

The objective is to find the optimal controller, characterised by a fuller structure consisting of the three polynomials X, Y and Z as they appear in Fig. 7. This structure is defined by the equation

$$X\,u = -\,Y\,s + Z\,e$$

The discrete process is governed by the development equation

$$As = B\,u + C \quad \text{with } (A,B) = 1 \text{ and } (A,C) = 1$$

in which the polynomial C depends on the initial conditions.
Given an input sequence defined by

$$Le = R \quad \text{with } (R,L) = 1$$

the objective is to find the polynomials X, Y and Z such that the error sequence defined by $\epsilon = e-y$, together with the control sequence, are finite and of minimal order on the error.

#### a) – Finding the optimal controller with $C(z^{-1}) = 0$

As a general rule,

$$\epsilon = \left(1 - \frac{BZ}{AX + BY}\right)\frac{R}{L} - \frac{CX}{AX + BY} \qquad (26)$$

$$u = \frac{AZ}{(AX+BY)}\frac{R}{L} - \frac{CX}{AX+BY} \qquad (27)$$

If $C(z^{-1}) = 0$, then to obtain finite $\epsilon$ and u simultaneously, with minimal $\epsilon$, the following must be true:

$$AX + BY = R^+$$

In addition, according to equation (26), the difference $R^+ - BZ$ must be divisible by L, and therefore

$$LH + BZ = R^+$$

This last equation shows that Z cannot contain L; otherwise this polynomial relationship would have no solution.

Consequently, according to (27), the sequence u cannot be finite unless A is divisible by L, which implies that $L_0 = 1$.

To summarise, in order to obtain the three polynomials X, Y, and Z of the optimal controller, it is necessary

a) to find the polynomials H and Z of minimal degree in H which are the solution of the polynomial equation

$$LH + BZ = R^+ \qquad (28)$$

b) to determine the polynomials X and Y which are any solution of the polynomial equation

$$AX + BY = R^+ \qquad (29)$$

The error and control sequences then become, respectively,

$$\epsilon = R^- H \qquad (30)$$

$$u = R^- Z A_0 \qquad (31)$$

b) - **Finding the optimal controller with $C(z^{-1}) \neq 0$**

On this hypothesis, the initial conditions of the process are unknown. The polynomial C can therefore be any polynomial. According to relationships (26) and (27), to obtain the sequences $u(z^{-1})$ and $\epsilon(z^{-1})$ the following condition must be present:

$$AX + BY = 1$$

Therefore the divisibility of the polynomial 1−BZ by L entails the second condition:

$$LH + BZ = 1$$

In addition, as before, since Z cannot contain L, the sequence $u(z^{-1})$ will be finite if A is divisible by L, and therefore

$$L_0 = 1$$

The above results may therefore be summarised as follows.

Given the condition $L_0 = 1$, the three polynomials X, Y, and Z forming the optimal controller are the solutions of the two polynomial equations

$$LH + BZ = 1 \tag{32}$$

$$AX + BY = 1 \tag{33}$$

The error and control sequences, which are both finite, are expressed in the form

$$\epsilon(z^{-1}) = RH - CX \tag{34}$$

$$u(z^{-1}) = RA_0Z - CY \tag{35}$$

It will be noted that the solution of minimal degree in either X or H must be chosen, depending on whether $\delta(C) > \delta(R)$ or $\delta(C) < \delta(R)$.

### 3.2.3 – SYNTHESIS OF DISCRETE CONTROLLERS MINIMISING A QUADRATIC CRITERION

For the set of problems described in the previous section, the specification of a polynomial error sequence resulted in responses of the type

$$\epsilon = MX \tag{36}$$

where the polynomial X was the solution of a polynomial equation of the type

$$AX + BY = C \tag{37}$$

The response, in minimal time, of the error $\epsilon$ therefore leads to the choice of the unique solution of a minimal degree in X. However, it is possible to optimise a quadratic criterion while accepting a deterioration of the response time. This idea will be developed in the following text.

#### 3.2.3.1 – Formulation of the problem

Let $X_0$, $Y_0$ be the solution of minimal degree in X of equation (37). Acording to the theory of polynomial equations summarised in Appendix 1, the set of solutions of this equation is written

$$X = X_0 - B\,\mu$$

$$Y = Y_0 + A\,\mu$$

where $\mu$ is any polynomial in $z^{-1}$.

The error sequence takes the following form:

$$\epsilon = MX$$

$$\epsilon = M(X_o - B\ \mu)$$

i.e.

$$\epsilon = \epsilon_o - N\ \mu$$

It is therefore necessary to choose the coefficients of a polynomial $\mu$ of degree n such that the criterion

$$J = |\epsilon|^2 \text{ is minimal.}$$

The error set will be finite but will no longer be of minimal order in $z^{-1}$.

3.2.3.2 - **Finding the optimal polynomial $\mu$**

Consider a polynomial P with real or complex coefficients of the general form

$$P = \alpha_o + \alpha_1\ z^{-1} + \ldots + \alpha_n\ z^{-n}$$

Let $\overline{P}$ and $<P>$ denote the following expressions:

$$\overline{P} = \overline{\alpha}_o + \alpha_1\ z + \ldots. + \alpha_n z^n$$

$$<P> = \alpha_o \text{ (constant term in } P(z^{-1}))$$

then the quadratic norm of the polynomial P defined by

$$||P||^2 = |\alpha_o|^2 + \ldots. + |\alpha_n|^2$$

may be written

$$||P||^2 = <\overline{P}\ P>$$

If the quadratic criterion to be minimised is expressed with these new notations, we obtain

$$J = <\overline{\epsilon}\ \epsilon> = <\overline{\epsilon}_o\ \epsilon_o> - <\overline{\epsilon}_o\ N\mu> - <\overline{N}\ \epsilon_o\overline{\mu}> + <\overline{N}\ N\ \mu\ \overline{\mu}>$$

$\mu$ is the optimal polynomial if and only if $\overline{\mu}$ leads to the same value of the criterion. Since the first two terms are independent of $\overline{\mu}$, the polynomial must be such that

$$<z^k\ \overline{N}\ N\ \mu> = <z^k\ \overline{N}\ \epsilon_o> \quad \forall k = 0,\ldots.,n$$

or by making the polynomial $\mu$ explicit:

$$\begin{bmatrix} \langle\bar{N}N\rangle & \langle\bar{N}Nz^{-1}\rangle & \cdots & \langle\bar{N}Nz^{-n}\rangle \\ \langle z\bar{N}N\rangle & \cdots & & \vdots \\ \vdots & & & \vdots \\ \langle z^{n}\bar{N}N & \cdots & & \vdots \end{bmatrix} \begin{bmatrix} \mu_o \\ \vdots \\ \mu_n \end{bmatrix} = \begin{bmatrix} \langle \bar{N}\epsilon_o\rangle \\ \langle z\bar{N}\epsilon_o\rangle \\ \vdots \\ \langle z^{n}\bar{N}\epsilon_o\rangle \end{bmatrix}$$

This matrix expression supplies the coefficients $\mu_o,\ldots, \mu_n$ of the optimal polynomial $\mu(z^{-1})$.

## 3.2.4 – SYNTHESIS OF CONTROLLERS FOR MULTIVARIABLE SYSTEMS

### 3.2.4.1 – Formulation of the problem

The purpose of this section is to generalise the results obtained in section 3.1.2.4 to multivariables systems. We are concerned with the process (see Fig. 8) described by the following relationship.

$$A_1 \; s = B_1 \; u + C_1 \qquad (38)$$

where

$A_1 \in R^{pxs}$

$B_1 \in R^{pxm}$

$C_1 \in R^{px1}$

with p and m being the number of outputs and control signals of the process respectively.
$R^{nxm}$ denotes the set of polynomial matrices of dimension nxm.
$A_1$ and $B_1$ represent left-hand prime polynomial matrices, denoted ($A_1$, $B_1$) lhp. The mathematical Appendix 1 shows the significance of this property.

The notation $(A,B)_{rhp}$ will be used to describe polynomial matrices which are right-hand prime such that

$$A_1^{-1} B_1 = B A^{-1} \quad \text{where} \quad \begin{matrix} A \in R^{mxm} \\ B \in R^{pxm} \end{matrix} \qquad (39)$$

(transfer matrices of the process)

In expression (38), $C_1$ denotes a polynomial matrix whose structure and parameters are a function of any assumed initial conditions:

* S is the output vector of the process of dimension p
* u is the control vector of the process of dimension m.

Let the reference signal e be defined thus:

$$L_1 e = R_1 \quad \text{where } (L_1, R_1)_{lhp} \quad R_1 \in R^{px1}, \quad L_1 \in R^{pxp}$$

The problem of optimalising consists in finding a law of the type

$$X_1\ u = -Y_1\ s + Z_1\ e$$

such that the control and error vectors are cancelled at the end of a finite period.

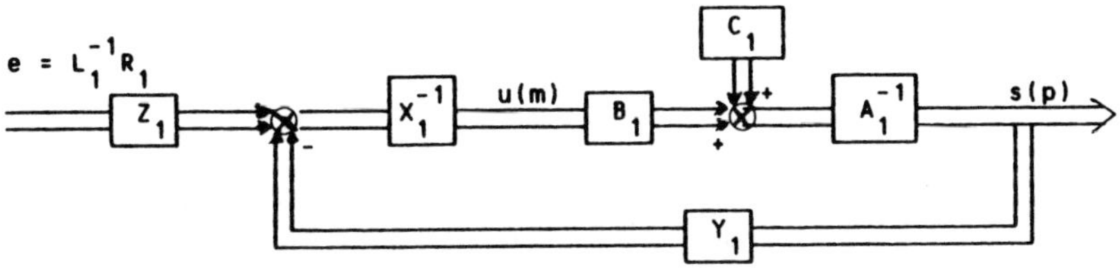


Fig. 8

### 3.2.4.2 - Finding the optimal controller

$\epsilon$ and u may be expressed thus:

$$\epsilon = e - s$$

$$\epsilon = [I-(A_1+B_1X_1^{-1}\ Y_1)^{-1}B_1X_1^{-1}Z_1]L_1^{-1}R_1-(A_1+B_1X_1^{-1}Y_1)^{-1}C_1 \quad (40)$$

$$U = (X_1+Y_1A_1^{-1}B_1)^{-1}\ (Z_1\ L_1^{-1}\ R_1 - Y_1\ A_1^{-1}\ C_1) \quad (41)$$

For the error and control vectors to be cancelled at the end of a finite period, the following conditions must be fulfilled:

1) $(X_1+Y_1A_1^{-1}B_1)^{-1}\ Y_1\ A_1^{-1} = M_1$ where $M_1 \in R^{mxp}$

2) $(A_1+B_1X_1^{-1}\ Y_1)^{-1} = M_2$ where $M_2 \in R^{pxp}$

3) $[I-(A_1+B_1X_1^{-1}\ Y_1)^{-1}B_1X_1^{-1}Z_1]L_1^{-1}R_1 = M_3$ where $M_3 \in R^{px1}$

4) $(X_1+Y_1A_1^{-1}B_1)^{-1}\ Z_1\ L_1^{-1}\ R_1 = M_4$ where $M_4 \in R^{mx1}$

It is known [Antsaklis (1979)] that there are polynomial matrices $X \in R^{pxp}$, $X_1 \in R^{mxn}$, $Y \in R^{mxp}$, and $Y_1 \in R^{nxp}$ related in the following way:

$$\left.\begin{aligned} X_1A + Y_1B &= I_m & AX_1 + YB_1 &= I_m \\ Y_1X - X_1Y &= 0_{mxp} & AY_1 - YA_1 &= 0_{mxp} \\ A_1B - B_1A &= 0_{pxm} & BX_1 - XB_1 &= 0_{pxm} \\ A_1X + B_1Y &= I_p & BY_1 + XA_1 &= I_p \end{aligned}\right] \quad (42)$$

These relationships show that $M_1 = Y$, $M_2 = X$, which satisfies conditions (1) and (2).

Given the above relationships, we may also write

$$M_3 = (I - BZ_1)\, L_1^{-1} R_1 \tag{43}$$

It is then evident that $L_1$ must be a right–hand divisor of $I-BZ_1$, in other words that the following equation must be true:

$$I - BZ_1 = P\, L_1 \tag{44}$$

In addition, taking the above relationships into account, $M_4$ can be written thus:

$$M_4 = A\, Z_1\, L_1^{-1}\, R_1 \tag{45}$$

It may therefore be deduced that $L_1$ must also be a right–hand divisor of $AZ_1$, so that

$$AZ_1 = QL_1 \tag{46}$$

The expression of $M_3$ multiplied by $A_1$ on the left gives

$$A_1 - A_1\, B\, Z_1 = A_1\, PL_1$$

or in other words

$$A_1 - B_1\, A\, Z_1 = A_1\, PL_1$$

The last two conditions (3) and (4) are therefore satisfied if

$$Z_1 = A^{-1}\, QL_1$$

where P and Q are the solution of the following matrix equation:

$$A_1P + B_1Q = A_1\, L_1^{-1}$$

The matrix $L_1$ must therefore be a right–hand divisor of $A_1$.

To summarise, the solution of the optimisation problem exists if $L_1$ is a right–hand divisor of $A_1$.

- First of all, the polynomials $X_1$ and $Y_1$ such that relationship (47) is complied with are found:

$$X_1 A + Y_1 B = I_m \tag{47}$$

- The polynomials P and Q complying with relationship (48) are then found:

$$A_1 P + B_1 Q = A_1 L_1^{-1} \tag{48}$$

- $Z_1$ is deduced from this by the following relationship:

$$Z_1 = A^{-1} Q L_1$$

Polynomials $X_1$, $Y_1$, and $Z_1$ determined in this way provide the optimal controller. Relationships (49) and (50) provide the final expression of the signals $\epsilon$ and u:

$$\epsilon = M_3 - M_2 C_1 = PR_1 - XC_1 \tag{49}$$

$$u = M_4 - M_1 C_1 = QR_1 - YC_1 \tag{50}$$

## *3.3 - POLE PLACEMENT CONTROLLERS*

This section describes conventional control strategies based on the placement of the poles (and zeros) of the closed loop system.

Initially (in 3.3.1) we consider the disturbance-free servo problem, consisting in the design of a controller such that the output of the system "follows the reference as well as possible".

The qualitative concept of "as well as possible" corresponds to the specification of system performance in dynamic and static terms, as regards both the output of the system and the control applied to the system. The performance will be expressed here in terms of placing poles, and zeros if necessary.

### 3.3.1 - DISTURBANCE-FREE SERVO PROBLEM

Since the control system has access to the reference e(t), the controller consists of two blocks:

- one block admitting $D_c(z^{-1})$ as the transfer function and corresponding to feedback;

- one block admitting $D_a(z^{-1})$ as the transfer function and corresponding to feedforward.

The block diagram of the closed loop system is shown in Fig. 8:

Fig. 8a: Digital control systems with feedforward and feedback. Case of the servo problem.

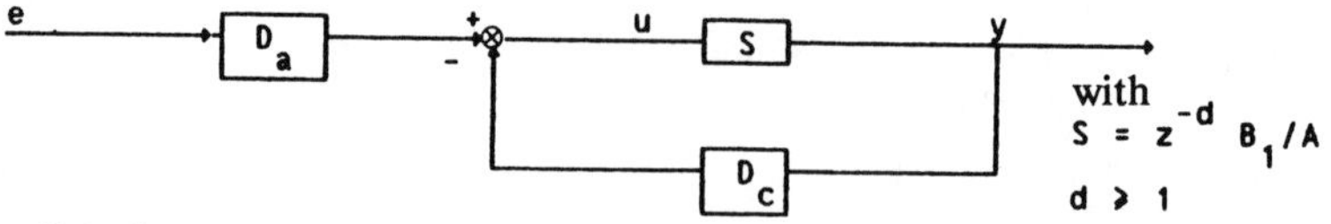


The digital control signal u(t) for the block diagram in Fig. 8 is calculated by means of the following expression:

$$u = D_a\, e - D_c\, y \tag{51}$$

The control law (51) associated with a strategy of pole and zero placement for the closed loop system will now be determined.

The strategy consists in specifying the desired behaviour of the closed loop system on the basis of the choice of the positioning of its poles and zeros. The control system is then determined in such a way that the closed loop system is stable and the transfer function relating the reference e(t) to the output y(t) is fixed by the following operator:

$$S_r = \frac{z^{-d}\, \gamma\, B_r}{A_r} \tag{52}$$

where $\gamma$ is a normalising gain and the polynomials $A_r$ and $B_r$ are assumed to be coprime, with $A_r(o) = 1$.

In the ideal case in which the transfer function of the system S is assumed to be exactly known, the stability of the closed loop system is provided by selecting a value of $A_r(z^{-1})$ such that its zeros are strictly outside the unit circle. For the sake of simplicity, it is assumed that the delay occurring in the model (52) is equal to the delay, d, of the system. In practice, it is sufficient to consider a model (52) with a delay $d' \geqslant d$. The normalising gain $\gamma$ must be made equal to $A_r(1)/B_r(1)$ so that the steady-state error is cancelled for a reference in step form.

First solution:

Letting

$$D_a = \gamma\, \frac{Z}{X}, \qquad D_c = \frac{Y}{X} \tag{53}$$

the output of the closed loop system may be written:

$$y = \frac{z^{-d}\, \gamma\, B_1 Z}{AX + z^{-d} B_1 Y}\, e \tag{54}$$

The problem of placing the poles and zeros of the closed loop system is therefore equivalent to the algebraic problem consisting in finding the polynomials X, Y and Z such that the following equality is respected:

$$\frac{z^{-d}\ \gamma\ B_1 Z}{AX + z^{-d} B_1 Y} = \frac{z^{-d}\ \gamma\ B_r}{A_r} \tag{55}$$

A problem of this type does not have a unique solution. Aström, Westerberg, and Wittenmark (1978) used this non-uniqueness to establish a general procedure for the calculation of the control law (see also Aström and Wittenmark (1979)).

Using equality (55), it is easily verified that, if the polynomial $B_r$ does not contain certain factors of $B_1$, then X must contain these factors; in other words, the controller cancels the open-loop zeros associated with these factors.

When the polynomial $B_1$ is factorised into $B_1^+\ B_1^-$, where

$B_1^+$ is the factor containing the stable and sufficiently well damped zeros of $B_1$, with $B_1^+(o) = 1$;

$B_1^-$ is the factor containing the unstable or poorly damped zeros of $B_1$, with $B_1^-(o) = b_o$;

and the controller is assumed to cancel the factor $B_1^+$, then

$$X = B_1^+ X_1 \tag{56}$$

$$B_r = B_1^- B_{r_1} \tag{57}$$

In general, the degree of the polynomial $AX_1 + z^{-d} B_1^- Y$ is higher than that of the polynomial $A_r$. This indicates that the polynomials Z and $AX_1 + z^{-d} B_1^- Y$ have common factors. It can be shown that these factors correspond to the poles of the observer, when the pole placement is realised by a state feedback (see, for example, Owens (1981)). Z is then factorised as:

$$Z = B_{r_1}\ Z_1 \tag{58}$$

where

$B_{r_1}$ corresponds to the desired closed-loop zeros;

$Z_1$ corresponds to the poles of the observer which may be chosen arbitrarily.

The control algorithm associated with the strategy of placing the poles and zeros of the closed loop system is summarised in Table 1 below:

1. Choice of the polynomials $A_r$, $B_{r_1}$, and $Z_1$
2. Factorisation of the polynomial $B_1$ into $B_1^+ B_1^-$
3. Solution of the polynomial equation

$$AX_1 + z^{-d} B_1^- Y = A_r Z_1 \tag{59}$$

with respect to the polynomials $X_1$ and Y
4. Calculation of the control signal using equation

$$u = \gamma \frac{B_{r_1} Z_1}{B_1^+ X_1} e - \frac{Y}{B_1^+ X_1} y \tag{60}$$

$$\text{with } \gamma = A_r(1)/ B_r(1) \tag{61}$$

Table 1: General pole-zero placement strategy for the servo problem.

Notes

1. The closed loop system is then written

$$B_1^+ Z_1 A_r[y(t) - y_r(t)] = 0 \tag{62}$$

where

$$y_r(t) \triangleq S_r e(t) \tag{63}$$

represents the output of the reference model with the transfer function $S_r$ defined in (52).

2. Since polynomials A and $B_1$ are assumed to be coprime, this is also true of A and $B_1^-$. Consequently the number of possible solutions of equation (59) is infinite. In practice, one of the following two minimal solutions will be chosen:

$$\delta(X_1) = \delta(B_1^-) + d-1 \qquad \delta(Y) = \text{Sup}[\delta(A)-1, \delta(A_r)+\delta(Z_1)-\delta(B_1^-)-d]$$

or

$$\delta(X_1) = \text{Sup}[\delta(B_1^-)+d-1, \delta(A_r)+\delta(Z_1)-\delta(A)], \delta(Y) = \delta(A)-1$$

3. Special cases

In the context of the self-tuning controllers, it is interesting to consider the two following special cases in which stage 2 of Table 1 can be eliminated, thus appreciably reducing the amount of calculation.

* Case in which the controller cancels all the process zeros.

Equations (56), (57), and (58) become

$$X = B_1 X_1 \tag{64}$$

$$B_r = B_{r_1} \tag{65}$$

and

$$AX_1 + z^{-d}Y = A_r Z_1 \tag{66}$$

This control strategy can only be used for minimum phase systems, since the controller compensates all the process zeros.

* Case in which the controller does not cancel the process zeros.

$$X = X_1 \tag{67}$$

$$B_r = B_1 B_{r_1} \tag{68}$$

and

$$AX_1 + z^{-d} B_1 Y = A_r Z_1 \tag{69}$$

By contrast with the previous special case, the use of this control strategy does not require the assumption of a system with minimum phase characteristics.

4. Interpretation of the control system as a system with a parallel reference model (Aström and Wittenmark (1979))

Using equations (56) and (59), the expression of u(t) (60) can be rewritten as

$$u = \frac{A}{B_1^+} \gamma \frac{B_{r_1}}{A_r} e - \frac{Y}{X} (y - y_r) \tag{70}$$

where $y_r$ is defined in (63).

Consequently the corresponding control system may be interpreted as a control system with a parallel reference model (see Fig. 9).

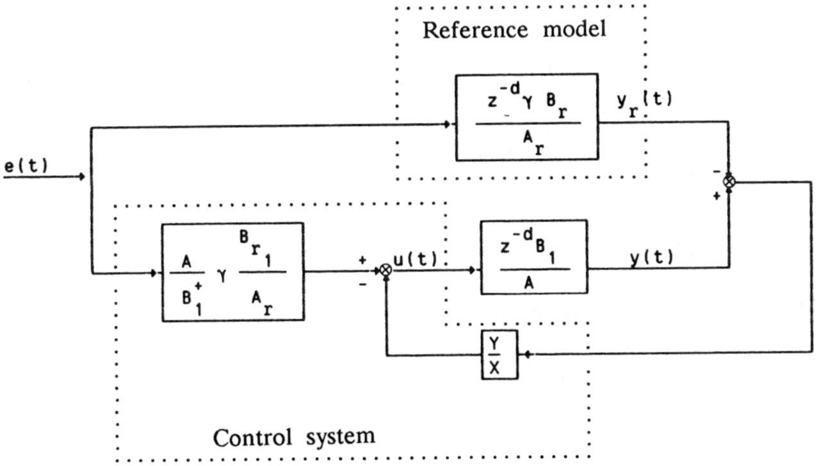


Fig. 9: Interpretation of the control system as a system with a reference model.

5. Example

Returning to the example in section 3.2.2.1, we find that

$$\frac{z^{-d}B_1}{A} = \frac{z^{-1}[e^{-1} + (1-2e^{-1})\, z^{-1}]}{(1-z^{-1})\ (1-e^{-1}\, z^{-1})}$$

Polynomials $A_r$, $B_{r_1}$, and $Z_1$ are chosen in such a way that

$$A_r = 1 - e^{-2}\, z^{-1}, \qquad B_{r_1} = Z_1 = 1$$

Consequently, polynomials ($X_1$, Y) which are the solution to (59) are defined as follows:

$$X_1 = 1, \qquad Y = 1+e^{-1}-e^{-2}-e^{-1}z^{-1}$$

and the control law (60) is given by

$$u = \frac{1 - e^{-2}}{e^{-1}+(1-2e^{-1})z^{-1}}\, e - \frac{1 + e^{-1}-e^{-2}-e^{-1}\, z^{-1}}{e^{-1} + (1-2e^{-1})z^{-1}}\, y$$

The output of the closed loop system is then written

$$(1-e^{-2}z^{-1})[e^{-1}+(1-2e^{-1})z^{-1}]\ [y(t) - y_r(t)] = 0$$

with

$$y_r(t) = \frac{z^{-1}(1-e^{-2})}{1-e^{-2}z^{-1}}\, e(t)$$

Second solution

The second solution is obtained by letting

$$D_a = \frac{\gamma B_{r_1}}{A_r}\, \frac{P}{X}, \qquad D_c = \frac{Y}{X} \tag{71}$$

with X being defined as in (56), and $X_1$ and Y being the solution of the following polynomial equation:

$$AX_1 + z^{-d}\, B_1^-\, Y = P \tag{72}$$

The closed loop system is then written

$$B_1^+\, P\ [y(t) - y_r(t)] = 0 \tag{73}$$

with $y_r(t)$ defined in (63).

Notes

Unlike (62), expression (73) shows that this second solution enables the servo error $y-y_r$ to be cancelled with a dynamic independent of the dynamic ($A_r$) of the reference model.

It should be noted that the polynomial P plays the same role as the product of the polynomials $Z_1$ and $A_r$ in the previous solution.

### 3.3.2 - SPECIAL CASE IN WHICH THE CONTROLLER CANCELS ALL THE PROCESS ZEROS (MINIMUM PHASE SYSTEM)

In this case,

$$X = B_1 X_1 \tag{74}$$

and

$$AX_1 + z^{-d} Y = P \tag{75}$$

It is then possible to interpret the control law in terms of d-step-ahead prediction of an auxiliary output $y^f(t)$ defined as the output y(t) filtered through the filter with transfer function $P(z^{-1})$.

Thus, taking the polynomial equation (75) into account,

$$y^f(t+d) \overset{\Delta}{=} P\ y(t+d) \tag{76}$$

$$= \frac{PB_1}{A} u(t) \tag{77}$$

$$= X\ u(t) + Y\ y(t) \tag{78}$$

It is easily verified that this expression (78) forms the optimal d-step-ahead predictor of $y^f(t)$, denoted $\hat{y}^f(t+d/t)$. Consequently the expression (51) for the control signal associated with the definitions (71) of $D_a$ and $D_c$ is also written

$$\hat{y}^f(t+d/t) = P\ y_r(t+d) \tag{79}$$

In other words, the corresponding control law is obtained by writing that the d-step-ahead prediction of the auxiliary output $y^f$ is equal to the desired output $Py_r$.

This control law was proposed by Landau and Lozano (1981) for the special case of a minimum phase system. Finally, it should be noted that, in the deterministic case, the d-step-ahead prediction error

$$\tilde{y}^f(t+d/t) \overset{\Delta}{=} y^f(t+d) - \hat{y}^f(t+d/t) \tag{80}$$

is zero.

### *3.4 – CONTROL LAWS BASED ON THE MINIMISATION OF A ONE STEP QUADRATIC CRITERION*

This section is concerned with the control of a single-variable system which is subject to a stochastic additive disturbance at the output level and which can be represented by an ARMAX model:

$$A\ y(t) = z^{-d}B_1\ u(t) + C\ n(t),\ C(o) = 1 \qquad (81)$$

where n(t) is a sequence of independent random variables (0, $\sigma$), and c is a stable polynomial.

Two control strategies will be considered:

- Control law with minimum output variance (Aström (1970)), called minimum variance control and obtained as its name indicates by minimising the output variance, i.e.

$$\underset{u(t)}{\text{Min}}\ \{E\ [y^2(t+d)\ /\ M_t]\} \qquad (82)$$

(E[.$M_t$] means conditional expectation [.] given the data set $M_t$ = {y(t), y(t−1),...., u(t−1), u(t−2),....} available at the time instant t).

- Control law with minimum generalised output variance (Clarke, Hastings and James (1971), Clarke and Gawthrop (1975, 1979)), called generalised minimum variance control and corresponding to the minimisation of the following one step quadratic criterion:

$$\underset{u(t)}{\text{Min}}\ \{E\{\{[\Gamma\ y(t+d) - Zy_r(t+d)]^2 + [\Lambda\ u(t)]^2\}/M_t\}\} \qquad (83)$$

where

$\Gamma$, Z and $\Lambda$ are polynomial weightings, with $\Gamma(o) = 1$;

$y_r(t)$ is the output of a reference model defined by means of the equation

$$y_r(t) \triangleq \frac{z^{-d_r}\gamma\ B_r}{A_r}\ e(t),\quad \gamma = \frac{A_r(1)}{B_r(1)}$$

$M_t$ represents the set of measurements available at the time instant t.

Since the control laws based on the minimisation of the one step quadratic criteria (82) and (83) are closely related to the solution of a problem of d-step-ahead prediction, the following section deals with the determination of a predictor of this type.

### 3.4.1 - OPTIMAL k-STEP-AHEAD PREDICTOR

Given the process y(t) modelled by equation (81) and the filtered output $y^f(t)$ defined by

$$y^f(t) \triangleq T\, y(t) \tag{84}$$

where T is a polynomial in $z^{-1}$ such that T(o) = 1, an attempt will be made to find the optimal k-step-ahead predictor (k ⩾ 1) of the process $y^f(t)$, denoted $\hat{y}^f(t+k/t)$, which minimises the cost function:

$$J = E\{[\tilde{y}^f(t+k/t)]^2\} \tag{85}$$

where

$$\tilde{y}^f(t+k/t) \triangleq y^f(t+k) - \hat{y}^f(t+k/t) \tag{86}$$

represents the k-step-ahead prediction error of $y^f$.

Equations (81) and (84) give

$$y^f(t+k) = \frac{B_1 T}{A}\, u(t+k-d) + \frac{CT}{A}\, n(t+k) \tag{87}$$

The last term of (87) may be separated into two parts:

- a predictable part which is a function of the signals available at the time instant t;

- a non-predictable part which is a function of the future values $\{n(\tau);\ \tau > t\}$ of the noise.

This separation into two parts may be carried out with the aid of the Euclid's algorithm which enables the division of polynomial CT by polynomial A up to the order k, thus:

$$\frac{CT}{A} = X_k + z^{-k}\,\frac{Y_k}{A} \tag{88}$$

or in an equivalent way:

$$CT = A\, X_k + z^{-k}\, Y_k \tag{89}$$

with

$$\delta(X_k) = k-1,\quad \delta(Y_K) = \mathrm{Sup}[\delta(A)-1,\ \delta(C)+\delta(T) - k] \tag{90}$$

and

$$C(o) = T(o) = 1 \rightarrow X_k(o) = 1 \tag{91}$$

By introducing (88) into (87) and deriving the expression of n(t) from (81), we then obtain

$$y^f(t+k) = X_k\ n(t+k) + \frac{Y_k}{C}\ y(t) + \frac{B_1X_k}{C}\ u(t+k-d) \qquad (92)$$

Let $\hat{y}^f$ denote any function of the measurements belonging to the set $M_t$ and calculate the cost function (85) for $\tilde{y}^f(t+k/t) = y^f(t+k) - \hat{y}^f$. Assuming that $k \leqslant d$, the term $X_k n(t+k)$ is then independent of the other terms of the second member of (92) and the cost J may be decomposed into the sum of two terms:

$$J = E\{[X_k\ n(t+k)]^2\} + [\frac{Y_k}{C}\ y(t) + \frac{B_1X_k}{C}\ u(t+k-d) - \hat{y}^f]^2 \qquad (93)$$

The minimisation of J with respect to $\hat{y}^f$ then gives the following value of the optimal k–step–ahead prediction of the signal $y^f$:

$$\hat{y}^f(t+k/t) = \frac{Y_k}{C}\ y(t) + \frac{B_1X_k}{C}\ u(t+k-d) \qquad (94)$$

To summarise, the process $y^f$ being modelled with the aid of the model (81) (84), the optimal k–step–ahead predictor of $y^f$ can be calculated in the two stages described in Table 2 below:

1. Find the polynomials $X_k$ and $Y_k$ which are the solutions of polynomial equation (89).
2. Calculate the optimal predictor by means of the recurrent equation

$$\hat{y}^f(t+k/t) = -\sum_{i=1}^{\delta(C)} c_i\hat{y}^f(t+k-i/t-i) + \sum_{i=0}^{\delta(Y_k)} y_i^{(k)}y(t-i)$$

$$+ \sum_{i=0}^{\delta(\beta_k)} \beta_i^{(k)}\ u(t+k-d-i) \qquad (95)$$

where

$$\beta_k(z^{-1}) \triangleq \sum_{i=0}^{\delta(\beta_k)} \beta_i^{(k)}\ z^{-i} \triangleq B_1\ X_k, \qquad \delta(\beta_k)=\delta(B_1)+k-1$$

$$Y_k(z^{-1}) \triangleq \sum_{i=0}^{\delta(Y_k)} y_i^{(k)}\ z^{-i} \qquad (96)$$

Table 2: Calculation of the optimal k–step–ahead predictor

Notes

1. It is easy to deduce from equations (92) and (94) the d–step–ahead prediction error:

$$\tilde{y}^f(t+k/t) = X_k\ n(t+k) \qquad (97)$$

and the variance of this prediction error is

$$E\{[\tilde{y}^f(t+k/t)]^2\} = \sum_{i=0}^{k-1} [x_i^{(k)}]^2 \sigma^2 \qquad (98)$$

In the special case of the optimal one-step-ahead predictor (k=1), we find that $\tilde{y}^f(t+1/t) = n(t+1)$, making it possible to verify that the white noise n(t) of model (81) corresponds to the one-step-ahead prediction error, in other words to the innovation process associated with the process $y^f$.

2. The optimal k-step-ahead predictor (in the sense of the minimisation of criterion (85) is linear in $\{y(\tau), u(\tau)\}$. According to (95), we may also conclude that this predictor is a dynamic system whose dynamic is fixed by the polynomial C which, by assumption, has all its zeros outside the unit circle. Consequently the k-step-ahead predictor is a stable system.

## 3.4.2 - MINIMUM VARIANCE REGULATOR

### 3.4.2.1 - Case of a minimum phase system

If T = 1 and k = d in the results of the previous section, then according to (92) we have

$$y(t+d) = X_d\, n(t+d) + \frac{Y_d}{C}\, y(t) + \frac{B_1 X_d}{C}\, u(t) \qquad (99)$$

where $X_d$ and $Y_d$ are solutions of the following polynomial equation:

$$C = AX_d + z^{-d}\, Y_d \qquad (100)$$

and

$$\delta(X_d)=d-1,\ \delta(Y_d) = \mathrm{Sup}\ (\delta(A)-1,\ \delta(C)-d) \qquad (101)$$

Because of the independence of $X_d\, n(t+d)$ with the two last terms of the second member of (99), the cost (82) to be minimised can be decomposed into the sum of two variances:

$$E\{y^2(t+d)/M_t\}=E\{[X_d n(t+d)]^2\} + E\{[\frac{B_1 X_d}{C}\, u(t) + \frac{Y_d}{C}\, y(t)]^2/M_t\} \qquad (102)$$

Since the first term of (102) is not a function of the variable u(t), the minimisation of this cost with respect to u(t) is obtained when the second term is zero, i.e.

$$\frac{B_1 X_d}{C}\, u(t) + \frac{Y_d}{C}\, y(t) = 0 \qquad (103)$$

The control law for the strategy of minimum output variance may therefore be calculated by the procedure described in Table 3:

| |
|---|
| 1. Find the polynomials $X_d$ and $Y_d$ which are solutions of polynomial equation (100) with (101)<br>2. Calculate the control signal u(t) from the expression<br>$$u(t) = -\frac{Y_d}{B_1 X_d} y(t) \qquad (104)$$ |

Table 3: Minimum output variance control law

Notes

1. If expression (104) for u(t) is introduced into (99), the closed loop system output can be written

$$y(t+d) = X_d \, n(t+d) \qquad (105)$$

and its variance is

$$E[y^2(t+d)] = \sum_{i=0}^{d-1} [x_i^{(d)}]^2 \sigma^2 \qquad (106)$$

Equation (105) shows that the output of the system controlled by a strategy of minimum output variance is a moving average process of order (d–1), equal to the d–step–ahead prediction error (see (97) with k = d). It is therefore possible to test the control optimality in the sense of criterion (82) by verifying that

$$E[y(t-\tau) \, y(t)] = 0 \qquad \text{for } \tau \geqslant d \qquad (107)$$

2. If expression (104) for u(t) is introduced into expression (94) of the k–step–ahead predictor, with k=d, T=1, we obtain

$$\hat{y}(t+d/t) = \frac{Y_d}{C} y(t) + \frac{B_1 X_d}{C} u(t) = 0 \qquad (108)$$

This means that the minimum output variance control law is such that the output d–step–ahead predictor coincides with the desired output, i.e. zero for the problem of stochastic regulation. The corresponding control law may therefore be obtained in the two following stages:

- Determination of the d–step–ahead predictor with d steps: $\hat{y}(t+d/t)$
- Calculation of u(t) in writing that this prediction is equal to the desired value, i.e. zero.

The minimum variance regulator is said to be of the "predictive type" because it implicitly calculates the d–step–ahead prediction of the system output.

3. According to (104), it may be concluded that the regulator associated with the strategy of minimum output variance cancels all the process zeros. This control strategy can therefore only be used for minimum phase systems.

According to (104) and (105), we have

$$u(t) = -\frac{Y_d}{B_1} n(t) \qquad (109)$$

If the system is non-minimum phase, the transfer function $1/B_1$ is unstable, and in this case the control signal u(t) no longer has a bounded variance.

### 3.4.2.2 - Case of a non-minimum phase system

Aström (1970) has proposed the following sub-optimal control law for a non-minimum phase system:

$$u(t) = -\frac{Y}{B_1^+ X} y(t) \qquad (110)$$

where X and Y are solutions of the polynomial equation

$$C = AX + z^{-d} B_1^- Y \qquad (111)$$

with

$$\delta(X) = \delta(B_1^- + d-1, \quad \delta(Y) = \mathrm{Sup}(\delta(A)-1, \ \delta(C)-\delta(B_1^-)-d) \qquad (112)$$

Therefore the control law (110) only cancels the stable part of $B_1$.

Subtracting (100) from (111), we have

$$A[X - X_d] = z^{-d} [Y_d - B_1^- Y] \qquad (113)$$

Since $z^{-d}$ and A are coprime, the polynomial X' may be defined such that

$$X - X_d = z^{-d} X' \qquad (114)$$

with $\delta(X') = \delta(B_1^-)-1$ if equation (111) is regular.

The closed-loop output of the system controlled by (110) can be written as:

$$y(t) = X\,n(t) \tag{115}$$

and according to (114) the variance of this output may be deduced:

$$E[y^2(t)] = \sum_{i=0}^{d-1} [x_i^{(d)}]^2 \sigma^2 + \sum_{i=0}^{\delta(X')} [x_i']^2 \sigma^2 \tag{116}$$

where the first sum corresponds to the strategy of minimum output variance (see (106)).

### 3.4.3 – GENERALISED MINIMUM VARIANCE CONTROLLER

This control law corresponds to the minimisation of criterion (83). Defining the polynomials $X_d$ and $Y_d$ as solutions of the equation

$$AX_d + z^{-d} Y_d = C\Gamma \tag{117}$$

and using the results of 3.4.1, with k = d and T = Γ, criterion (83) can be written

$$E\{[X_d n(t+d)]^2\} + \{\frac{Y_d}{C} y(t) + \frac{B_1 X_d}{C} u(t) - Zy_r(t+d)\}^2 + \{\Lambda u(t)\}^2 \tag{118}$$

Minimising (118) with respect to u(t) gives

$$b_o[\frac{Y_d}{C} y(t) + \frac{B_1 X_d}{C} u(t) - Zy_r(t+d)] + \lambda_o \Lambda\, u(t) = 0 \tag{119}$$

where $b_o$ is the constant coefficient of the expansion of $B_1X_d/C$. (Noting that $a_o = c_o = \gamma_o = 1$, we find that $b_o = b_{1o}$). The control signal can therefore be expressed as:

$$u(t) = \frac{CZ}{X} y_r(t+d) - \frac{Y}{X} y(t) \tag{120}$$

with

$$Y = Y_d,\ X = B_1 X_d + C\Lambda^M,\quad \Lambda^M = \frac{\lambda_o \Lambda}{b_{1o}} \tag{121}$$

If the value of u(t) obtained from (120–121) is carried into the equation of the model (81), the output of the closed loop system is written

$$y(t) = \frac{B_1 Z\, y_r(t) + X\, n(t)}{B_1\, \Gamma + A\, \Lambda^M} \tag{122}$$

From this expression for y(t), it may be concluded that the poles of the closed loop system are functions of the weightings Γ and $\Lambda^M$.

In order to calculate the steady-state servo error, a normalisation gain $\eta$ is introduced into the weighting polynomial Z, thus:

$$Z = \eta Z_1 \tag{123}$$

The gain $\eta$ is determined in such a way that the static gain of the transfer function relating $y_r(t)$ to y(t) is equal to unity; therefore,

$$\eta = \frac{\Gamma(1)}{Z_1(1)} + \frac{A(1)\ \Lambda^M(1)}{B_1(1)\ Z_1(1)} \tag{124}$$

The calculation of the control law corresponding to the strategy of minimisation of criterion (83) is summarised in Table 4.

| |
|---|
| 1. Choice of weightings Γ, Z and Λ<br>Determination of the normalisation gain $\eta$ by means of formula (124) |
| 2. Determination of polynomials $X_d$ and $Y_d$ which are solutions of equation (117) |
| 3. Calculation of u(t) using equations (120), (121) |

Table 4: Minimum generalised output variance control law

Notes

1. Deterministic case

In the case of a deterministic model obtained by making n(t) equal to zero in (81), criterion (83) becomes

$$\min_{u(t)} \{[\Gamma\ y(t+d) - Zy_r(t+d)]^2 + [\Lambda\ u(t)]^2\} \tag{125}$$

The corresponding control law may be derived in the same way as before by assuming that

$$C = 1 \tag{126}$$

The resulting algorithm is shown in Table 5:

| |
|---|
| 1. Identical to (1) in Table 4<br>2. Determination of polynomials $X_d$ and $Y_d$ which are solutions of the equation<br>$AX_d + z^{-d} Y_d = \Gamma$ (127)<br>3. Calculation of u(t) using the equations<br>$u(t) = \frac{Z}{X} y_r(t+d) - \frac{Y}{X} y(t)$ (128)<br>with<br>$Y = Y_d, \quad X = B_1 X_d + \Lambda^M, \quad \Lambda^M = \frac{\lambda_o \Lambda}{b_{1o}}$ (129) |

Table 5: Deterministic case

Applying control law (128), the output of the closed loop system is

$$y(t) = \frac{B_1 Z}{B_1 \Gamma + A\Lambda^M} y_r(t) \qquad (130)$$

Therefore we may conclude, as in the stochastic case, that the poles of the closed loop system are functions of the weightings $\Gamma$ and $\Lambda^M$. The control law described in Table 5 may therefore be used to place the poles of the closed loop system. The corresponding algorithm is described in Table 6:

| |
|---|
| 1. Choice of a polynomial P characterising the desired pole placement<br>2. Determination of weightings $\Gamma$ and $\Lambda^M$ which are solutions of the equation<br>$B_1 \Gamma + A\Lambda^M = P$ (131)<br>3. Determination of polynomials $X_d$ and $Y_d$ which are solutions of equation (127)<br>4. Calculation of the control signal using equations (128) and (129) |

Table 6: Pole placement algorithm using the weightings of a one step quadratic criterion.

Table 7 shows the expression of the transfer function $y(z)/y_r(z)$ obtained by applying the control procedure described in Table 6 for different choices of the polynomials P and $\Lambda^M$.

| P | $\Lambda^M$ | $y/y_r$ | Cancellation of the open-loop zeros ($B_1$) | Placement of supplementary zeros (Z) |
|---|---|---|---|---|
| ∀ | ≠0 | $B_1Z/P$ | NO | YES |
| $P = \Gamma$ | 0 | $Z/P$ | YES | YES |
| $P=P_oZ/P_o$ ∀ | ≠0 | $B_1/P_o$ | NO | NO |
| $P=P_o$ $Z=\Gamma$ | 0 | $1/P_o$ | YES | NO |

Table 7: Expression of $y/y_r$ for different choices of P and $\Lambda^M$

It should be noted that the control system cancels the open-loop zeros if $\Lambda^M = 0$. Therefore this value $\Lambda^M$ can only be envisaged in the case of a minimum phase system.

2. Choice of a weighting $\Lambda = 0$

Expressions (120) and (122) become

$$u(t) = \frac{CZ}{B_1X_d} y_r(t+d) - \frac{Y_d}{B_1X_d} y(t) \tag{132}$$

$$y(t) = \frac{Z}{\Gamma} y_r(t) + \frac{X_d}{\Gamma} n(t) \tag{133}$$

As indicated above, the controller defined by (132) cancels the open-loop zeros, making it necessary for the controlled system to have minimum phase characteristics.

If in addition $y_r = 0$ (as in the case of the pure stochastic realisation problem), the choice $\Gamma = 1$ leads to the minimum output variance control law.

3. Choice of weighting $\Lambda = (1-z^{-1})\Lambda'$

The normalisation gain $\eta$ becomes independent of the system parameters A and $B_1$ and is equal to $\Gamma(1)/Z_1(1)$ (see equation (124)).

4. It must be noted that, for the calculation of u(t) by formula (120), the output of the reference model $y_r$ must be known at the instant (t+d), or alternatively the reference e must be known at the instant $(t+d-d_r)$. Consequently, for $d_r < d$, the reference e must be known in advance (pointing problem) or predicted from observations (tracking problem).

5. Interpretation as a minimum output variance strategy

It will be demonstrated that the minimisation of criterion (83) is equivalent to the minimisation of the following auxiliary output variance:

$$\Phi(t+d) \overset{\Delta}{=} \Gamma\, y(t+d) - Z\, y_r(t+d) + \Lambda^M u(t) \tag{134}$$

Taking expression (118) into account, the auxiliary output $\Phi$ can also be written as:

$$\Phi(t+d) = \frac{Y_d}{C}\, y(t) + \frac{B_1 X_d}{C}\, u(t) - Z\, y_r(t+d) + \Lambda^M u(t) + X_d n(t+d) \tag{135}$$

Therefore

$$\underset{u(t)}{\mathrm{Min}^2} \{E\{\Phi^2(t+d)/M_t\}\} = \underset{u(t)}{\mathrm{Min}} \{[\frac{Y_d}{C}\, y(t) + \frac{B_1 X_d}{C}\, u(t) - Z y_r(t+d) + \Lambda^M$$
$$+ \Lambda^M u(t)]^2 + E\{[X_d\, n(t+d)]^2\}\} \tag{136}$$

and the minimisation of (136) obtained by cancelling the partial derivative with respect to u(t) leads to:

$$\hat{\Phi}(t+d/t) = \frac{Y_d}{C}\, y(t) - Z\, y_r(t+d) + [\frac{B_1 X_d}{C} + \Lambda^M]\, u(t) = 0 \tag{137}$$

which is identical to equation (119).

To summarise:

$$\underset{u(t)}{\mathrm{Min}} \{E\{\{[\Gamma\, y(t+d) - Z\, y_t(t+d)]^2 + [\Lambda\, u(t)]^2\}/M_t\}\}$$

$$\updownarrow \qquad\qquad \updownarrow$$

$$\underset{u(t)}{\mathrm{Min}} \{E[\Phi^2(t+d)/M_t]\} \longleftrightarrow \hat{\Phi}(t+d/t) = 0$$

The minimisation of criterion (83) therefore provides a control law identical to that obtained by the strategy of the minimum output variance applied to the auxiliary output $\Phi$. Consequently it is possible to interpret this control law in terms of the d–step–ahead prediction of $\Phi$, u(t) being calculated with the aid of the equation

$$\hat{\Phi}(t+d/t) = 0 \tag{138}$$

This version of the control law will be used in Chapter 6 to develop a direct adaptive control algorithm.

It should be noted that

$$\tilde{\Phi}(t+d/t) \triangleq \Phi(t+d) - \hat{\Phi}(t+d/t) = X_d \ n(t+d) \tag{139}$$

Since the polynomial $X_d$ is of degree (d−1), the same tests of optimality as those described by equation (107) apply in replacing y by $\Phi$.

6. Another expression of u(t)

Since the reference signal $y_r$ is assumed to be known by anticipation or prediction, according to the definition of $\Phi$ (134), we have:

$$\hat{\Phi}(t+d/t) = \hat{y}^f(t+d/t) - Z\ y_r(t+d) + \Lambda^M u(t) \tag{140}$$

where the filtered signal $y^f$ is defined by

$$y^f(t) \triangleq \Gamma\ y(t) \tag{141}$$

Applying the results of section 3.4.1 with T = $\Gamma$ and k = d, formula (94) gives the expression of the d-step-ahead predictor of the signal $y^f$:

$$\hat{y}^f(t+d/t) = \frac{Y_d}{C} y(t) + \frac{B_1 X_d}{C} u(t) \tag{142}$$

Equation (119), used to calculate the control signal u(t), may therefore also be written

$$b_{1o}[\hat{y}^f(t+d/t) - Z\ y_r(t+d)] + \lambda_o\ \Lambda\ u(t) = 0 \tag{143}$$

so that u(t) may alternatively be written

$$\boxed{u(t) = \frac{1}{\Lambda^M} [Z\ y_r(t+d) - \hat{y}^f(t+d/t)]} \tag{144}$$

It should be noted, however, that the prediction $\hat{y}^f$ appearing in the second member of (144) is itself a function of u(t) according to (142).

As will be seen in Chapter 6, this second way of writing the control law as a function of $\hat{y}^f$ has certain advantages over the first expression of u(t) as a function of $\Phi$ (equation 137). It has been used by Clarke and Gawthrop (1979 a and b, 1981) to develop a second form of direct adaptive control algorithm.

### 3.4.4 – NUMERICAL EXAMPLE

Consider the controlled system represented by the following ARMAX model:

$$A\ y(t) = z^{-d}\ B_1\ u(t) + C\ n(t) \qquad (145)$$

where

$$A(z^{-1}) = 1-0.5\ z^{-1}, \quad B_1(z^{-1}) = C(z^{-1}) = 1, \quad d = 2 \qquad (146)$$

$\sigma$ designates the standard deviation of the noise n(t).

### 3.4.4.1 – Study of the two step predictor of the output y(t)

The application of the results of 3.4.1 with T = 1 and k = d gives

$$C\ \hat{y}(t+d/t) = Y_d\ y(t) + B_1\ X_d\ u(t) \qquad (147)$$

where $X_d$ and $Y_d$ are solutions of the polynomial equation

$$C = AX_d + z^{-d}\ Y_d \qquad (148)$$

where

$$\delta(X_d) = d-1, \quad \delta(Y_d) = \text{Sup}\ [\delta(A)-1,\ \delta(C)-d] \qquad (149)$$

Equalising the terms with the same power in the two members of equation (148), we obtain

$$X_2 \triangleq x_o^{(2)} + x_1^{(2)}\ z^{-1} = 1+0.5\ z^{-1}, \quad Y_2 \triangleq y_o^{(2)} = 0.25 \qquad (150)$$

Consequently the equation for the two step predictor is as follows:

$$\boxed{\hat{y}(t+2/t) = 0.25\ y(t) + (1 + 0.5\ z^{-1})\ u(t)} \qquad (151)$$

The two-step prediction error $\tilde{y}(t+2/t)$ is given by formula (97) and is as follows:

$$\tilde{y}(t+2/t) = (1 + 0.5\ z^{-1})\ n(t+2)$$

and the variance of this prediction error is

$$E[\tilde{y}^2(t+2/t)] = 1.25\ \sigma^2$$

The block diagram for this predictor is given below (Fig. 10).

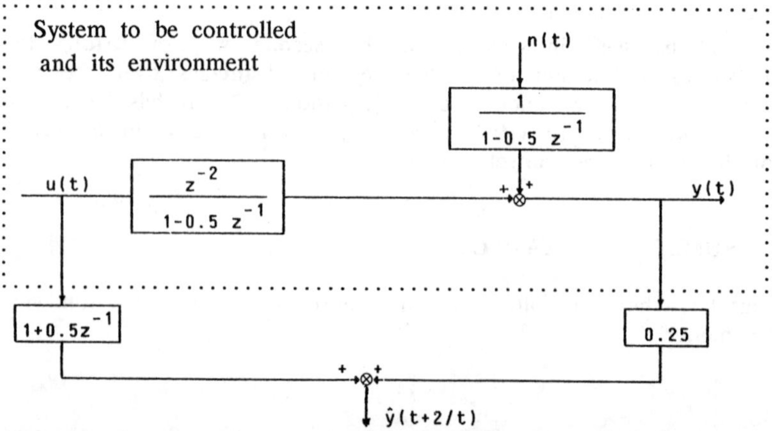


Fig. 10: Block diagram of the 2–step predictor.

3.4.4.2 – **Calculation of the control law for minimum generalised output variance**

This section is concerned with the closed loop system controlled by a control law corresponding to the minimisation of the following one-step quadratic criterion:

$$\underset{u(t)}{\text{Min}}\ \{E\{\{[y(t+d) - \eta y_r(t+d)]^2 + \lambda\ u^2(t)\}/M_t\}\} \tag{152}$$

where $y_r(t)$ is the reference signal to be followed and $\lambda \geqslant 0$.

The application of formulae (120) and (121) with

$\Gamma = 1$, $Z = \eta$, and $\Lambda = \sqrt{\lambda}$

gives

$$u(t) = \frac{\eta C}{B_1X_2 + \dfrac{\lambda}{b_{1o}} C}\ y_r(t+2) - \frac{Y_2}{B_1X_2 + \dfrac{\lambda}{b_{1o}} C}\ y(t) \tag{153}$$

therefore

$$\boxed{u(t) = \frac{\eta}{1+\lambda+0.5\ z^{-1}}\ y_r(t+2) - \frac{0.25}{1+\lambda+0.5\ z^{-1}}\ y(t)} \tag{154}$$

As indicated in 3.4.3 (Note 4), the calculation of u(t) by formula (154) requires that the reference $y_r(t)$ be known in advance or that a two-step prediction of it be carried out.

The block diagram for this control law is shown in Fig. 11.

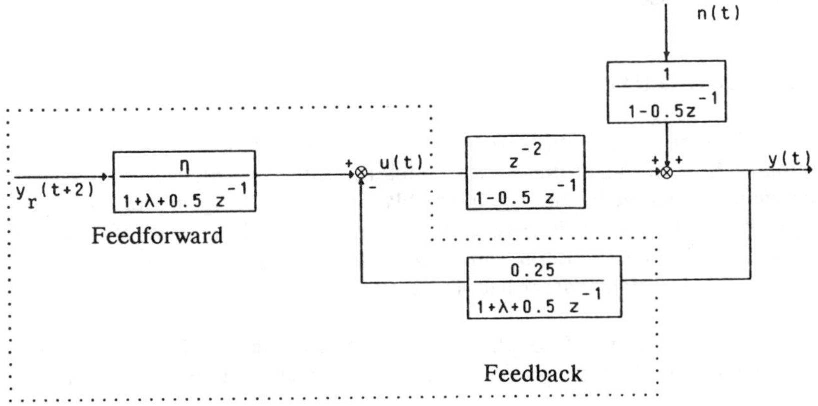


Fig. 11: Block diagram of the control system.

The output of the closed loop system can then be calculated by using formula (122); thus, for $t \geqslant 0$:

$$\boxed{y(t) = \frac{\eta}{1+\lambda-0.5\lambda z^{-1}} \; y_r(t) + \frac{1 + \lambda + 0.5\, z^{-1}}{1+\lambda-0.5\lambda z^{-1}} \; n(t)} \qquad (155)$$

The normalisation gain $\eta$ may be calculated in such a way that the static gain of the transfer function relative to the reference $y_r$ is equal to unity. Application of formula (124) gives

$$\eta = 1 + \frac{\frac{\lambda}{b_{1o}} A(1)}{B_1(1)}$$

$$\boxed{\eta = 1 + 0.5\,\lambda} \qquad (156)$$

In response to the stochastic disturbance n(t), the variance of the servo error for the above control signal is as follows:

$$E(\epsilon^2(t)) = \frac{1.25 + 2.5\,\lambda + \lambda^2}{1 + 2\,\lambda + 0.75\,\lambda^2}\,\sigma^2$$

and therefore

$$\text{for } \lambda = 1 \qquad E(\epsilon^2(t)) = \frac{19}{15}\,\sigma^2$$

$$\text{for } \lambda = 0 \qquad E(\epsilon^2(t)) = 1.25\,\sigma^2$$

For $\lambda = 0$, the variance of the servo error is minimised, and the controlled system behaves as a system with minimal response time. For $\lambda = 1$, the servo performance is degraded in terms of response time and error variance. However, the amplitude of the control signal is less in this last case.

## *3.5 – CONCLUSION*

Whether the objective is a minimal response time or the minimisation of quadratic criteria (deterministic or stochastic), the use of polynomial methods leads inevitably to the solution of polynomial equations in the single-variable case. The flexibility of this new approach is also useful in the synthesis of the controllers for multivariable systems described in section 3.2.4.

The synthesis of digital controllers for **stochastic** linear systems, touched on in the last section, will form the subject of Chapter Five.

### *REFERENCES*

K.J. Aström, Introduction to stochastic control theory, Academic Press, v. 70, New York, 1970

K.J. Aström, B. Westerberg, and B. Wittenmark, Self tuning controllers based on pole-placement design - Report TFRT 3148, Dept. of Automatic Control, Lund Institute of Technol., Sweden, May 1978

K.J. Aström and B. Wittenmark, Self tuning controllers based on pole-zero placement - Report TFRT 7180, Dept. of Automatic Control, Lund Institute of Technol., Sweden, Oct. 1979, published in IEE Proc., v. 127, pt. D, No. 3, pp. 120-130, May 1980

D.W. Clarke and P.J. Gawthrop, Self tuning controller, Proc. IEE, v. 122, No. 9, pp. 929-934, Sept. 1975

D.W. Clarke and P.J. Gawthrop, Self tuning control, Proc. IEE, v. 126, No. 6, pp. 633-640, June 1979

D.W. Clarke and P.J. Gawthrop, Implementation and application of microprocessor-based self tuners - Proc. 5th IFAC Symposium on Identification and System Parameter Estimation, Darmstadt, pp. 197-208, Sept. 1979, published in Automatica, v. 17, No. 1, pp. 233-244, 1981

D.W. Clarke and R. Hastings James, Design of digital controllers for randomly disturbed systems - Proc. IEE, v. 118, No. 10, pp. 1503-1506, Oct. 1971

I.D. Landau and R. Lozano, Unification of discrete time explicit model reference adaptive control design, Automatica, v. 17, No. 4, pp. 593-611, 1981

D.H. Owens, Multivariable and optimal systems, Academic Press, 1981

# CHAPTER 4

# PRINCIPLES OF INTERNAL MODEL CONTROL

## 4.1 – INTRODUCTION AND DEFINITIONS

The ergonomist analysing the work of a human operator normally assumes that the operator constructs an operating image of the system or process that he is controlling. A driver is aware of the reactions of his vehicle. The operator of an industrial process has an idea of the response times and gains of the various parts of the process.

This **operating image** is always abstract and can be reproduced by a mathematical model. This abstract representation of the process made by the operator is called the **internal model**.

A process is generally a complex system in which the different variables interact. Its operation is always subject to **constraints** and to **disturbances** arising from the environment.

The internal model represents the process with a degree of accuracy which, although varying according to the desired performance, is always imperfect. This difference of behaviour between the process and its internal model must always be taken into account in the development of algorithms for the control of the process. This point is examined further below.

A coherent solution to the problems posed by the control of processes can only be obtained by a global approach which takes into account all the variables and their interactions, allowing for the constraints of application and operation, and which provides a range of functions covering all the modes of operation of the process.

These imperatives make it necessary to define a realistic, powerful, general–purpose control method which is suitable for **multivariable and non–linear processes**.

The resources available for achieving this aim are as follows.

- **Equipment**: rapid digital computers having sufficient memory to store information obtained from past experience;

- **Methodology**: methods capable of correctly modelling complex processes and algorithms capable of using these models in real time.

If the human operator is replaced by a digital computer provided with an internal model of the process, the model will be available at all times to enable a command or a strategy to be selected.

This approach leads to the development of control algorithms which are more or less heuristic and therefore fundamentally different from the conventional approaches to control, in which a controller acts on the variation between the real output and the desired output of the process.

The limitations of the conventional approach are well known. Digital controllers are generally very sensitive to the quality of the model and variations in the process.

The **robustness** of a controller or a control law designates its ability to tolerate divergences between the model and the process. The robustness is generally measured by sensitivity studies and by the application of stability criteria.

**The principles of internal model control** correspond to the specifications described above.

First of all, we shall show (see 4.2) how it is possible to move from a conventional concept of regulation to internal model control for the compensation of a **measured** variation between the desired and the observed behaviours of the process.

We shall then establish (see 4.3) the basic principles of internal model control, making it possible to compensate the **measured and predicted** variations between the desired and estimated behaviours of the process.

Since any representation may be used for the internal model, the example taken (see 4.4) is the case of a multivariable system represented in state form.

## *4.2 - TRANSFERRING FROM CONVENTIONAL REGULATION TO INTERNAL MODEL CONTROL*

### 4.2.1 - PRINCIPLE

Consider the conventional closed loop system (Fig. 1) of a process represented by its sampled transfer function $s(z)=Z(B_0(p)\ S(p))$, a controller D(z), a signal e(z), and a stationary disturbance p(z) at the output of the process.

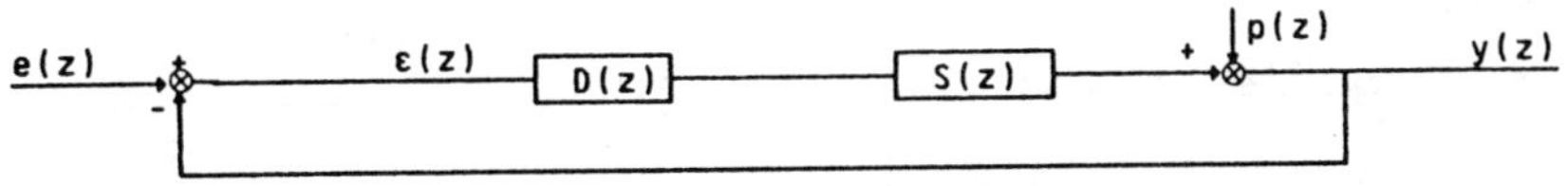


Conventional digital control loop

Fig. 1

Assuming that a model of the process, $S_M(z)$, is available, it is now necessary to connect this into the loop.

Fig. 2 below shows how $S_M(z)$ is brought into the loop without modifying the control structure.

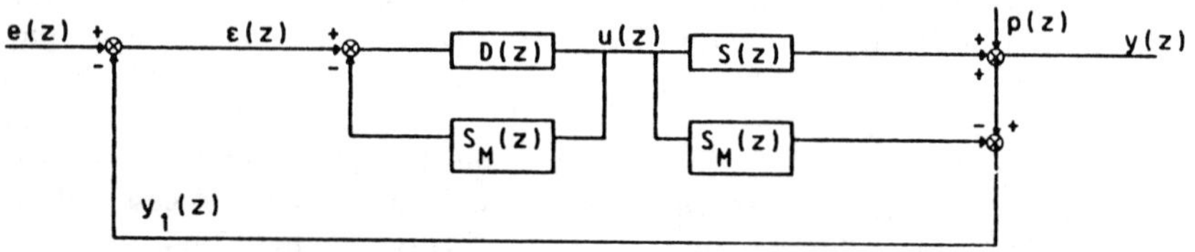


Introduction of the internal model into the control structure

Fig. 2

If the control signals u(z) of the two cases are compared, we find that

– in the case of Fig. 1:

$$u(z) = D(z)\ [e(z) - y(z)] \tag{1}$$

– in the case of Fig. 2:

$$u(z) = D(z)\ [\epsilon(z) - S_M(z)\ u(z)] \tag{2}$$

$$\epsilon(z) = e(z) - y_1(z)$$

$$y_1(z) = y(z) - S_M(z)\ u(z)$$

Expressions (1) and (2) are therefore equal, there being an equivalence between the two representations.

In the conventional control system, it is assumed that the model of the process is perfect, and this is used to determine the controller or control law.

In the second system, the variation between the process and its representation is taken into account. If the loop $[D(z), S_M(z)]$ is considered to be a controller $D_1(z)$, the control structure shown in Fig. 3 is obtained.

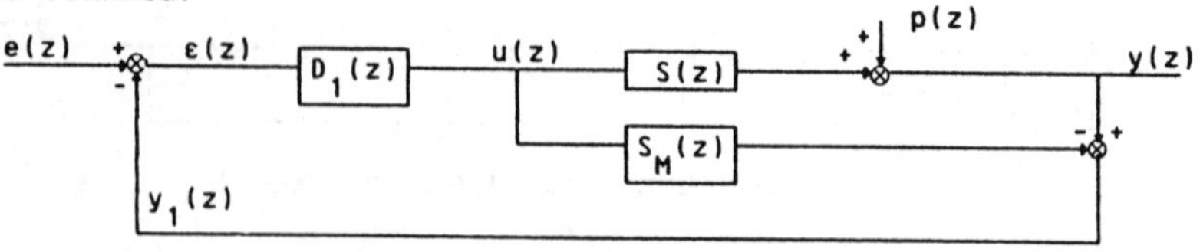


Fig. 3: Structure of control by internal model

Thus the conventional control system has been transformed progressively into an internal model control system by introducing the model of the process into the control loop in an explicit way. The properties of this control structure will now be examined.

### 4.2.2 - PROPERTIES

#### 4.2.2.1 - Calculation of the expressions of the control signals u(z) for feedback, $y_1(z)$, and output, y(z)

At this stage of the study, the constraints present at the level of the control signal u(z), namely saturation of amplitude and/or speed, will not be taken into account; the method of allowing for these will be described in 4.3.3.

The feedback signal $y_1(z)$ is written

$$y_1(z) = S(z)\ u(z) + p(z) - S_M(z)\ u(z) \tag{3}$$

where

$$u(z) = D_1(z)\ [e(z) - y_1(z)]$$

$$y_1(z) = [S(z) - S_M(z)]\ D_1(z)\ [e(z) - y_1(z)] + p(z) \tag{4}$$

Equation (4) shows that the inclusion of $y_1(z)$ in the loop causes the variation between the process and the internal model to intervene directly. Therefore,

$$y\ (z) = \frac{p(z) + [S(z) - S_M(z)]\ D_1(z)\ e(z)}{1 + [S(z) - S_M(z)]\ D_1(z)} \tag{5}$$

a) - **The ideal case**

With a perfectly matching internal model, $S_M(z) = S(z)$, then $y_1(z) = p(z)$

The loop structure enables the disturbance p(z) to be estimated.

If the disturbance is zero, $p(z) = 0 \rightarrow y_1(z) = 0$, we return to an open loop structure; but since $D_1(z)$ contains the model of the process, we return to the conventional control structure, thus:

Control structure in the ideal case

Fig. 4

$$y(z) = S(z)\ D_1(z)\ e(z) \tag{6}$$

$$D_1(z) = \frac{D(z)}{1 + S(z)\ D(z)}$$

and therefore

$$y(z) = \frac{S(z)\ D(z)}{1 + S(z)\ D(z)}\ e(z) \tag{7}$$

which is the expression for the conventional structure in Fig. 1.

Consequently, in the ideal case, the internal model approach contains the conventional approach.

b – **General case**

To summarise, we may say that control by internal model contains conventional control, that it enables the disturbances to be estimated, and that it uses as a feedback signal the variation between the behaviour of the process and that of its internal model.

The expressions of u(z) and y(z) will now be defined:

$$\epsilon(z) = \frac{e(z) - p(z)}{1 + [S(z) - S_M(z)]\ D_1(z)} \tag{8}$$

$$u(z) = D_1(z) \left[\frac{e(z) - p(z)}{1 + [S(z) - S_M(z)]D_1(z)}\right] \tag{9}$$

$$y(z) = S(z)\ D_1(z) \left[\frac{e(z) - p(z)}{1 + [S(z) - S_M(z)]D_1(z)}\right] \tag{10}$$

4.2.2.2 – **Stability analysis**

The stability of the control law and that of the loop as a whole will be studied in succession. The stability of a linear sampled system is given by the position of the roots of the characteristic equation in relation to the unit circle. For the control signal u(z), the characteristic equation is written

$$\frac{1}{D_1(z)} + S(z) - S_M(z) = 0 \tag{11}$$

and for the output the characteristic equation is written

$$\frac{1}{S(Z)\ D_1(Z)} + \frac{S(z) - S_M(z)}{S(z)} = 0 \tag{12}$$

a) In the ideal case, $S(z) = S_M(z)$, equation (11) implies that

$$\frac{1}{D_1(z)} = 0$$ and therefore the control structure is stable;

and equation (12) implies that

$$\frac{1}{S(z)D_1(z)} = 0$$ and therefore the process is stable.

In order to create an internal model control system, it is necesary for the process S(z) to be asymptotically stable and for the control law $D_1(z)$ to be equally stable.

If the process is naturally unstable or is integrative, it must first be stabilised by a local loop.

b) - In the general case, the characteristic equations (8), (9) and (10) cause the variation between the process and the internal model, $S(z) - S_M(z)$, to come into action. This demonstrates the relationship between the concepts of stability and robustness.

**Compromise between stability and robustness**: If the gain $K_D$ of the regulator and the gain $K_S$ of the variation between the real system and the model are included, as follows:

$$D(z) = K_D\ D^*(z)$$

$$S(z) - S_M(z) = K_S.(S(z) - S_M(z))^*$$

then the characteristic equation (11) becomes

$$1 + K_D.K_S.\ D^*(z)\ (S(z) - S_M(z))^* = 0 \qquad (13)$$

The gain $K_D$ of the regulator is associated with the performance of the regulation loop. As $K_D$ becomes greater, the dynamic performance of the regulation loop improves.

The gain, $K_S$, indicates the quality of the model, and is therefore connected with the concept of robustness.

The conditions of stability of the loop determine the value of the product $K_DK_S$. It follows that, for given stability conditions, the robustness is inversely proportional to the performance.

### 4.2.2.3 - Study of the conventional regime

The application of the final value theorem to the expression of the output y(z) (equation (10)) gives $P_o$ and $e_o$ for a disturbance and a variation of the ramp reference input respectively.

$$\lim_{z \to 1} (z-1) \left[ \frac{D_1(z) \, . \, S(z)}{1+[S(z)-S_M(z)]D_1(z)} \left(\frac{e_o - p_o z}{z-1}\right) + \frac{p_o\, z}{z-1} \right]$$

i.e.

$$y(\infty) = \frac{D_1(1) \, . \, S(1)}{1+[S(1)-S_M(1)]D_1(1)} [e_o - p_o] + p_o$$

The first factor must be equal to 1 for there to be no error in the conventional regime; i.e.

$$D_1(1)\ S(1) = 1 + D_1(1)\ S(1) - S_M(1)\ D_1(1)$$

and therefore

$$D_1(1) = \frac{1}{S_M(1)} \tag{15}$$

In conclusion, there will be no error in the conventional regime if the gain of the control law is equal to the inverse of the gain of the internal model, regardless of the mismatch between the internal model and the process.

#### 4.2.2.4 – Application of a regulator using the internal model control approach

By comparison with the structure presented hitherto, the application leads to the introduction of two additional operators, namely a generator of reference trajectories, R(z) at the level of the reference signal and a robustness filter F(z) in the feedback loop.

Fig. 4 shows the application of an internal model control structure.

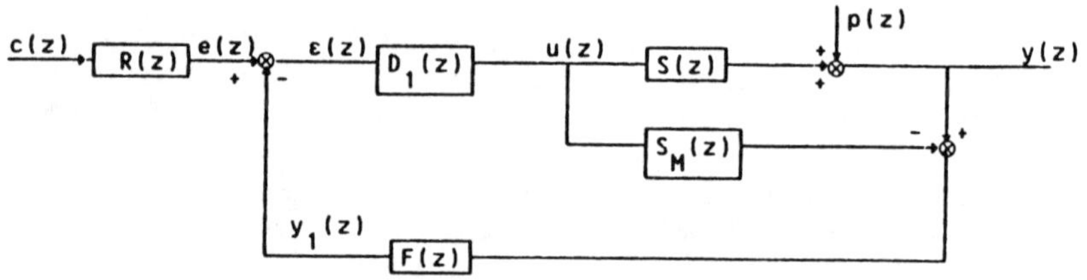


Application of control by internal model

Fig. 5

Examining the robustness, we find that

$$y(z) = p(z) + \frac{D_1(z) \, . \, S(z)}{1+[S(z)-S_M(z)]\ F(z).D_1(z)} [e(z)-F(z)p(z)] \tag{16}$$

Comparison of (16) and (10) shows that the filter F(z) acts on the difference $S(z)-S_M(z)$ as well as on the disturbance p(z):

$$u(z) = \frac{D_1(z) \quad [e(z)-F(z)p(z)]}{1 + [S(z)-S_M(z)] \; F(z) \; D_1(z)} \qquad (17)$$

The analysis of the characteristic equation demonstrates the effect of F(z); the stability depends on the zeros of

$$\frac{1}{D_1(z)} + (S(z) - S_M(z)) \; F(z) = 0 \qquad (18)$$

## 4.2.3 - CHOICE OF THE CONTROL LAW: CALCULATION OF D(z)

In the ideal case, the structure is of the open loop type (Fig. 4). A simplistic approach would be to say that, in order for y(z) to resemble e(z) as closely as possible, it is sufficient to choose D(z), the inverse of the process, as the control law.

In the general case, the process is only known through its internal model; consequently the inverse of the internal model of the process is chosen as the control law.

In this case, according to equation (10),

$$y(z) = S(z) \; . \; S_M^{-1}(z) \left[ \frac{e(z) - p(z)}{1+[S(z)-S_M(z)] \; S_M^{-1}(z)} \right] + p(z)$$

and therefore

$$y(z) = e(z)$$

Theoretically, even if the internal model is not a perfect match, the output reproduces the input on condition that $S_M(z)$ can be inverted.

$S_M(z)$ cannot always be inverted, and it must also be stable. If $S_M(z)$ contains unstable zeros (non-minimal phase characteristics) or time delays, it is not possible to invert it. To resolve this difficulty, $S_M(z)$ is separated into two operators, namely $S_{M+}(z)$ which has all its singularities inside the unit circle and therefore can be inverted, and $S_{M-}(z)$ which has its singularities outside the unit circle.

$$S_M(z) = S_{M-}(z) \; . \; S_{M+}(z)$$

and therefore

$$D_1(z) = \frac{1}{S_{M+}(z)}$$

The control law is found by inverting the stable part of the internal model. In these conditions, the closed loop becomes:

$$y(z) = p(z) + \frac{S(z) \; / \; S_{M+}(z)}{1 + (S(z) - S_M(z))/S_{M+}(z)} \; [e(z) - p(z)]$$

i.e.

$$y(z) = p(z) + \frac{S(z)}{S_{M+}(z) + S(z) - S_M(z)} \; [e(z) - p(z)]$$

In the ideal case, $S(z) = S_M(z) = S_{M-}(z) \,.\, S_{M+}(z)$

and therefore

$$y(z) = p(z) + S_{M-}(z) \; [e(z) - p(z)]$$

$$y(z) = S_{M-}(z) \; e(z) + [1 - S_{M-}(z)] \; p(z) \qquad (19)$$

It can be seen from equation (19) that it would be impossible for example to compensate time delays or non-minimal phase characteristics.

### 4.2.4 - CONSEQUENCES

The use of the internal model as described in section 4.2, while satisfactory in that it takes into account the difference in behaviour between the process and its model, does not fully exploit the possibilities of the model.

In fact, only the difference between the process and the model at the present instant n is taken into account; this is defined as **state resetting**. The procedure described in 4.2 acts as a controller, in other words it reacts to the measured difference, whereas the model makes it possible to react to the predicted difference, as will be seen later (see 4 and 5).

## *4.3 - PRINCIPLES OF INTERNAL MODEL CONTROL*

Internal model control is based on three principles which may be applied in different ways depending on the means available and the objectives.

### 4.3.1 - INTERNAL MODEL

It is useful to have a model of the process which will be used not to govern the regulator with a time lag but to predict in real time, for each sampling period, the future effect of a control sequence.

#### 4.3.1.1 - Nature of the model

A model of a process is available if it is possible to predict the behaviour of the process when it is subjected to a known input. This definition and this mode of application will be strictly adhered to.

The model may be in the form of knowledge or representation. It may be represented mathematically in any form provided that it can be simulated.

Different representations may be used, such as:
- finite difference equations,
- state representation,
- weighting sequences (Volterra),
- a file in a data-processing model,
- consultation of experts.

#### 4.3.1.2 - State resetting

For each sampling period, the state of the model must be reset according to the state of the process; in other words, the variables used to calculate the future outputs of the model must be reset to match the corresponding real system variables. Thus the real system described by equation (20):

$$S_o(n+1) = a_1 S_o(n) + a_2 S_o(n-1) + b_1 e_o(n) + b_2 e_o(n-1) \tag{20}$$

has a corresponding model whose equation may be written

$$S_M(n+1) = f_M (S_o(n-i), e_o(n-i)) + b_1 e_M(n)$$

where

$S_M(n+1)$ is specified by the reference model (see 4.3.2)

$f_M (S_o(n-i), e_o(n-i))$ is calculated from present and past measurements

$e_M(n)$ is to be determined.

#### 4.3.1.3 - Structure resetting

Generally, the model is only of limited validity. The problem of updating the model may be solved in different ways. There are three possible approaches:

- Adaptation: a frequently used technique.

In this case, the structural parameters of the model depend in a predetermined way on an external measurement which is validated with caution. An example is the development of the effectiveness of actions according to the "load" of an industrial process. If the development law is known, it is possible to take into account the non-linear deviations of variables and thus to avoid the restriction of purely local control methods.

– Self–tuning

This technique has the objective of identifying the structural parameters of the process on line. Although this is necessary in certain cases, it still presents theoretical and practical problems, the reliability of instrumentation being a critical point in some cases.

– Reliance on robustness

This technique resolves the problem by developing the robustness of the control laws. We may call this an example of passive adaptation.

#### 4.3.1.4 – **Simulation**

The behaviour of the internal model, reset according to the real system at each instant, enables predictive simulation to be achieved. In the general case where the model is complex and where the operator connecting the control signals to the state is not invertible, it is necessary to perform iterative tests (scenarios) of a number of control sequences in order to make the behaviour of the system conform to a predetermined objective.

At any instant n, the use of the internal model is described by the diagram below:

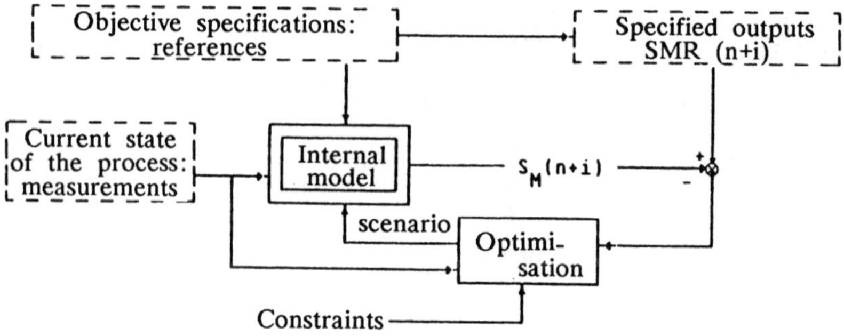


Use of the internal model

Fig. 6

### 4.3.2 – REFERENCE TRAJECTORY

The desired future output of the real system process must be specified. The reference trajectory is thus initialised in the measured or estimated state of the real process, and must tend towards the known future reference. Assuming, for example, that the reference is constant, the following diagram will describe the system.

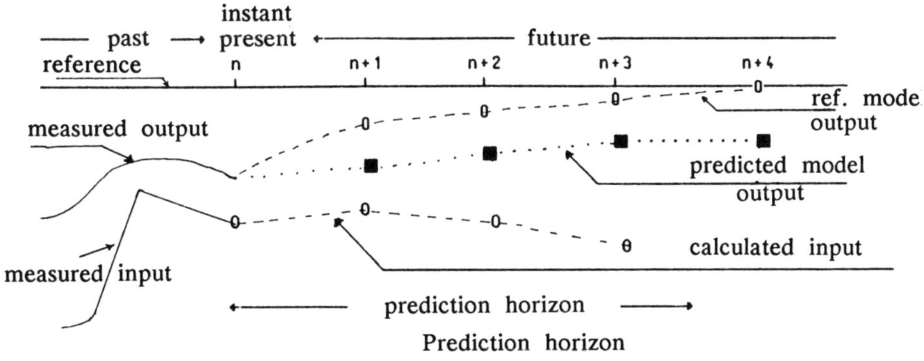


Fig. 7

This trajectory may be established in different ways: it may be plotted or calculated by a stationary operator or may depend on a procedure which may make use of information of various types issued by the process or by the operator.

For example, in the case of a multivariable process, it is possible to specify as many different reference trajectories as there are output variables. The reference trajectory may depend on the value of the output variable. In the case of zone references, where the output variables must belong to a certain domain, the process may have more output variables than reference variables.

### 4.3.2.1 – **Response time of the reference model**

The response time of the reference system $T_R$ determines the behaviour of the system in closed loop form. The order of magnitude of this response time is given by the dynamics of the process in open loop form.

In the case of multivariable processes, each output may have a different corresponding reference trajectory.

### 4.3.2.2 – **Coincidence horizon**

The coincidence between the predicted outputs of the process and the desired outputs can only be ensured at certain points of the reference trajectory. These points constitute the coincidence horizon.

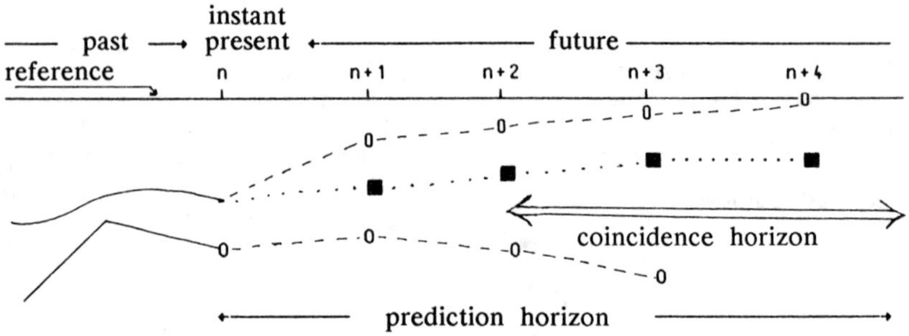


Coincidence horizon

Fig. 8

For example, the diagram can be used to find inputs e(n), e(n+1),... e(n+4) which make the behaviour of the process and the reference model identical at instants (n+2), (n+3), (n+4).

For theoretical and practical reasons concerned with controllability, it is useful to be able to shift the coincidence points on the prediction horizon. In the case of stable systems with time delays, non-minimal phase angles, or any other singularities, it is possible to establish a horizon which is sufficiently distant to make the outputs and control signals stable.

Seeking the coincidence at several points rather than at a single point results in:

- better specification of the dynamic behaviour of the process;
- better allowance for the constraints, by formulating the problem in terms of the minimisation of the distance between two functions rather than in terms of the solution of equations.

### 4.3.3 - CONTROL ALGORITHM

Let a process be represented by an internal model such that

$$S_M(n+1) = H(X(n), E(n))$$

where X(n) is a vector whose components are updated according to the state of the real system;

E(n) is a control vector to be determined;

and let SMR(n+i) be the outputs determined by the reference trajectory which specifies the objective.

The control algorithm is then required to solve the problem of determining the vector E(n) such that the objective is achieved while the constraints are observed.

The objective is to make the predicted outputs and the reference model

outputs coincide at all points on the reference horizon. Therefore the distance between the predicted real system outputs and the reference model outputs must be minimised, subject to the constraints which affect

- the control variables:

$$e_m < E(n) < e_M$$

$$|E(n) - E(n-1)| < \delta_e$$

- the intermediate variables:

$$U(n) = F(E(n))$$

$$u_m < U(n) < u_M$$

When formulated in this way, the problem can be solved by many methods, such as linear porgramming, quadratic control, optimal control with direct or iterative solutions, heuristic control, etc. The choice depends essentially on the data processing resources, the compromise made between precision and calculation time, and the characteristics of portability, simplicity, etc.

## 4.3.4 - STABILITY AND ROBUSTNESS

### 4.3.4.1 - **Definition**

It has been shown that an appropriate choice of the coincidence horizon makes it possible to find a strategy for the stable control of an asymptotically stable process.

With this approach, the problems of stability become noticeably easier to deal with; the order and complexity of the process no longer affect the stability. The essential problem is now that of robustness, in the unadapted case.

The control strategy has been determined with a model $M_o$ of the process $P_o$, and nominally provides the desired coincidence between the outputs of the process and the reference trajectory. The performances of the controlled system are therefore fixed: let $C_o$ be a criterion, which may be, for example:

- the response time, $T_o$, to a reference step input;
- the variance of the deviation when the process is subjected to a given noise.

If the process varies and becomes $P_o + \Delta P$, with the model $M_o$ remaining constant, the performance is affected and the quadratic criterion becomes $C_o + \Delta C$.

For simplicity, it will be assumed that the modification $\Delta P$ is due to a variation of a scalar parameter K (for example, the gain of the process). The robustness is defined as the following ratio:

$$R = \Delta K / \Delta C$$

#### 4.3.4.2 – The robustness–performance compromise

Generally, there is an incompatibility between the robustness and the nominal performance. The principal tuning factor is the response time of the reference trajectory: if this is short, the nominal performance is high, but the robustness is low; if it is long, the controlled system will be slow but robust.

The user can choose the appropriate compromise according to the objectives of the control system.

The *a priori* calculation of the robustness is often impossible in the general case of practical applications. In a theoretical case, it can be shown that if the reference trajectory has a decrement $\alpha_R$ and if q is the ratio of the gains of the internal model and the process, then the closed loop system follows a trajectory of decrement $\alpha$ given by

$$\alpha = 1 - q(1 - \alpha_R)$$

for $\alpha_R = .9$, the difference in gain which would result in unacceptable behaviour would be $0 = 1 - q_L(1-.9)$, $q_L = 10$. This example shows that $\alpha$ (performance) and q (robustness) vary in an inverse relationship.

## *4.4 – EXAMPLE OF APPLICATION IN THE CASE OF A SYSTEM DESCRIBED BY ITS STATE EQUATIONS*

### 4.4.1 – DEFINITION OF THE CONTROL PROBLEM

* Multivariable linear system of which a discrete state representation is:

$$\begin{aligned} x(k+1) &= F\, x(k) + G\, u(k) \\ y(k+1) &= H\, x(k+1) \end{aligned} \qquad (21)$$

where

$$x \in R^n,\ y \in R^p,\ u \in R^m$$

k and k+1 indicate two consecutive sampling instants.

* Model of the same order as (21) with coefficients which may vary

from those of the matrices of (21):

$$x_M(k+1) = F_M \ x_M(k) + G_M \ u(k)$$
$$y_M(k+1) = H_M \ x_M(k+1) \tag{22}$$

* Desired outputs (or references): $y_d(k)$, $k = 1,...$
where
$y_d \in R^p$

The control problem consists in the calculation of the values of the vector u(k) at each instant such that the outputs y(k) of the system are equal to the required outputs $y_d(k)$ at all points of the prediction horizon of length N.

The following assumptions will be made:

- the system (21) and model (22) are stable (the eigenvalues of the matrices F and $F_M$ have a modulus of less than 1);
- if necessary, the transfer matrices between the inputs and outputs of the real system (21) and model (22) can be inverted (requiring that p=m). These conditions, assumed in order to simplify certain calculations, may be relaxed by conditions of attainability.

## 4.4.2. - CONTROL STRATEGY

### 4.4.2.1 - Reference trajectory

If an output $y_i$ is not equal to the corresponding reference $y_{di}$, the adjustment is carried out according to a procedure determined by a reference trajectory.

Reference trajectories are generated, for example, by a first order system (reference model) with unit gain and with a control time constant of $\alpha_i$:

$$y_R(k+1) = \alpha \ y_R(k) + [I - \alpha] \ y_d(k) \tag{23}$$

where
$\alpha = \text{diag}\ (\alpha_i)$

The reference model is stable ($|\alpha_i| < 1 \ \forall = 1,...,p$)

### 4.4.2.2 - Resetting the reference trajectories

At each instant k, the reference trajectories are initialised at a value $y_R(k)$:

$$\bar{y}_R\ (k) = \beta_R \ y(k) + (I - \beta_R) \ y_R(k) \tag{24}$$

This initialisation permits different possibilities between no resetting ($\beta^{Ri} = 0$) and total resetting to the outputs of the system ($\beta^{Ri} = 1$). The future reference trajectories calculated at the instant k are given by

$$y_R(k+l) = \alpha^l y_R(k) + (I-\alpha) \sum_{j=1}^{l} \alpha^{j-1} y_d(k+l-j) \tag{25}$$

4.4.2.3 – **Calculation of future outputs**

At each instant k, the future (or predicted) outputs of the system are calculated by means of the model (22) and reset to the outputs of the real system as follows:

$$y_p(k+l) = y_M(k+l) + \beta_M[y(k) - y_M(k)] \tag{26}$$

Note that when l=0, (26) becomes

$$y_p(k) = \beta_M\, y(k) + (I-\beta_M)\, y_M(k)$$

which represents a loop similar to loop (24) of the reference trajectories. In particular, the different possibilities between no resetting of the predicted outputs and total resetting to the outputs of the system are given by the choice of $\beta_M$.

The predicted outputs corresponding to given control signals u(k+j) are then

$$\begin{aligned} y_p(k+l) = {} & H_M F_M^l x_M(k) + H_M \sum_{j=1}^{l} F_M^{j-1} G_M\, u(k+l-j) \\ & + \beta_M [y(k) - H_M x_M(k)] \end{aligned} \tag{27}$$

4.4.2.4 – **Calculation of control signals**

At each instant k, the values of the future control signals u(k+l), l⩾0, are calculated to ensure the equality (or coincidence) of the predicted outputs and the reference trajectories at a future instant k+L.

Subsequently only the values u(k) are accepted and sent to the system. It is necessary to solve the equation

$$y_p(k+L) = y_R(k+L)$$

which may be written by using (27)

$$\begin{aligned} & H_M \sum_{j=1}^{L} F_M^{j-1} G_M\, u(k+L-j) \\ & = y_R(k+L) + H_M(\beta_M - F_M^L)\, x_M(k) - \beta_M\, y(k) \end{aligned} \tag{28}$$

Equation (28) is a vectorial relationship containing L unkown vectors (u(k),..., u(k+L−1)). There is therefore an infinite number of solutions. This freedom may be used to comply with any constraints or to optimise a criterion. It will be assumed that the future control signals are related to each other by

$$u(k+l) = \Phi_l \quad u(k) \qquad l = 0,1,\ldots, L-1 \tag{29}$$

with

$$\Phi_o = I$$

Two examples of the choice of the matrices $\Phi_L$ will be given:

a) – **over-sampling of the future control signals**

The future control signals are assumed to be constant between the instants k and k+L−1. This enables the calculation of the control signals to be simplified. In this case,

$$\Phi_l = I \qquad\qquad l = 0,\ldots,L-1 \tag{30}$$

b) – **control signal with minimal energy**

Assuming that the transfer matrix of the model (22) is invertible, the minimisation of the control vectors on the horizon $l \in [0, L-1]$ is performed by

$$\Phi_l = G_M^T \, (F_M^{L-l-1})^T \, H_M^T \, [G_M^T \, (F_M^{L-1})^T \, H_M^T]^{-1} \tag{31}$$

Applying (29), (28) becomes

$$(H_M \sum_{j=1}^{L} F_M^{j-1} \, G_M \, \Phi_{L-j}) \; u(k) = y_R(k+L) + H_M(\beta_M - F_M^L) \; x_M(k)$$
$$- \beta_M \; y(k) \tag{32}$$

which, assuming that m=p, has at least one solution.

The control strategy thus defined will be denoted the strategy of the "L-th point following". In case (a) above, the strategy of the "following block", with a block representing the length of the over-sampling of the future control signals, will be referred to. When l=1, the strategy of the "following point" will be referred to.

## 4.4.3 – TRANSFER MATRIX OF THE CONTROL SYSTEM

In order to calculate the properties of convergence, stability and robustness of the control method, the discrete closed loop transfer matrix of the controlled system will be calculated by calculating the Z transform of (32).

Let U(z), Y(z) and $Y_d(z)$ be the z transforms of the sequences u(k), y(k) and $y_d(k)$ respectively, with

$$S(z) = Z[s(k)] = \sum_{k=0}^{\infty} s(k) z^{-k}$$

According to (24) and (25),

$$Z[y_R(k+L)] = [zI-(I-\beta_R)\alpha]^{-1}[z\alpha^L\beta_R y(z) + \alpha^L(I-\beta_R)(I-\alpha)y_d(z)]$$

$$+ (I-\alpha) \sum_{j=1}^{L} \alpha^{j-1} z^{L-j} y_d(z) \qquad (33)$$

According to the definition of $x_M(k)$ (see section 4.3.2), we can write directly from (22):

$$Z[x_M(k)] = (zI - F_M)^{-1} G_M U(z) \qquad (34)$$

By introducing (33) and (34) into (13), we obtain a relationship in the following form:

$$A_1(z) U(z) = A_2(z) . Y(z) + A_3(z) Y_d(z) \qquad (35)$$

where

$$A_1(z) = H_M \sum_{j=1}^{L} F_M^{j-1} G_M \Phi_{L-j} - H_M(\beta_M-F_M^L)(zI - F_M)^{-1} G_M$$

$$A_2(z) = z[zI-(I-\beta_R)\alpha]^{-1} \alpha^L \beta_R - \beta_M \qquad (36)$$

$$A_3(z) = (I-\alpha)\{[zI-(I-\beta_R)\alpha]^{-1} \alpha^L(I-\beta_R) + \sum_{j=1}^{L} \alpha^{j-1} z^{L-j}\}$$

The equations provide a relationship between U(z) and Y(z):

$$Y(z) = H(zI - F)^{-1} G u(z) \qquad (37)$$

Equations (35) and (37) can be used to write

$$U(z) = [A_1(z)-A_2(z)H(zI-F)^{-1}G]^{-1} A_3(z) Y_d(z) \qquad (38)$$

where $A_1$, $A_2$ and $A_3$ are functions of the coefficients of the internal model and of the reference model (equation (36)).

* Transfer matrix of the controlled system

Assuming that $[H(zI-F]^{-1}G]^{-1}$ exists (a condition of the controllability of the system),

$$[A_1(z)[H(zI-F)^{-1}G]^{-1} - A_2(z)]y(z) = A_3(z) . Y_d(z) \qquad (39)$$

and the closed loop transfer matrix is given by

$$T_{BF}(z) = [A_1(z)[H(zI-F)^{-1}G]^{-1} - A_2(z)]^{-1} A_3(z)$$

or

$$T_{BF}(z) = [H(zI-F)^{-1}G][A_1(z) - A_2(z)H(zI-F)^{-1}G]^{-1} A_3(z) \tag{40}$$

where it is possible to recognise

- the open loop transfer matrix of the system:

$$T_{BO}(z) = H(zI-F)^{-1}G$$

- the closed loop characteristic matrix of the system:

$$D(z) = A_1(z) - A_2(z) T_{BO}(z)$$

- a matrix connected with the reference model: $A_3(z)$

Equation (36) can be used to illustrate the control system in the following form (Fig. 9):

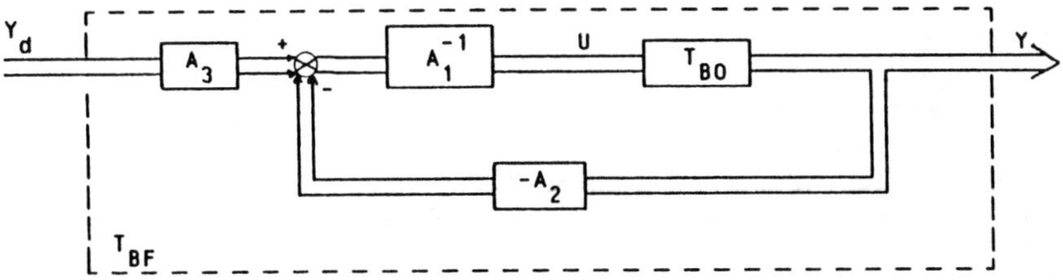


Outline diagram of control system

Fig. 9

By definition, if $\beta_R = \beta_M = 0$, there is no looping on the outputs, as may be seen in Fig. 9, since $A_2 = 0$.

### 4.4.4 - STUDY OF CONVERGENCE

The control method (or algorithm) is said to converge if the outputs y(k) tend towards the references $y_d(k)$, while k tends towards infinity (absence of bias). This is expressed by

$$T_{BF}(z) \Big|_{z=1} = I$$

or according to (39)

$$A_1(1) - [A_2(1) + A_3(1)]\ [H(I-F)^{-1}G] = 0 \quad (41)$$

which represents the convergence equation of the method. The limits $A_2(1)$ and $A_3(1)$ exist if the matrix $[I-\alpha+\beta_R\alpha]$ can be inverted.

**4.4.4.1 – The case in which $[I-\alpha+\beta_R\alpha]$ is singular**

Since this matrix is diagonal,

$$I - \alpha + \beta_R\ \alpha = 0$$

i.e. (42)

$$\beta_R = I - \alpha^{-1}$$

In this case, equation (41) cannot be used. Multiply (39) by $[zI-[I-\beta_R)\alpha]$ and take the limit when z tends towards 1. This leaves (see (36))

$$-\alpha^L\ B_R\ \tilde{y} = (I-\alpha)\ \alpha^L(I-\beta_R)\tilde{Y}_d$$

where $\tilde{y}$ and $\tilde{y}_d$ represent the limits of $y(k)$ and $y_d(k)$ when k tends towards infinity,

and, using (42),

$$-\alpha^L(I-\alpha^{-1})\ \tilde{y} = (I-\alpha)\ \alpha^{L-1}\ \tilde{y}_d$$

i.e.

$$\tilde{y} = \tilde{y}_d \qquad \forall\ \alpha \neq I$$

Note that the matrices of the system and of the model, as well as $\beta_R$ and $\beta_M$, no longer appear in the convergence equation.

<u>Property 1</u>:

If the reference trajectories are reset with $\beta_{R_i} = 1 - 1/\alpha_i$ ($\forall$ i = 1, m), the method converges regardless of the model of the system and the resetting of this model, in the absence of any constraint on the control variables.

4.4.4.2 – **Case in which $[I - \alpha + \beta_R\ \alpha]$ can be inverted**

In this case, the convergence equation is correct if

* $\beta_M = I$, for any $H_M$, $F_M$, $G_M$.

  corresponding to the resetting of the predicted outputs to the system outputs at each instant;

* $H_M\ (I-F_M)^{-1}G_M = H(I-F)^{-1}G$, for any $\beta_M$

  signifying that the matrix of the static gains of the model (22) is equal to the matrix of the static gains of the system.

### 4.4.5 - STUDY OF THE DYNAMIC PERFORMANCES

When convergence has been established (asymptotic property of the conventional regime), the transient regime of the controlled system can be examined in order to evaluate the dynamic performance of the system, particularly in terms of stability and robustness.

It is assumed that the reference trajectories and the predicted outputs are reset to the system outputs:

$$\beta_R = \beta_M = I$$

and $T_R(z)$ is the transfer matrix of the reference model:

$$T_R(z) = (I-\alpha)\ (z-\alpha)^{-1} \tag{43}$$

The closed loop transfer matrix of the system can be written in the form

$$T_{BF}(z) = T_{BO}(z)\ D^{-1}(z)(z^L-\alpha^L)\ T_R(z) \tag{44}$$

where $T_{BO}(Z)$ is the open loop transfer matrix of the system and $D(z) = A_1(Z) - A_2(z)\ T_{BO}(z)$ is the characteristic matrix of the system in closed loop configuration.

The controlled system can be illustrated by the diagram in Fig. 10:

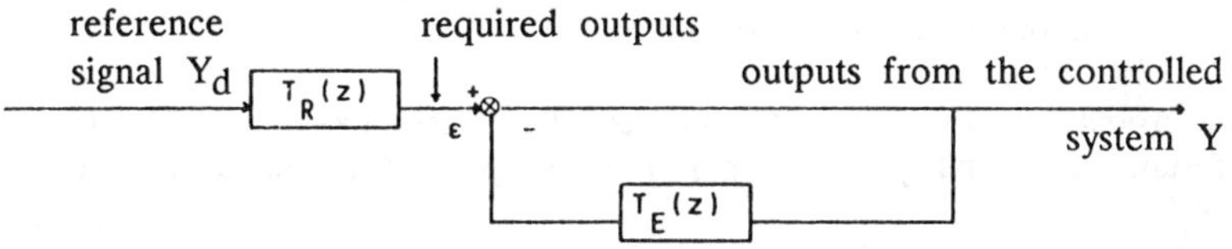


Diagram of the controlled system

Fig. 10

with

$$T_E(z) = [D(z) - (z^L-\alpha^L)\ T_{BO}(z)]\ (z^L-\alpha^L)^{-1}\ T_{BO}{}^{-1}(z) \tag{45}$$

$\epsilon(z) = T_E(z)\ (Yz)$, representing the dynamic error between the required outputs and the actual outputs of the system.

The ideal case, $T_E(z) = 0$, will be present if and only if

$$L = 1 \text{ and } T_M(z) = T_{BO}(z)$$

Therefore the following properties are present:

Property 2: In the case of the "following point" strategy, the outputs of the controlled system are identical to the required outputs if the model (22)is an exact representation of the system.

Property 3: In the case of the strategy of the "L-th point following", the outputs of the controlled system are different from the required outputs, regardless of the quality of the model of the system.

Property 4: The strategy of the "following block" is equivalent to the strategy of the "following point" with a model whose observation equations are different ($H_M \rightarrow \tilde{H}_M$) and with a different reference model ($\alpha \rightarrow \alpha^L$).

where

$$\tilde{T}_M(z) = \tilde{H}_M(zI-F_M)^{-1}\, G_M$$

$$\tilde{H}_M = H_M(I-F_M^L)\,(I-F_M)^{-1}$$

The mismatch of the model may be considered acceptable if, for example, the controlled system is stable. The examination of the roots of $D(z) = 0$ will then make it possible to find the acceptable limits of mismatch.

The difference between the outputs of the system and the required outputs is explained by the freedom allowed to the predicted outputs of the system between the instants k+1 and k+L−1, which makes it possible to have control values different from those which provide a coincidence over the whole prediction horizon.

The roots of the characteristic equation are examined in order to ascertain the stability (or margin of stability) of the controlled system.

In the case of the "L-th point following" strategy, L is a parameter which may be defined in such a way as to provide the requisite margin of stability.

## *4.5 - CONCLUSION*

These control methods are essentially of a different type from the "conventional" regulation systems in which the regulator has a fixed analytic structure which is defined previously. The internal model control methods have the following principal characteristics:

* They are generally applicable: their limitation is in controllability more than in the complexity of the process. The process may be of the single- or multi-variable, linear or non-linear type. Any type of representation may be used.

* They are realistic: the control signals, internal variables, and reference signals all take into account operating constraints which may be as complex as their expressions.

* They are robust: they tolerate mismatching and have generally better performance and robustness characteristics than those of methods not using an internal model as a prediction device.

* They are powerful: they enable a systematic approach to be made to the problem of the operation of a process, similar to an aided design stage. The performance of the system in closed loop configuration is specified (decrement $\alpha_R$ in the case of a first order reference model) without the need to develop a regulator structure to control a number of parameters which increases with the complexity of the process.

On the other hand, they require more investment, in the form of the development of a model of the process, and the use of more powerful computing equipment. The conventional techniques, by contrast, require less knowledge of the process and less equipment, and are therefore more economical. Their range of application and performance are more restricted.

The system designer must therefore have access to a variety of methods which he must be capable of selecting in accordance with the complexity of the problems, the performance expected, and the available equipment.

## *REFERENCES*

ADERSA GERBIOS - IDCOM, Conduite algorithmique des processus industriels de fabrication, Paris, 1976

G. Garcia, Morari, Internal Modern Control - a unifying review and some new results, Ind. Eng. Chem. Processes Des. Des., v. 21, 308-323, 1982

A. Rault, D. Jaume, and M. Verge, Commande par modèle interne: présentations et applications, 1984

J. Richalet, A. Rault, and R. Pouliquen, Identification des processus par la méthode du modèle, Gordon et Breach, 1971

J. Richalet, A. Rault, J.L. Testud, and J. Papon, Algorithmic control of industrial processes - 4th IFAC Symposium on Identification and System Parameters Estimation, Tbilisi (USSR), 1976

J. Richalet, A. Rault, J.L. Testud, and J. Papon, Model predictive heuristic control: application to industrial processes, Automatica, v. 14, 413–429, 1978.

# CHAPTER 5

# DISCRETE OPTIMAL CONTROL OF LINEAR STOCHASTIC SYSTEMS

## 5.1 - INTRODUCTION

This chapter takes up the problem of optimal control with minimisation of a quadratic criterion, using different approaches.

Historically, this problem was initially tackled in the frequency domain, using the Wiener-Hopf techniques. The application of these techniques to problems of closed loop control had a major disadvantage in that it was impossible to deal with the case of an unstable system and/or one with non-minimal phase characteristics.

The subsequent approach using state space methods replaced the transfer functions approach because of its universality. The analytic and algebraic procedures used to solve the problems of synthesis are now well known and unified. They result in a state feedback, which, assuming that this is not measurable, requires an observer whose nature, in the deterministic case, can be determined arbitrarily by the designer. However, if additional inputs or delays or the presence of coloured noise are taken into account, this leads to a further complexity of the representation which does not appear in the case of transfer functions, owing to the flexibility of representation by block diagrams. These factors have led many authors to propose new and more effective methods of synthesis based on the techniques of Wiener. This is why, after a period of eclipse due to the success of the state space approach, there is now a revival of the transfer function approach. Complete and well-structured theories are appearing. They make use of advanced mathematical techniques and provide a conceptual framework for optimal control which differs radically from the state approach. The proposed methods are generally based on the algebraic theory of polynomial matrices. They have shed new light on the subject and are described in the first section of this chapter.

The second section deals with the time domain formulation without going into any great detail (the subject is already well documented). Finally, the last section establishes a connection between the time and frequency approaches for the stationary case with an infinite horizon.

At the end of the book, Appendix 1 brings together the principal mathematical features of the polynomial matrix approaches used in the text.

## *5.2 – OPTIMAL CONTROL BY THE TECHNIQUES OF POLYNOMIAL ALGEBRA*

The procedure which will be used enables the deterministic or stochastic cases to be dealt with separately. In Chapter 3, this technique was used for the synthesis of digital controllers to optimise a minimum time criterion. In this chapter, only the stochastic case, which is the most general one, will be considered.

### 5.2.1 – SINGLE-INPUT, SINGLE-OUTPUT CASE

The formulation of the optimisation problem for a given system depends partly on the control structure adopted and partly on the quadratic criterion which is to be minimised. In the case of a control structure in open loop form, the problem is formulated directly; in other words, the transfer function of the controller is considered to be the unknown quantity in the algebraic calculations. In this case, the solution is found without difficulty.

In the case of the closed loop control structure, the problem is much more complicated. The most widely used approach consists in the transformation of the initial structure into an equivalent open loop structure so that the canonical system of Wiener can be used. In the case of an unstable process and/or one with non-minimal phase characteristics, the use of this method inevitably led to systems which were unstable on return to the closed loop configuration. Moreover, this procedure very often produced unfeasible controllers.

Recently, Youla, Bongiorno and Jabr (1976) have introduced the concept of parametrisation of the set of transfer functions which provide asymptotic stability of the system, taking into account its closed loop structure. Expressed in other terms, the idea consists in the definition of a transformation so that the optimisation procedure intrinsically checks the conditions of coarseness. We shall now establish the basic lemma which defines the subset of closed loop transfer functions which ensure the stability of the loop.

Consider the following standard structure (Fig. 1):

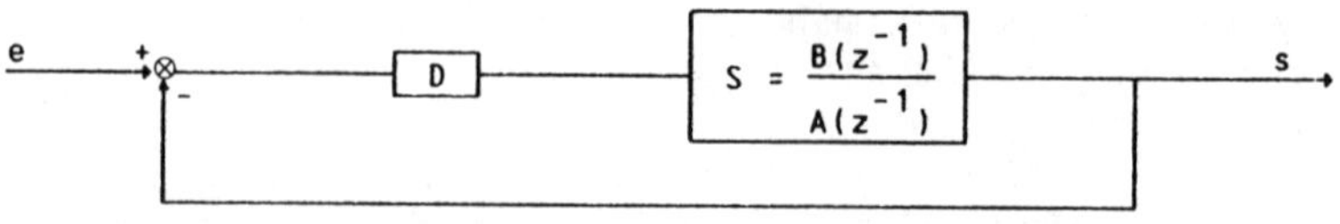


Fig. 1

where S and D represent the $z^{-1}$ transfer functions of the process and of

the controller for the system operating in the servo mode.

The closed loop transfer function G relating the input to the output is written

$$G = \frac{DS}{1 + DS}$$

Lemma:

The set of closed loop transfer functions for a process S which guarantees the asymptotic stability of the loop and avoids compensation for poles and zeros outside the unit circle is written

$$G = BV + \mu\, AB \tag{1}$$

where U and V are any solution of the polynomial equation

$$AU + BV = 1 \tag{2}$$

and $\mu$ is any rational but stable fraction.

Proof

Let D be the transfer function of the controller

$$D = \frac{Y}{X}$$

We know (see 3.1.1) that the controllers which guarantee the asymptotic stability and avoid compensation for poles and zeros outside the unit circle are such that the polynomials X and Y are the solution of the polynomial equation

$$AX + BY = \Delta \qquad \text{where } (A,B) = 1 \tag{3}$$

in which $\Delta$ is a stable polynomial.

The set of the solutions X and Y of this equation, which exists because (A,B) = 1, is given by

$$\begin{aligned} X &= \Delta U - TB \\ Y &= \Delta V + TA \end{aligned} \tag{4}$$

where $\Delta U$ and $\Delta V$ represent a particular solution of equation (3) and T represents any polynomial in $z^{-1}$.

The set of closed loop transfer functions having the desired properties is written

$$G = \frac{BY}{AX + BY} = \frac{BY}{\Delta} \qquad \text{according to (3)} \qquad (5)$$

i.e.

$$G = BV + \frac{T}{\Delta} AB \qquad \text{according to (4)}$$

The latter expression is in the required form (1).

Before approaching the actual synthesis of optimal controllers, some very important notations which will be used in this chapter must be explained.

* If $E(z^{-1})$ is a rational fraction in $z^{-1}$, $\bar{E}$ denotes the rational fraction in which the operator $z^{-1}$ is replaced by the operator Z.

* $E^-$ denotes the domain defined by

$$\{z^{-1} / |z^{-1}| \leqslant 1\}$$

which forms the unstable domain outside the unit circle, including the boundary.

* $F^+$ and $F^-$ represent the set of polynomials whose roots belong to $E^+$ and $E^-$ respectively.

* For a polynomial $P(z^{-1})$, $P^+$ and $P^-$ denote the stable and unstable parts of P respectively, so that

$$P = P^+ P^-$$

* $\delta(P)$ represents the degree of the polynomial $P(z^{-1})$.

* The notation $\tilde{P}$ designates the expression

$$\tilde{P} = z^{-\delta(P)} \bar{P}$$

* $P^*$ is a polynomial which has all its roots inside the unit circle and is such that $P^* \overline{P^*} = P \bar{P}$. This polynomial exists if P has no roots on the unit circle.

* (A,B) denotes the GCD of the polynomials A and B.

* The three following expressions are equivalent:

$$\sum_{k=0}^{\infty} \epsilon^2(k) = \frac{1}{2\pi j} \oint_{\Gamma} \epsilon \,\bar{\epsilon} \,\frac{dz}{z} \triangleq \langle \epsilon \;\bar{\epsilon} \rangle$$

where $\Gamma$ is formed by the unit circle.

### 5.2.1.1 – Optimal control in the open loop configuration

a) – **Formulation of the problem**

Consider the system shown in Fig. 2:

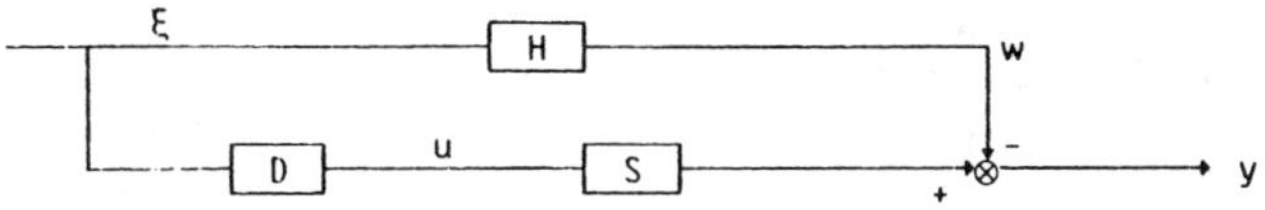


Fig. 2

where

* $S = B(z^{-1})/A(z^{-1})$ represents the transfer function of the process assumed to be stable such that $(A,B) = 1$;

* w denotes an additive noise on the output characterised by a white noise sequence $\xi(0,1)$ passing through a stable filter $H = R(z^{-1})/L(z^{-1})$;

* y represents the output of the system.

The objective is to find the optimal stable controller $D(z^{-1})$ which minimises the effect of a disturbance on the system, while using an open loop control system applying the knowledge of $\xi$.

Two criteria will be used in succession to qualify the performance of the system:

1) a quadratic criterion on the output with an infinite horizon (denoted 1.a):

$$J = E\{ \sum_{k=0}^{\infty} y^2(k) \}$$

2) a quadratic criterion on the output weighted by the control signal (denoted 1.b)

$$J = E\{ \sum_{k=0}^{\infty} y^2(k) + \eta \sum_{k=0}^{\infty} u^2(k) \} \quad \text{where } \eta \;\; 0$$

b) – **Finding the optimal corrector minimising criterion 1.a**

In the rest of this study, the following notation will be used:

$$A_o = \frac{A}{(A,L)} \qquad\qquad (6)$$

$$L_o = \frac{L}{(A,L)}$$

We shall now state the theorem which defines the optimal controller.

Theorem 1.a

The optimal controller minimising the quadratic criterion on the output is given by the following rational fraction:

$$D(z^{-1}) = \frac{A_o Y_1}{L_o B^*} \tag{7}$$

where $X_1$ and $Y_1$ are the solution of minimum degree in $X_1$ of the polynomial equation

$$X_1 L + B^- Y_1 = \tilde{B}^- R \text{ (1)} \tag{8}$$

This solution exists because the polynomials L and $B^-$ are prime polynomials with respect to each other.

Proof

According to Parseval's theorem, the criterion can be written

$$J = E \{ \frac{1}{2\pi j} \oint (y \, \bar{y}) \frac{dz}{Z} \} = \langle y \, \bar{y} \rangle$$

with

$$y = \frac{B}{A} u - \frac{R}{L} \xi \quad \text{and} \quad u = D\xi$$

Given that $B^* = B^+ \tilde{B}^-$, and since $\Phi_{\xi\xi} = 1$, it is easily seen that

$$J = \langle y \, \bar{y} \rangle = \langle y_o \bar{y}_o \rangle$$

where

$$y_o = \frac{B^*}{A} D - \frac{\tilde{B}^- R}{B^- L} \quad \{1\}$$

Using equation (8), the second member of the last relationship can be transformed to give

$$y_o = \frac{B^* D}{A} - \left( \frac{X_1}{B^-} + \frac{Y_1}{L} \right)$$

where $X_1$ and $Y_1$ are interrelated by equation (8).

By letting

$$W = \frac{B^* D}{A} - \frac{Y_1}{L} \tag{9}$$

{1} The notation $\tilde{B}^-$ is used henceforth for the expression $\tilde{B}^- = [B^-]$

we obtain

$$J = \langle y_o \, \bar{y}_o \rangle = \langle W \, \bar{W} \rangle + \langle W \, \overline{\left[\frac{X_1}{B^-}\right]} \rangle + \langle \bar{W} \, \frac{X_1}{B^-} \rangle + \langle \frac{X_1}{B^-} \, \overline{\left[\frac{X_1}{B^-}\right]} \rangle \qquad (10)$$

If we choose the solution in $X_1$ and $Y_1$ such that $\delta(X_1) < \delta(B^-)$, i.e. the solution of minimal degree in $X_1$, we may make the following basic statements:

– Since a stable controller D is being sought, relationship (9) shows that W(z) is a stable rational fraction since A and L are of this nature according to the hypothesis. In addition, all the poles of the rational fraction

$$\overline{\left[\frac{X_1}{B^-}\right]} (z)$$

are stable. Consequently, the expression

$$\frac{W(z)}{z} \left[ \overline{\left[\frac{X_1}{B^-}\right]} (z) \right]$$

has all its poles inside the unit circle, and its numerator is a polynomial in z of a degree less by at least two than that of the denominator. This last constraint implies that

$$\sum_{\mathrm{Res}} \frac{W}{z} \overline{\left[\frac{X_1}{B^-}\right]} = 0 \text{ for all the poles}$$

It may be deduced from this that

$$\langle W \overline{\left[\frac{X_1}{B^-}\right]} \rangle = 0$$

– Similarly, it will be noted that all the poles of $\bar{W}(z)$ are unstable. The condition $\delta(X) < \delta(B^-)$ implies that

$$\frac{X_1}{B^-} = \frac{z \, N(z)}{D(z)}$$

Consequently all the poles of the rational fraction in z, $\bar{W}X_1/zB^-$, are unstable, and therefore

$$<\overline{W} \frac{X_1}{B^-}> = 0$$

Returning to equation (10), it follows that the minimum of J will be obtained when W = 0, which implies equation (7) and completes the proof of the theorem.

* It is verified that the optimal controller is stable, since L is stable by hypothesis and B* is stable by definition.

* The value of the criterion at the optimum level is as follows:

$$J = < \left[\frac{X_1}{B^-}\right] \overline{\left[\frac{X_1}{B^-}\right]} >$$

* For processes such that $B^- = 1$, it is necessary to specify that $X_1=0$ and $Y_1=R$, thus obtaining the obvious solution D = AR/LB, and therefore J = 0.

c) **Finding the optimal corrector to minimise criterion 1.b**

Before stating the theorem, some new values will be introduced. C denotes a stable polynomial such that

$$\eta \, A \, \bar{A} + B \, \bar{B} = C \, \bar{C} \quad \text{where } C \in F^+ \tag{11}$$

The notations l, m, and h are used as follows:

$$\begin{aligned} l &= h_1 - \delta(C) \\ m &= h_1 - \delta(B) \end{aligned} \tag{12}$$

where

$$h_1 = \max(\delta(A), \delta(B), \delta(C))$$

<u>Theorem 1.b</u>

The optimal controller to minimise the quadratic criterion with a weighting on the control signal (1b) is given by

$$D = \frac{A_o Y_1}{L \, C} \tag{13}$$

where $X_1$ and $Y_1^o$ are the solution of minimal degree in $X_1$ of the polynomial equation

$$X_1 \, L + z^{-1} \, \tilde{C} \, Y_1 = z^{-m} \, \tilde{B} \, R \tag{14}$$

<u>Proof</u>

The criterion J is written

$$J = E\{ \frac{1}{2\pi j} \oint_{\Gamma} (y\,\bar{y} + \eta\, u\, \bar{u}) \frac{dz}{Z} \}$$

$$J = E\{ \frac{1}{2\pi j} \oint_{\Gamma} \left[ D\,\bar{D}\,(\eta + \frac{B\bar{B}}{A\bar{A}}) - \frac{R\bar{B}}{L\bar{A}}\bar{D} - \frac{\bar{R}B}{\bar{L}A} D + \frac{R\bar{R}}{L\bar{L}} \right] \frac{dz}{Z} \}$$

By letting

$$\epsilon_o = D \frac{C}{A} - \frac{\tilde{B}\, Z^{-m}}{\tilde{C}\, Z^{-1}} \frac{R}{L}$$

it may be seen that

$$J = <\epsilon_o \bar{\epsilon}_o> + \eta <\frac{RA}{LC}, \overline{\frac{RA}{LC}}>$$

As before, $\epsilon_o$ is decomposed in the form

$$\epsilon_o = D \frac{C}{A} - ( \frac{X_1}{Z^{-1}\,\tilde{C}} + \frac{Y_1}{L}) \qquad (15)$$

where $X_1$ and $Y_1$ are related by equation (14).

In order to minimise J, therefore, it is simply necessary to minimise $<\epsilon_o, \bar{\epsilon}_o)$. Using the same proof as for theorem 1a, and specifying the polynomial $X_1$ of minimal degree, in other words such that

$$\delta(X_1) < \delta(Z^{-1}\,\tilde{C}),$$

the optimal solution is given by

$$D \frac{C}{A} - \frac{Y_1}{L} = 0$$

which implies equation (13) and completes the proof of the theorem.

Notes

1. If $\eta=0$, it is sufficient to replace C with B* according to (11) to obtain the result found in the previous section.

2. The solution of the polynomial equation(14) exists. Since L and C are stable, the polynomials L and $Z^{-1}\,\tilde{C}$ are prime polynomials with respect to each other.

### 5.2.1.2 - Discrete optimal closed loop control

**a) - Formulation of the problem**

The aim is to find a discrete controller denoted by $D(Z^{-1})$ which minimises a quadratic criterion of performance relating to a process subjected to a disturbance input which is random in nature (see Fig. 3). This may represent a measurement noise or a state noise related to the output of the system:

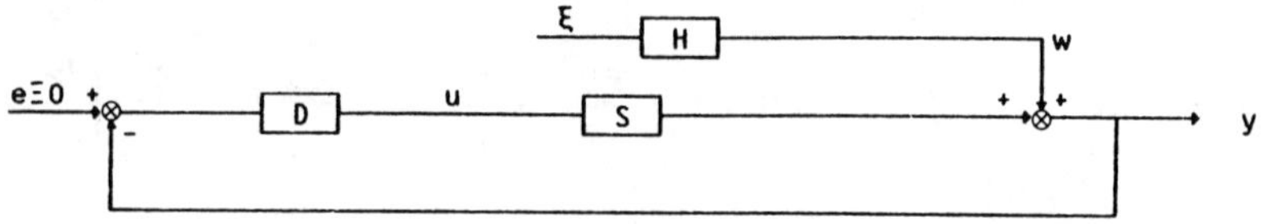


Fig. 3

* $S = B(z^{-1})/A(z^{-1})$ represents the transfer function, which is not necessarily stable, of the process with $(A,B) = 1$.

* w denotes an aditive noise at the output characterised by a white noise sequence $\xi(0,1)$ passing through a stable filter $H = R/L$;

* y represents the output of the process.

The objective is to find the optimal controller D which minimises the criterion

$$J = E\{\sum_{k=o}^{\infty} y^2(k)\}$$

The criterion

$$J = E\{\sum_{k=o}^{\infty} y^2(k) + \eta\, u^2(k)\}$$

appears below as a special case.

b) - **Finding the optimal criterion**

Let $L_o$ be defined as above, being assumed to belong to $F^+$

Theorem 2

The optimal corrector which minimises the quadratic criterion relating to the output in the closed loop is unique and is given by

$$D = \frac{A_o^+ Y_1}{L_o B^+ X_1} \tag{16}$$

where $X_1$ and $Y_1$ is the solution of minimal degree in $X_1$ of the polynomial equation

$$L A_o^- X_1 + B^- Y_1 = \tilde{B}^- R^* \tilde{A}_o^- \tag{17}$$

Proof

It has been shown at the beginning of this chapter that the set of closed loop transfer functions which guarantee the asymptotic stability of the loop and avoid the compensation of poles and zeros outside the unit circle is written in the form of (1), which is reproduced here:

$$G = BV + \mu\ AB \tag{18}$$

where U and V are the solution of AU+BV = 1.

Referring to Fig. 3, the output y is written as a function of G=DS/1+DS:

$$y = (1-G)\ \frac{R}{L}\ \xi$$

in other words, $y = A(U-\mu B)\ R\xi/L$, since the optimal controller in this subclass is to be found.

The problem can therefore be represented by the following diagram (Fig. 4):

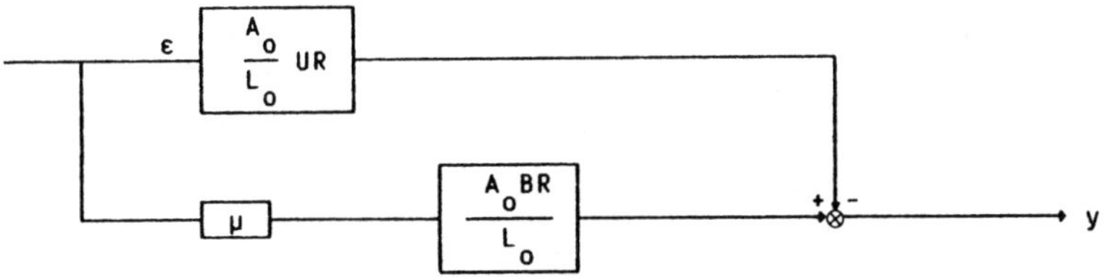


Fig. 4

By this transformation, the closed loop problem is changed into an open loop structure similar to that in Fig. 1. The optimal $\mu$ which minimises the quadratic criterion is to be found.

The direct application of the results of Theorem 1a gives

$$\mu = \frac{Z_1}{B^* \ R^* \ A_o^*} \tag{19}$$

where $X_1$ and $Z_1$ is the solution of minimal degree in $X_1$ of the polynomial

equation

$$L_o\ X_1 + B^-\ Z_1 = \tilde{B}^-\ R^*\ A_o^*\ U \tag{20}$$

At this stage, the problem is solved, but the procedure for obtaining the optimal solution is not convenient in this form. The following formulation is to be preferred:

Multiply equation (20) by A and, assuming that

$$A_o^+\ Y_1 = A\ \ Z_1 + \ B^*\ R^*\ A_o^*\ V \tag{21}$$

and taking (2) into account, we return to equation (17):

$$L\, A_o^- X_1 + B^- Y_1 = \tilde{B}^- R^* \tilde{A}_o^-$$

The closed loop transfer function G then becomes, taking (19) into account,

$$G = BV + \mu\, AB = BV + \frac{Z_1\, AB}{B^* R^* A_o^*}$$

and therefore, taking (21) into account,

$$G = \frac{B\, A_o^+ Y_1}{B^* R^* A_o^*} = \frac{B^- Y_1}{\tilde{B}^- R^* \tilde{A}_o^-}$$

The optimal controller is immediately obtained from the relationship

$$D = \frac{G}{(1-G)\, S}$$

i.e.

$$D = \frac{A_o^+ Y_1}{L_o B^+ X_1}$$

Thus we have returned to equation (16) of Theorem (2).

Notes

1. The polynomial equation (17) has solutions. Since L is stable, and since the polynomials A and B are prime polynomials with respect to each other,

$$(L\, A_o^- ,\; B^-) = 1$$

c) – **Applications**

Consider the closed loop system of Fig. 5:

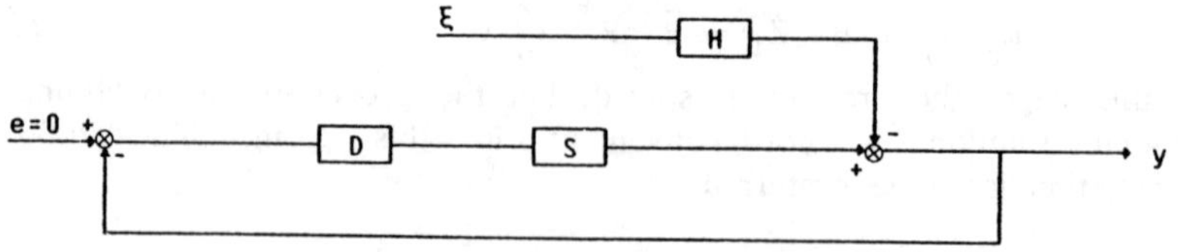


Fig. 5

1) The process S is stable, has time delays and minimum phase characteristics:

$$S = \frac{z^{-k} B_1}{A} \qquad \text{where } (A, z^{-k} B_1) = 1$$

$$H = \frac{C}{A} \qquad \text{where } A, B_1, C \in F^+$$

The application of Theorem 2 immediately leads to the optimal controller which minimises

$$J = E\{ \sum_{k=o}^{\infty} y^2(k) \}$$

$$D = \frac{Y_1}{B_1 X_1} \qquad \text{since } A_o = L_o = 1 \text{ and } B^+ = B_1 \tag{22}$$

where $X_1$ and $Y_1$ are the solution of minimal degree in $Y_1$ of the polynomial equation

$$A X_1 + z^{-k} Y_1 = C \tag{23}$$

This result is the minimum variance law proposed by Aström (1970) which is used in adaptive control techniques.

2) – If the process has non-minimum phase characteristics, then $B_1{}^+$ and $B_1{}^-$ denote the stable and unstable parts of $B_1$ respectively.

The $z^{-1}$ transfer function of the optimal controller which minimises the quadratic criterion is expressed as follows:

$$D = \frac{Y_1}{B_1^+ X_1} \tag{24}$$

where $X_1$ and $Y_1$ are the solution of minimal degree in $X_1$ of the polynomial equation

$$A X_1 + z^{-k} B_1^- Y_1 = C \tilde{B}_1^-$$

This is the minimum variance law for systems with non-minimum phase characteristics proposed by Peterka (1972).

**5.2.1.3 – Discrete optimal control in a closed loop (with 2 sources of disturbance)**

The preceding problem will now be extended to the system shown in Fig. 6, where two different types of disturbance (state disturbance and measurement noise) act on the system.

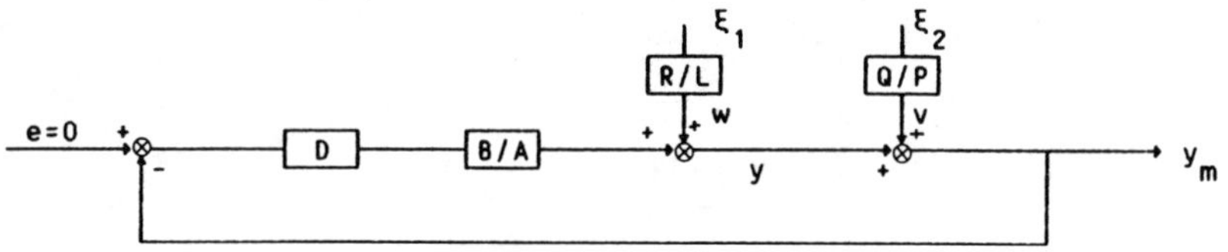


Fig. 6

The following notation is used:

* D is the $z^{-1}$ transfer function of the controller to be determined.

* $S = B(z^{-1})/A(z^{-1})$, the transfer function of the dynamic of the process.

* w is a disturbance with a mean value of zero characterised by its spectral power density

$$\Phi_{ww} = \frac{R(z^{-1})\ R(z)}{L(z^{-1})\ L(z)}$$

It is generated by a white noise $\xi_1$ passing through a filter R/L with $\Phi_{\xi_1 \xi_1} = 1$.

* v is the noise of the sensor with a mean value of zero characterised by its spectral power density

$$\Phi_{vv} = \frac{Q(z^{-1})\ Q(z)}{P(z^{-1})\ P(z)}$$

It is generated by a white noise $\xi_2$ passing through a filter Q/P with $\Phi_{\xi_2 \xi_2} = \alpha$.

The objective is to find an optimal controller D such that a linear combination of variance of the control signal and the output is minimised, namely:

$$J = E\left\{ \sum_{k=o}^{\infty} y^2(k) + \eta\, u^2(k) \right\}$$

Before the theorem is formulated, some definitions which are important for understanding the theorem will be given.

Definitions

$$A_o = \frac{A}{(A,L)}$$
$$L_o = \frac{L}{(A,L)}$$

- $C_1(z^{-1})$ denotes a **stable** polynomial in $z^{-1}$ such that

$$C_1 \bar{C}_1 = \eta A \bar{A} + B \bar{B}$$

- $C_2(z^{-1})$ denotes a **stable** polynomial in $z^{-1}$ such that

$$C_2 \bar{C}_2 = R \bar{R} P \bar{P} + \alpha Q \bar{Q} L \bar{L}$$

$$\left.\begin{aligned} l &= h_1 - \delta(C_1) \\ m &= h_1 - \delta(B) \\ n &= h_1 - \delta(A) \end{aligned}\right] \quad \text{where } h_1 = \max(\delta(A),\ \delta(B),\ \delta(C_1))$$

$$\left.\begin{aligned} i &= h_2 - \delta(C_2) \\ j &= h_2 - \delta(QL) \\ k &= h_2 - \delta(RP) \end{aligned}\right] \quad \text{where } h_2 = \max(\delta(QL),\ \delta(RP),\ \delta(C_2))$$

The theorem establishes optimal control based on the following hypotheses:
(A,B) = 1, (R,L) = 1, and (Q,P) = 1

Theorem 3

Optimal control exists for the preceding problem if

* $(L_0)^{-1}$ is analytic in $E^-$

* $(A,P^-) = 1$

The optimal controller is unique and is given by

$$D = \frac{PY_1}{X_1}$$

where $X_1$, $Y_1$ and $Z_1$ form the solution of the system of polynomial equations

$$AX_1 + BP\,Y_1 = C_1\,C_2\,A_0^*$$
$$\tilde{C}_1\,\tilde{C}_2\,z^{-(l+i)}\,Y_1 + L_o\,A^+\,Z_1 = A_o^*\,R\tilde{R}\tilde{P}\tilde{B}z^{-(m+k)}$$

such that

$$\delta(Z_1) < \delta(A^- \tilde{C}_1 \tilde{C}_2 z^{-(1+i)})$$

The proof of this theorem, which is of the same nature as the previous theorems, is not given here, but the reader will find it in Castel and Reboulet (1973).

Note

This theorem provides the basis for the solution of some particularly significant special cases. Corollaries 3a and 3b give details of these, and the theoretical results are illustrated with practical examples.

Corollary 3a

Given the following hypotheses:

1) $\alpha = 0$ (a single noise source)

2) $(A,B) = (R,L) = 1$

the optimal solution of the above problem exists if $(L_0)^{-1}$ is analytic in $E^-$.

The optimal controller is unique and is given by

$$D = \frac{Y}{X}$$

where X, Y, and Z form the solution of the following system of polynomial equations:

$$AX + BY = C_1 R^* A_o^*$$

$$z^{-1} \tilde{C}_1 Y + A L_o Z = R^* A_o^* \tilde{B} z^{-m}$$

with $\delta(Z) < \delta(z^{-1} \tilde{C}_1)$

This is the result obtained by Kucera (1979) which will now be illustrated by the following example:

Example of application

We shall consider a discrete first-order system subjected to a control signal u, with its output disturbed by a coloured measurement noise.

The discrete state equations governing the development of the process are as follows:

$$x(k+1) = ax(k) + b\,u(k)$$

$$w(k+1) = d\,w(k) + \xi(k+1) \quad \text{where } E(\xi^2(k)) = 1$$

$$y(k) = x(k) + w(k)$$

and this can be represented by the following diagram (Fig. 7).

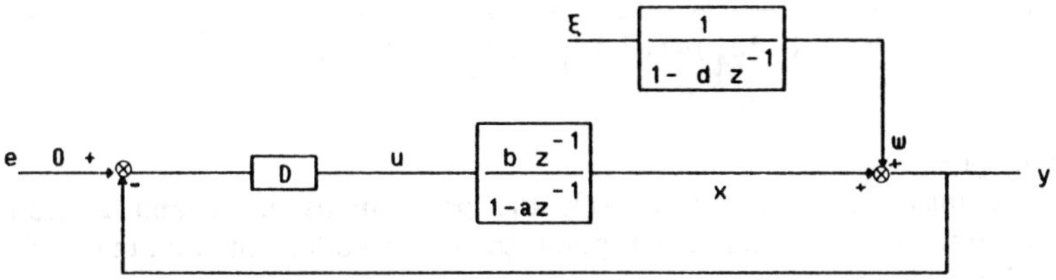


Fig. 7

The objective is to find the controller $D(z^{-1})$ which minimises the criterion

$$J = E\{\sum_{k=o}^{\infty} y^2(k) + \eta\, u^2(k)\} \qquad n \geqslant 0$$

In order to apply the results of corollary 3a, it is necessary to perform the factorisation which leads to the stable polynomial $C_1$ such that

$$C_1\,\bar{C}_1 = \eta\, A\bar{A} + B\bar{B}$$

This condition is equivalent to finding the parameters $\nu_1$ and $c_1$ of $C_1 = \nu_1(1-c_1\, z^{-1})$ such that

$$b^2\, z^{-1} + \eta(1-a\, z^{-1})(z^{-1}- a) = \nu_1^2(1-c_1\, z^{-1})(z^{-1}-c_1)$$

Therefore it may be seen that $\nu_1$ and $c_1$ are solutions of the following system:

$$a\,\eta = c_1\,\nu_1^2$$

$$b^2 + \eta(1+a^2) = \nu_1^2\,(1+c_1^2)$$

and it may be noted that

$$\nu_1^2 = \frac{ab^2}{(a-c_1)\quad(1-a\,c_1)}$$

<u>First case</u>

We shall initially consider the case of a stable process $|a| < 1$. Corollary 3a leads to X, Y, Z as the solution of minimal degree in Z of the polynomial system

$$X + b\,z^{-1}\,Y = (1 - c_1\,z^{-1})\,\nu_1$$

$$(1-d\,z^{-1})Z + \nu_1(z^{-1} - c_1)Y_1 = b$$

on condition that $Y = (1-az^{-1})Y_1$.
The optimal controller is therefore given by the relationship

$$D = \frac{Y}{X} = \frac{d(a - c_1)(1-a c_1)(1-a z^{-1})}{b[a(1-dc_1)+(a^2 dc_1 + dc_1 - ad - ac_1) z^{-1}]}$$

Second case

For unstable systems ($|a|>1$), the polynomials X, Y and Z leading to the optimal controller correspond to the solution of minimal degree in z of the following polynomial system:

$$(1-a z^{-1}) X + b z^{-1} Y = \nu_1 (1-c_1 z^{-1})(1 - \frac{z^{-1}}{a})$$

$$(1-d z^{-1})(1- a z^{-1}) Z + \nu_1 (z^{-1}-c_1) Y = b (1 - \frac{z^{-1}}{a})$$

The optimal controller therefore has the following expression for a rational fraction in $z^{-1}$ :

$$D = \frac{Y}{X} = \frac{a-c_1}{b} \frac{ad-1 + a^2 - a^2 dc_1 - ad(a-c_1) z^{-1}}{a^2(1-dc_1) - (a^2 d - 2adc_1 + c_1) z^{-1}}$$

This example shows that the simplicity of application of the method is independent of the stability of the system.

Corollary 3b

This corollary is important because, on the stated hypotheses, we obtain the standard case of a system where the additive noise on the states of the process is not coloured. An example will be given to illustrate this statement.

HYPOTHESES

1) $(A,B) = (R,L) = (Q,P) = (A,P) = 1$

2) $\delta(L) > \delta(\tilde{R}) - p_1 - b_1$

where $p_1$ and $b_1$ are the numbers of roots of the polynomials P and B respectively.

3) $L_0$ is a factor of $A_0^* \; R \; \tilde{R} \; \tilde{P}$

On these three hypotheses, the optimal solution exists if $(L_0)^{-1}$ is analytic in $E^-$.
The following rational fraction in $z^{-1}$ defines the optimal controller:

$$D = \frac{PY_1}{X}$$

in which the polynomials X and $Y_1$ are the solution of minimal degree in $Y_1$ of the equation

$$AX + BPY_1 = C_1\{C_2A_o^*/L_o\}$$

Application

We shall consider a discrete first-order system disturbed by a white noise, with the measurement being disturbed by a coloured noise. The discrete equations governing the system are as follows:

$$x(k+1) = ax(k) + bu(k) + \xi_1(k)$$

$$y(k) = x(k)$$

$$y_m(k) = y(k) + v(k)$$

where

$$v(k+1) = pv(k) + \xi_2(k+1)$$

$$E(\xi_1^2(k)) = 1$$

$$E(\xi_2^2(k)) = \alpha$$

and the noises $\xi_1$ and $\xi_2$ are independent: $E(\xi_1(k), \xi_2(k)) = 0$. The system may be represented as follows (Fig. 8):

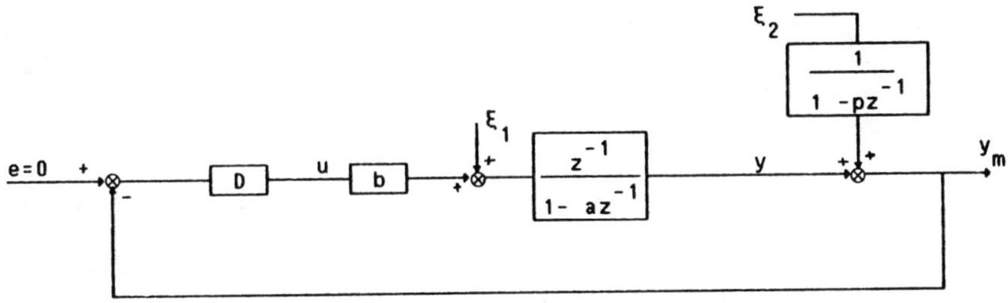


Fig. 8

The objective is to find the optimal controller D which minimises

$$J = E\left[\sum_{k=o}^{\infty} y^2(k) + \eta\, u^2(k)\right]$$

The first factorisation is a matter of finding $C_2 = \nu_2(1-c_2\, z^{-1})$, the solution of the equation

$$C_2 \bar{C}_2 = R\bar{R}\ P\bar{P} + \alpha\ Q\bar{Q}\ L\bar{L}$$

The parameters $c_2$ and $\nu_2$ must therefore comply with the following polynomial equality:

$$\nu_2^2 (1-c_2 z^{-1})(z^{-1}-c_2) = (1-pz^{-1})(z^{-1}-p) + \alpha(1-az^{-1})(z^{-1}-a)$$

The parameters $c_1$ and $\nu_1$ must also comply with the following equality:

$$\nu_1^2 (1-c_1 z^{-1})(z^{-1}-c_1) = b^2 z^{-1} + \eta(1-az^{-1})(z^{-1}-a)$$

The optimal controller is then expressed as follows for the $z^{-1}$ transfer function:

$$D = \frac{(1-pz^{-1})\ Y_1}{X}$$

where the polynomials X and $Y_1$ are the solution of minimal degree in $Y_1$

of the equation

$$(1-az^{-1})\ X + bz^{-1}(1-pz^{-1})\ Y_1 = \nu_1 \nu_2 (1-c_1 z^{-1})(1-c_2 z^{-1})$$

i.e.

$$Y_1 = \frac{(a-c_1)(a-c_2)}{b(a-p)} \quad \text{and } X = 1 + \left[\frac{ap-p(c_1+c_2)+c_1 c_2}{p-a}\right] z^{-1}$$

## 5.2.2 – THE MULTIVARIABLE CASE

The polynomial matrix approach enables us to generalise the above set of results to the multivariable case. The principal properties of these matrices are stated in Appendix 1 so that the reader can familiarise himself with them.

Certain important notations which will be used in this section will now be introduced.

* If $E(z^{-1})$ is a matrix of rational fractions in $z^{-1}$, then $\bar{E}$ is used to denote the matrix $E^T(z)$.

* If $p(z^{-1})$ represents a polynomial matrix in $z^{-1}$,
  - $\delta(P)$ denotes the highest degree of the polynomials forming P;
  - $\tilde{P}$ denotes the polynomial such that

$$\tilde{P} = z^{-\delta(P)}\ \bar{P}$$

* $(P,Q)_{lhp}$ and $(P,Q)_{rhp}$ denote two polynomial matrices which are left-hand prime and right-hand prime respectively.

### 5.2.2.1 – Optimal stochastic control of a multivariable process in open loop form

**a) Formulation of the problem**

The block diagram in Fig. 9 represents the system which is to be optimised by an open loop control signal:

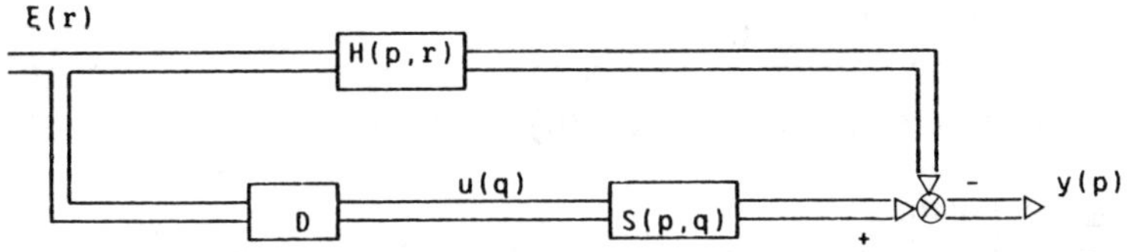


Fig. 9

* S represents the transfer matrix of the process consisting of rational fractions in $z^{-1}$. This has q inputs and p outputs. This transfer matrix may be written in the form

$$S = \underset{(p\times q)}{B} \; \underset{(q\times q)}{A^{-1}}$$

where B and A represent two right-hand prime polynomial matrices. The process is also assumed to be stable.

* w denotes a disturbance vector of dimension p characterised by a white noise vector ξ of dimension r passing through the filter $H = RL^{-1}$ which is assumed to be stable with

$$\Phi_{\xi\xi} = \Psi$$

The aim is to find the discrete controller D in the form of a rational fraction in $z^{-1}$ which generates, from the measurement of the disturbance ξ, a control vector which minimises the effect of this disturbance on the process. The criterion to be minimised may take either of the two following forms:

a) $$J_a = E\{\sum_{k=0}^{\infty} y^T(k)\Psi_1 y(k)\}$$ where $\Psi_1$ is a scalar matrix defined as positive

b) $$J_b = E\{\sum_{k=0}^{\infty} y^T(k)\Psi_1 y(k) + u^T(k)\Phi_1 u(k)\}$$

The second criterion has a term to weight the control signals in order to limit their amplitude. The scalar matrices $\Psi_1$ and $\Phi_1$ are defined as positive.

**b) – Finding the optimal corrector to minimise the criterion $J_a$**

Before establishing the basic theorem, two new definitions will be introduced:

– B* is used to denote the polynomial matrix, having a right–hand inverse denoted $B_d^{*-1}$ which is assumed to be stable, such that

$$\bar{B}\ \Psi_1\ \bar{B} = \overline{B^*}\,B^* \tag{26}$$

– m denotes the difference $\delta(B) - \delta(B^*)$.

Important notation

In expressions denoted $\tilde{K}^*$ or $\tilde{K}^{-1}$, the operation ~ affects the values $K^*$ or $K^{-1}$.

Theorem 4a

The optimal controller minimising the criterion $J_a$ is given by the following $z^{-1}$ rational fraction transfer matrix:

$$D = A\ B_d^{*-1} Y_1 L^{-1} \tag{27}$$

where $X_1$ and $Y_1$ are the solution of the polynomial matrix equation

$$X_1 L + z^{-m}\ \tilde{B}^* Y_1 = \tilde{B}\ \Psi_1 R \tag{28}$$

such that the matrix $z^m(B_d^{*-1})\ X_1$ is strictly proper (29)

Proof

According to Parseval's theorem,

$$J_a = \langle \bar{y}\ \Psi_1\ y \rangle \tag{30}$$

where

$$y = (BA^{-1}D - RL^{-1})\ \xi$$

Choosing a sequence $y_o$ such that

$$y_o = B^* A^{-1} D - \overline{B_d^{*-1}}\ \bar{B}\ \Psi_1 R\ L^{-1}$$

it will be shown that

$$\min J_a = \min \langle \bar{y}_o\ y_o \rangle \equiv \min \mathrm{tr} \langle y_o\ \bar{y}_o \rangle$$

By decomposing expression (30) it is easily seen that

$$\min J_a = \min \langle \bar{y}_o y_o \rangle \quad \text{if}\ \overline{B^*}\ \overline{B_d^{*-1}}\ \bar{B} = \bar{B}$$

To prove the last equality, the matrix B will be decomposed in the Hermite form. There is a matrix $U_{pxq}$ such that

$$B_{pxq} = [K_{pxr} \ 0] \, U_{qxq}$$

There is also a matrix $K^*_{rxr}$ having a stable inverse such that

$$\bar{K} \, K = \overline{K^*} \, K^*$$

Taking this last equality into account, it can be deduced that

$$B^* = [K^* \ 0\} \, U$$

Finally, since this matrix has a right-hand inverse denoted $B_d^{*-1}$,
we obtain

$$B_d^{*-1} = U^{-1} \begin{bmatrix} K^{*-1} \\ T \end{bmatrix}$$

where T is any stable rational matrix.

These different results immediately show that the following relationship is true:

$$\overline{B^*} \; \overline{B_d^{*-1}} \; \bar{B} = \bar{B}$$

We can now rewrite the value $y_o$ defined above in the following form:

$$y_o = B^* A^{-1} D - z^m \, (\tilde{B}_d^{*-1}) \, \tilde{B} \Psi_1 R \, L^{-1}$$

and this expression can be decomposed to

$$y_o = W - z^m \, (\tilde{B}_d^{*-1}) X$$

where

$$W = B^* \, A^{-1} \, D - YL^{-1}$$

The values X and Y appearing in these two relationships are connected by the matrix relationship (28).

At this stage of the study, we return to a procedure identical to the single-input, single-output case described in section 5.1.1.1. The minimum of the criterion is obtained when W = 0 (implying (27)) on condition that the matrix

$z^m (B_d^{*-1}) X$ is proper.

This conclusion completes the proof of Theorem 4a.

It now remains to demonstrate the existence of the solutions of the matrix equation (28) which comply with condition (29).

Equation (28) is written

$$XL + \tilde{U} z^{-m} \begin{bmatrix} \tilde{K}^* \\ 0 \end{bmatrix} Y = \tilde{U} \begin{bmatrix} \tilde{K} \\ 0 \end{bmatrix} \Psi_1 R$$

By letting

$$\tilde{U}^{-1} X = \begin{bmatrix} X_{11} \\ X_{12} \end{bmatrix}$$

the previous equation becomes

$$X_{11} L + z^{-m} \tilde{K}^* Y = \tilde{K} \Psi_1 R$$

In this equation, the square matrices L and $\tilde{K}^*$ can be decomposed into the diagonal Schmit form in which the polynomial invariants of L are stable and those of $\tilde{K}^*$ are unstable. Therefore the set of polynomial equations to which these decompositions lead has an infinite number of solutions. The solution which makes the matrix $(z^{-m} \tilde{K}^*)X_{11}$ strictly proper is unique.

Consequently, for any

$$X_1 = \begin{bmatrix} X_{11} \\ 0 \end{bmatrix}$$

the above result implies that $z^{-m} (\tilde{B}_d^{*-1}) X_1$ is strictly proper. It may therefore be asserted that $Y_1$ exists and is unique.

The notation $\tilde{B}_d^{*-1}$ signifies $(\tilde{B}_d^{*-1})$.

---

<u>Note</u>

1. Although $Y_1$ is unique, there is an infinite number of right-hand inverses $B_d^{*-1}$ of $B^*$ unless $B^*$ is a square matrix. In this case the controller is unique.

c) – **Finding the optimal corrector minimising the criterion $J_b$**

Before the theorem defining the optimal controller is stated, the new equation to be substituted for relationship (26) will be defined.

Assuming that

$$\text{rank} \begin{bmatrix} \Phi_1 & A \\ \Psi_1 & B \end{bmatrix} = q$$

then it is known that there is a square polynomial matrix of rank q, having a stable inverse, such that

$$\bar{A} \Psi_1 A + \bar{B} \Psi_1 B = \bar{C} C$$

l, m, and $h_1$ denote the following:

$$l = h_1 - \delta(C) \qquad \text{where } h_1 = \max(\delta(A), \delta(B), \delta(C))$$
$$m = h_1 - \delta(B)$$

Theorem 4b

The optimal criterion minimising the criterion $J_b$ is unique and is given by the relationship

$$D = A\, C^{-1}\, Y_1\, L^{-1}$$

$X_1$ and $Y_1$ are the solution of the polynomial matrix equation

$$X_1\, L + z^{-1}\, \tilde{C}\, y_1 = z^{-m}\, \tilde{B}\, \Psi_1\, R \tag{35}$$

which satisfies $(z^{-1}\, \tilde{C})^{-1}\, X_1$ strictly proper. (36)

Proof

According to Parseval's theorem, the criterion is written

$$J_b = E\{\frac{1}{2\pi j} \oint_\Gamma \Big[\bar{\epsilon}(\bar{D}\, \Phi_1 D + \bar{D}\, \overline{A^{-1}}\, \bar{B}\Psi_1 B\, A^{-1} D)\epsilon - \bar{\epsilon}(\bar{D}\, \overline{A^{-1}}\bar{B}\, \Psi_1 R\, L^{-1})\epsilon$$
$$-\bar{\epsilon}(\overline{L^{-1}}\, \bar{R}\Psi_1 B\, A^{-1} D)\,\epsilon + \bar{\epsilon}(\overline{L^{-1}}\, \bar{R}\, \Psi_1 R L^{-1})\epsilon\Big] \frac{dz}{z}\}$$

This may be put into the form

$$J = J_1 + J_2$$

where

$$J_1 = \langle \bar{E}_o\, E_o \rangle$$

and

$$E_o = C\, A^{-1} D - (z^{-1}\, \tilde{C})^{-1}\, z^{-m}\, \tilde{B}\, \Psi_1\, R L^{-1}$$

Since the term $J_2$ is independent of the controller D, the minimisation of J can be reduced to that of $J_1$, i.e.

$$\min J_o = \min \langle \bar{E}_o\, E_o \rangle \equiv \min \operatorname{tr} \langle E_o\, \bar{E}_o \rangle$$

As before, $E_o$ can be rewritten in the form

$$E_o = C\, A^{-1} D - Y_1 L^{-1} - (z^{-1}\, \tilde{C})^{-1}\, X_1 = W - (z^{-1}\, \tilde{C})^{-1}\, X_1$$

on condition that the polynomials $X_1$ and $Y_1$ are connected by matrix equation (35).

Returning to the procedure formulated for the single-variable case (see 5.1.1.1) it will be seen immediately that the minimum of this criterion will be attained when W = 0 if condition (36) is satisfied.

The condition W = 0 is identical to that of the optimal controller of Theorem 4b, and this completes the proof of the theorem.

**5.2.2.3 – Optimal stochastic closed loop control of a multivariable process**

**a) – Formulation of the problem**

We shall consider the multivariable process of the stochastic type controlled in the closed loop configuration and illustrated in Fig. 10.

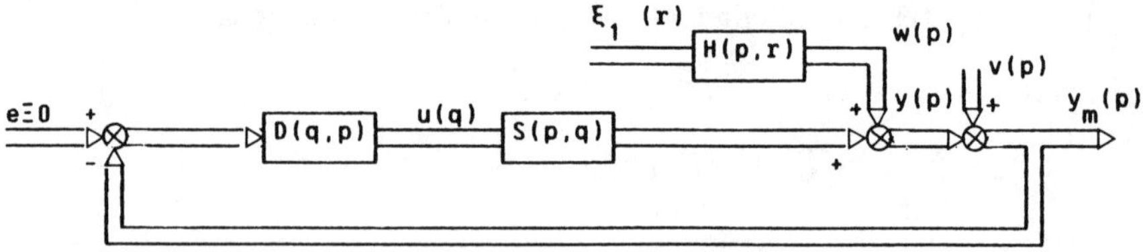


Fig. 10

* S denotes the transfer matrix characterising the dynamic of the process which can be modelled as follows:
   - either by a right-hand prime representation:

$$S = \underset{(p\times q)}{B}\ \underset{(q\times q)}{A^{-1}}$$

   - or by a left-hand prime representation:

$$S = \underset{(p\times p)}{A_1^{-1}}\ \underset{(p\times q)}{B_1}$$

* D represents the transfer matrix of the controller which is the objective of the optimal synthesis.

* w denotes a disturbance vector such that

$$w = L_1^{-1}\ R_1 \xi_1 \qquad \text{where} \quad \Phi_{\xi_1 \xi_1} = \Psi$$

where

$$\underset{(p\times p)}{L_1^{-1}}\ \underset{(p\times r)}{R_1} \quad \text{is the left-hand prime representation of } H(p,r)$$

and $\xi_1$ is a disturbance vector whose components are white.

* v denotes a white noise vector having a spectral power density of

$$\Phi_{vv} = \Phi$$

* the signals w and v are assumed to be independent.

On the basis of the above hypotheses, a transfer matrix will be found which characterises an optimal controller such that

- the quadratic criterion

$$J = E\left\{ \sum_{k=0}^{\infty} u^T(k)\, \Phi_1 u(k) + y^T(k) \Psi_1 y(k) \right\}$$

is minimal;

- the internal stability of the system, taking into account its closed loop structure, is assured; and

- the convergence of the criterion is guaranteed.

A number of definitions must now be given before the basic theorem can be stated.

b) - **Definitions**

1) If the rank of the matrix

$\begin{bmatrix} \Phi_1 & A \\ \Psi_1 & B \end{bmatrix}$ is equal to q, there is a polynomial square matrix $C(z^{-1})$ of rank q having a stable inverse such that

$$\bar{A}\, \Phi_1 A + B \bar{\Psi}_1 B = \bar{C}\, C$$

2) If the rank of the matrix $[L_1 \;\; \Phi \;\; R_1 \;\; \Psi]$ is equal to p, there is a polynomial square matrix of rank p denoted $C'(z^{-1})$ having a stable inverse such that

$$L_1 \Phi_1 \bar{L}_1 + R_1 \Psi \bar{R}_1 = C' \bar{C'}$$

3) l, m, n, and $h_1$ have the following meanings:

$$l = h_1 - \delta(C)$$
$$m = h_1 - \delta(B) \qquad \text{where } h_2 = \max(\delta(A_1), \delta(B), \delta(C))$$
$$n = h_1 - \delta(A)$$

4) i, j, k, and $h_2$ have the following meanings:

$$i = h_2 - \delta(C')$$
$$j = h_2 - \delta(L_1) \qquad \text{where } h_2 = \max(\delta(L_1), \delta(R_1), \delta(C'))$$
$$k = h_2 - \delta(R_1)$$

5) The matrices $A_{10}$, $L_{10}$, $A_{20}$ and $L_{20}$ model the right-hand prime and left-hand prime representations respectively such that

$$\underset{(p,p)}{A_1}\ \underset{(p,p)}{L_1^{-L}} = \underset{(p,p)}{A_{10}}\ \underset{(p,p)}{L_{10}^{-1}} \qquad \text{with } (A_{10}, L_{10}) \text{ rhp}$$

$$= L_{20}^{-1}\ A_{20} \qquad \text{with } (A_{20}, L_{20}) \text{ lhp}$$

6) $D_1$ denotes a polynomial matrix having a stable inverse such that

$$A_{20}\ C'\ \bar{C}'\ \bar{A}_{20} = D_1 \bar{D}_1$$

7) The matrices $D_o$, $L_o$, $A_2$, $D_2$, $C_o$, $B_2$ model the prime representations such that

$$\underset{(p\times p)}{D_o}\ \underset{(p\times p)}{L_o^{-1}} = \underset{(p,p)}{L_{20}^{-1}}\ \underset{(p,p)}{D_1} \qquad \text{with } (D_o, L_o) \text{ rhp}$$

$$\underset{(p,p)}{D_2}\ \underset{(p,p)}{A_2^{-1}} = \underset{(p,p)}{A_1^{-1}}\ \underset{(p,p)}{D_o} \qquad \text{with } (A_2, D_2) \text{ rhp}$$

$$\underset{(p,p)}{C_o^{-1}}\ \underset{(p,q)}{B_2} = \underset{(p,q)}{B}\ \underset{(q,q)}{C^{-1}} \qquad \text{with } (C_o, B_2) \text{ lhp}$$

The basic theorem can now be stated.

c) - **Theorem 5**

Let the process S be strictly causal so that $\delta(\tilde{B}) < \delta(B)$.

Let C and C' be the polynomial matrices obtained by the preceding factorisations (definitions 1 and 2).

The convergence of the criterion is then ensured if each of the matrices $L_{20}$ and $L_{10}$ has a stable inverse.

The optimal controller exists and is unique. Its transfer matrix is given by the matrix relationship

$$D = A\ C^{-1}\ Y_1\ (X_1\ A_2)^{-1}\ C_o \tag{37}$$

where $X_1$, $Y_1$, and $Z_1$ represent the solution of the following system of polynomial equations:

$$B_2\ Y_1 + X_1\ A_2 = C_o\ D_2 \tag{38}$$

$$z^{-1}\ \tilde{C}\ Y_1 - Z\ L_o\ A_2 = z^{-m}\ \tilde{B}\ \Psi_1\ D_2 \tag{39}$$

such that the matrix $(z^{-1}\ \tilde{C})^{-1}\ Z$ is strictly proper. (40)

Note

1. In many practical cases, equation (38) is sufficient for the unique determination of the solution. The purpose of the following corollary is to define the precise context in which this result is applicable.

   All the results shown here have been given without proof. The interested reader will find the details of the proofs concerned in Reboulet and Castel (1984).

d) – **Corollary 5**

HYPOTHESES

The two following hypotheses are added to those of Theorem (5):

$L_o = I_p$

$\delta(A_2) \geqslant \delta(D_2)$

The optimal controller exists and is unique, and is given by

$$D = A\ C^{-1}\ Y_1\ (X_1\ A_2)^{-1}\ C_o \qquad (41)$$

where $X_1$ and $Y_1$ form the solution of the polynomial equation

$$B_2\ Y_1 + X_1\ A_2 = C_o\ D_2 \qquad (42)$$

such that the matrix $Y_1\ A_2^{-1}$ is strictly proper.

This set of results will now be illustrated in the following application.

e) – **Application**

We shall consider the stochastic multivariable process with one input and two outputs represented in Fig. 11, and we shall attempt to find the

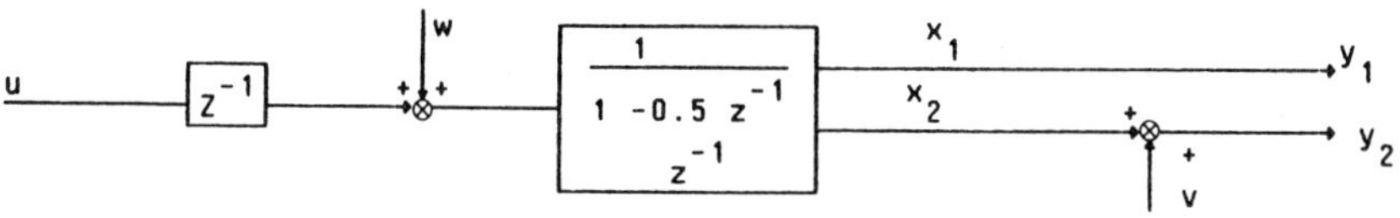


Fig. 11

optimal controller minimising the criterion

$$J = \sum_{k=0}^{\infty} u^2(k) + 7\ x_1^2(k) + 7\ x_2^2(k))$$

The disturbances w and v are white noises with $\Phi_{ww} = \Phi_{vv} = 1$.

<u>First stage: Modelling</u>

The first stage is to put the process into the right–hand prime and left–hand prime form of representation. We may write

$$S = \begin{bmatrix} z^{-1} \\ z^{-2}(1-0.5z^{-1} \end{bmatrix} (1 - 0.5\ z^{-1})^{-1} = BA^{-1}$$

or

$$S = \begin{bmatrix} 1 - 0.5z^{-1} & 0 \\ 0 & 1 \end{bmatrix}^{-1} \begin{bmatrix} z^{-1} \\ z^{-2} \end{bmatrix} = A_1^{-1} B_1$$

while the left-hand prime representation of H is as follows:

$$H = \begin{bmatrix} 1 - 0.5z^{-1} & 0 \\ 0 & 1 \end{bmatrix}^{-1} \begin{bmatrix} 1 \\ z^{-1} \end{bmatrix} = L_1^{-1} R_1$$

<u>Second stage: Solving the factorisation equations</u>

The matrices conforming to the factorisation equations must be found in accordance with the definitions.

* According to definition (1), a matrix must be found such that

$$\bar{A}\,\Phi_1 A + B\bar{\Psi}_1 B = \bar{C}\,C = -4z^{-1} + 17 + 4z$$

where

$$\Phi_1 = 1 \quad \text{and} \quad \Psi_1 = \begin{bmatrix} 7 & 0 \\ 0 & 7 \end{bmatrix}$$

Therefore $C = (4 - z^{-1})$.

* According to definition (2), the matrix C' complying with the following condition is found:

$$L_1 \Phi\, \bar{L}_1 + R_1 \Psi \bar{R}_1 = C'\overline{C'} = \begin{bmatrix} 1 & Z \\ z^{-1} & 2 \end{bmatrix}$$

and therefore

$$C' = \begin{bmatrix} 1 & 0 \\ z^{-1} & 1 \end{bmatrix} \quad \text{since } \Phi = \begin{bmatrix} 0 & 0 \\ 0 & 1 \end{bmatrix} \text{ and } \Psi = 1$$

* Definition (5) implies that

$$A_{10} = L_{10} = L_{20} = A_{20} = I_p$$

* Definition (6) is used to find the matrix $D_1$ which is the solution of the equation

$$A_{20}\; C'\; \bar{C}'\; \overline{A_{20}} = D_1\; \bar{D}_1$$

and therefore $D_1 = C'$.

* Definitions (7) lead to the matrices $D_o$, $L_0$ $(A_2, D_2)$ rhp and $(C_o, B_2)$ lhp, as follows:

$$D_o = C' \qquad\qquad L_o = I_p$$

$$\underbrace{\begin{bmatrix} 1-0.5z^{-1} & 0 \\ 0 & 1 \end{bmatrix}^{-1}}_{A_1^{-1}} \underbrace{\begin{bmatrix} 1 & 0 \\ z^{-1} & 1 \end{bmatrix}}_{D_o} = \underbrace{\begin{bmatrix} 1 & 0 \\ 0 & 1 \end{bmatrix}}_{D_2} \underbrace{\begin{bmatrix} 1-0.5z^{-1} & 0 \\ -z^{-1}(1-0.5z^{-1}) & 1 \end{bmatrix}^{-1}}_{A_2^{-1}}$$

$$\underbrace{\begin{bmatrix} z^{-1} \\ z^{-2}(1-0.5z^{-1}) \end{bmatrix}}_{B} \underbrace{\begin{bmatrix} 4-z^{-1} \end{bmatrix}^{-1}}_{C^{-1}} = \underbrace{\begin{bmatrix} 4-z^{-1} & 0 \\ 0 & 4-z^{-1} \end{bmatrix}^{-1}}_{C_o^{-1}} \underbrace{\begin{bmatrix} z^{-1} \\ z^{-2}(1-0.5z^{-1}) \end{bmatrix}}_{B_2}$$

<u>Third stage: Application of Corollary 5</u>

The conditions for application of the corollary are present, since $L_o = I_p$ and $\delta(A_2) \geqslant \delta(D_2)$.

The optimal controller is therefore given by the following relationship (41):

$$D = A\ C^{-1}\ Y_1\ (X_1\ A_2)^{-1}\ C_o$$

where $X_1$ and $Y_1$ correspond to the solution of minimal degree in $Y_1$ of the matrix equation

$$B_2\ Y_1 + X_1\ A_2 = C_o\ D_2$$

i.e.

$$\begin{bmatrix} z^{-1} \\ z^{-2}(1-0.5z^{-1}) \end{bmatrix} \begin{bmatrix} Y_{11} & Y_{12} \end{bmatrix} + \begin{bmatrix} X_{11} & X_{12} \\ X_{21} & X_{22} \end{bmatrix} \begin{bmatrix} 1-0.5z^{-1} & 0 \\ -z^{-1}(1-0.5z^{-1}) & 1 \end{bmatrix}$$

$$= \begin{bmatrix} 4-z^{-1} & 0 \\ 0 & 4-z^{-1} \end{bmatrix}$$

By solving this system, it will be immediately apparent that

$$Y_{12} = X_{12} = 0$$

$$X_{22} = 4 - z^{-1}$$

To obtain the other polynomials of the matrices $X_1$ and $Y_1$, it is necessary to solve the system

$$z^{-1}\ Y_{11} + X_{11}\ (1 - 0.5\ z^{-1}) = (4 - z^{-1})$$

$$z^{-2}\ Y_{11} + Y_{21} = z^{-1}\ (4-z^{-1})$$

which leads to

$$Y_{11} = 1 \qquad X_{11} = 4 \qquad X_{21} = 4\ z^{-1}(1-0.5\ z^{-1})$$

and to the controller

$$D = \begin{bmatrix} \frac{1}{4} & 0 \end{bmatrix}$$

The right-hand prime polynomial matrices $A_2$ and $D_2$ become

$$A_2 = (1 - 0.5\ z^{-1}) \qquad D_2 = \begin{bmatrix} 1 \\ z^{-1}(1 - 0.5\ z^{-1}) \end{bmatrix}$$

It is then necessary to find the polynomial matrices $X_1$ and $Y_1$ forming the solution of the matrix equation

$$\begin{bmatrix} z^{-1} \\ z^{-2}(1-0.5z^{-1}) \end{bmatrix} Y_1 + X_1\ (1-0.5z^{-1}) = \begin{bmatrix} 4 - z^{-1} \\ z^{-1}(4-z^{-1})(1-0.5z^{-1}) \end{bmatrix}$$

which results in

$$Y_1 = 1$$

$$X_1 = \begin{bmatrix} 4 \\ 4z^{-1}(1-0.5z^{-1}) \end{bmatrix}$$

The family of stable inverse matrices to the left of $D_o$ is expressed by

$$(D_o)_g^{-1} = (1 - \alpha\ z^{-1},\ \alpha)$$

where $\alpha$ is an analytic fraction in $E^-$.

The transfer matrix of the optimal controller is then written thus:

$$D = \begin{bmatrix} \dfrac{1 - \alpha z^{-1}}{4} & \dfrac{\alpha}{4(1-0.5z^{-1})} \end{bmatrix}$$

### 5.2.3 - CONCLUSION

It is interesting to note that finding the optimal solution for each of the problems dealt with only requires algebraic manipulation of polynomials, which are remarkably suitable for use in digital computer programs. In fact, efficient algorithms already exist for the solution of both the problem of factorisation and the problem of solving matrix systems of polynomial equations.

The procedure used here enables both the deterministic and the stochastic case to be dealt with equally well, although only the latter case was used to illustrate the methodology. Finally, it should be noted that, although the flexibility of representation provided by the transfer function approach makes it simple to deal with the case of additional inputs, the cases of coloured noise and delays, and singular cases by the state space method (with zero weighting on the control signal, and no measurement noise), this method is not capable of dealing with the case of criteria on a finite horizon, and is only applicable to the control of stationary systems.

## *5.3 - THE STATE VECTOR MODELLING APPROACH*

In parallel with the polynomial approach using the transfer function representation, the earliest form of control, that of discrete linear stochastic systems, has been based on state vector modelling. The purpose of this section is to bring together the principal results obtained and to make a highly instructive comparison with the polynomial method.

### 5.3.1 - NOTES ON THE PRINCIPAL RESULTS OBTAINED IN DISCRETE OPTIMAL STOCHASTIC CONTROL

First of all, the general formulation of the problem of discrete optimal stochastic control and its solution will be stated. Certain modifications of this initial formulation, necessary for the solution of more complicated problems, will then be detailed. The expressions obtained will often be algebraically equivalent to those found for the continuous case and described in Chapter 2. In the following text, therefore, reference is always made to the discrete version of the LQG (linear quadratic Gaussian) controller and the differences will be pointed out as they arise.

#### 5.3.1.1 - Formulation of the problem

Consider a discrete linear system described by the difference equation

$$x_{k+1} = F_k x_k + G_k u_k + w_k \qquad k = 0,1,\dots,N-1 \tag{43}$$

in which $F_k$ denotes a square nxn matrix, $G_k$ a nxq matrix, $x_k$ and $w_k$ vectors of dimension n and $u_k$ a vector of dimension q. The state noise $w_k$ is a pseudo-white stationary discrete process with a zero mean, i.e.

$$E(w_k) = 0 \qquad \forall k$$

$$E(w_k,\ w_l^T) = M_k\ \delta_{k-l} \qquad k \text{ and } l \in \{0,\dots,N-1\} \tag{44}$$

In addition, the initial state conditions $x_0$ have known stochastic values as follows:

$$E(x_0) = \bar{x}_0 \qquad E((x_0-\bar{x}_0)(x_0-\bar{x}_0)^T = X_0$$

The state is observed through the measurement equation:

$$y_k = H_k x_k + v_k \tag{45}$$

where $H_k$ is a matrix of dimensions pxn, and $y_k$ and $v_k$ are vectors of dimension p. The measurement noise $v_k$ is a stationary process with a zero mean characterised by

$$E(v_k) = 0 \qquad E(v_k \, v_l^T) = N_k \, \delta_{k-l} \qquad k, l \in \{0, 1, \ldots, N-1\}$$

In addition, it is assumed that the state and measurement noises are not intercorrelated, i.e.

$$D(w_k \, v_l^T) = 0 \qquad \forall k, l$$

The objective is to find the control law $u_k = f(y^k, u^{k-1})$ which minimises the stochastic quadratic criterion

$$J = E\,\{ \sum_{k=1}^{N} x_k^T \, Q_k \, x_k + \sum_{k=o}^{N-1} u_k^T \, R_k \, u_k) \tag{46}$$

It is assumed that the matrices $Q_k$, $R_k$, $M_k$ and $N_k$ are defined as positive and it is known that, in this case, the control signal $u_o$, $u_1, \ldots, u_{N-1}$ which minimises the criterion is found in two stages.

First stage

The control signal is expressed as a function of the state reconstructed optimally from information available at the instant k, denoted by $\hat{x}_k/I_k$.

The formulae are then those for discrete optimal linear control. Starting with the terminal instant N, the successive control signals are obtained by

$$u_k = -L_k \, \hat{x}_k/I_k \qquad k = N-1,\ N-2, \ldots, 0 \tag{47}$$

where the control gain $L_k$ is a function of the criterion matrix $S_k$

$$L_k = (R_k + G_k^T \, S_{k+1} \, G_k)^{-1} \, G_k^T \, S_{k+1} \, F_k \tag{48}$$

The criterion matrix is obtained from the recursive equation

$$S_k = (F_k - G_k L_k)^T \, S_{k+1} (F_k - G_k L_k) + L_k^T \, R_k L_k + Q_k \tag{49}$$

where $S_k$ is initialised to $S_N = Q_N$

Second stage

The formulae used to find $\hat{x}_k/I_k$ are conventional formulae for the non-stationary Kalman filter. Two formulations are possible, however, depending on whether or not the available information $I_k$ contains the observation at the instant k, in other words $y_k$. In the second case,

$$K_k = \{y_0, y_1, \ldots, y_{k-1}, u_0, u_1, \ldots, u_{k-1}\} = y^{k-1}$$

On this hypothesis, the estimate of the state is developed from the initial estimated state according to the following recursive equation:

$$\hat{x}_0 = E(x_0) = \bar{x}_0$$

$$\hat{x}_{k+1/y^k} = F_k \hat{x}_{k/y^{k-1}} + G_k u_k + K_k (y_k \; H_k \hat{x}_{k/y^{k-1}}) \quad (50)$$

The estimation gain $K_k$ is obtained from

$$K_k = F_k P_{k/y^{k-1}} H_k^T (H_k P_{k/y^{k-1}} H_k^T + N_k)^{-1} \quad (51)$$

and the matrix $P_K$ is developed from the equation

$$P_{k+1/y^k} = (F_k - K_k H_k) P_{m/y^{k-1}} (F_k - K_k H_k)^T + K_k N_k K_k^T + M_k \quad (52)$$

on the basis of the initialisation $P_0 = X_0$.

Relationships (51) and (52) demonstrate the formal duality of these equations with the control equations of the first stage.

Important notes

1. By a similar method to that used in the continuous case, the problem of designing an optimal stochastic controller is divided into two sub-problems whose formulation is dual. The first important difference concerns the structure of the matrices to be inverted which are used in these two stages, namely

$$R_k + G_k^T S_{k+1} G_k \quad \text{and} \quad H_k P_{k/y^{k-1}} H_k^T + N_k \quad .$$

The condition $R_k > 0$ or $N_K > 0$ is no longer necessary for the calculation of the inverses. It remains sufficient since the added terms

$$(G_k^T S_{k+1} G_k) \quad \text{and} \quad H_k P_{k/y^{k-1}} H_k^T$$

remain positive or zero if they were such on initialisation.

2. A second note concerns the form of the state estimation equations. The most usual formulation of Kalman filters concerns the estimation of the state **after** the k-th measurement. The result is written $\hat{x}_k/I_k$, where $I_k$ contains $y_k$.

Fig. 12 shows the variance between the two systems of equations (in a scalar case):

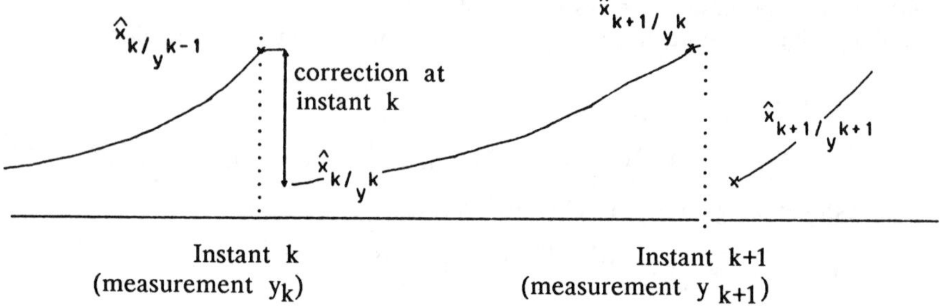


Fig. 12

In the formulation shown above, the control signal at instant k+1 depends on the best prediction of the state at instant k+1. This prediction consists of the correction at instant k $(K_k(y_k-H_k\ x_{k}/y^k)$ followed by the prediction for the interval k, k+1, thus:

$$\Delta x_k = F_k x_k + G_k u_k$$

If the measurement at instant k+1 is assumed to be available for the calculation of the control signal at the same instant, the whole system can be displaced towards the right-hand side of the sampling instants. The equations then take on the following standard form:

$$\hat{x}_{k+1/_y k} = F_k \hat{x}_{k/_y k} + G_k u_k \qquad \text{prediction equation}$$

$$\hat{x}_{k+1/_y k+1} = \hat{x}_{k+1/_y k} + \bar{K}_{k+1}\ (y_{k+1} - H_{k+1}\hat{x}_{k+1/_y k})$$

By letting $K_k = F_k\ \bar{K}_k$, the estimation equations are modified as follows, with the introduction of the covariance $P_{k/y^k}$

$$\bar{K}_k = P_{k/_y k-1}\ H_k^T (H_k P_{k/_y k-1} H_k^T + N_k)^{-1}$$

$$P_{k+1/_y k} = F_k\ P_{k/_y k}\ F_k^T + M_k$$

$$P_{k+1/_y k+1} = (I - \bar{K}_{k+1}\ H_{k+1}) P_{k+1/_y k} (I - \bar{K}_{k+1} H_{k+1})^T$$

$$+ \bar{K}_{k+1}\ M_{k+1}\ \bar{K}_{k+1}^T$$

It should be noted that the last equation (which in general provides the covariance of the estimation error even when the gain $\bar{K}_{k+1}$ is not optimal) is written as follows when $\bar{K}_{k+1}$ is optimal:

$$P_{k+1/y^{k+1}} = (I - \bar{K}_{k+1} \; H_{k+1})P_{k+1/y^{k}}$$

This formulation is more simple, but has the disadvantage of not being quadratic, which may cause practical problems of propagation of the symmetry and positivity of the covariance matrices.

Finally, it should be noted that the equations of the propagation of the covariances and gains are independent of the measurements $y_k$ and may be calculated *a priori*.

### 5.3.1.2 – Stationary case – fixed parameters

The stationary case (matrices F, G, H, M, N not indexed) with fixed parameters is very important in practice; moreover, there is an equivalence with the solutions obtained directly from the operational representation of the system (by the Wiener method or the polynomial approach). The following formulations are written if the system contains no unobservable or uncontrollable modes which are unstable or at the limit of stability:

* in the case of the control signal:

$$L = (R + G^TSG)^{-1} \; G^TSF$$

$$S = (F - GL)^TS(F - GL) + L^TRL + Q$$

The elimination of L between the two equations gives the equivalent of the continuous Riccati equation:

$$S = F^TSF + Q - F^T \; SG(R + G^TSG)^{-1} \; G^TSF$$

* in the case of the estimation:

The prediction covariance matrix P is, by duality, the solution of a similar equation:

$$P = FPF^T + M - FPH^T(N \quad HPH^T)^{-1} \; HPF^T$$

The covariance matrices of the corrected state $\bar{P}$ and the gain $\bar{K}$ are obtained directly from the following equations:

$$\bar{P} = P - PH^T(HPH^T + N^{-1} \; HP$$

$$\bar{K} = PH^T(HPH^T + N)^{-1}$$

Regardless of the structure of the filter adopted ("predictive" or "corrective"), the fact of assuming fixed parameters entails the possibility of the determination of a digital controller by solving the control and estimation equations with respect to the input–output relation where the measurements act as the input and the control signals act as the output (it should be remembered that we are discussing the case of a controller).

Therefore,

*in the case of the "predictor",

$$\hat{x}_{k+1/_y k} = F\,\hat{x}_{k/_y k-1} - GL\,\hat{x}_{k/_y k-1} + K(y_k - H\,\hat{x}_{k/_y k-1})$$

$$u_k = -L\,\hat{x}_{k/_y k-1}$$

and therefore

$$u = -L(Iz - (F-GL-KH))^{-1}Ky$$

* in the case of the "estimator",

$$\hat{x}_{k+1/_y k+1} = F\hat{x}_{k/_y k} - GL\,\hat{x}_{k/_y k}\bar{K}(y_{k+1} - H(F\hat{x}_{k/_y k} - GL\hat{x}_{k/_y k}))$$

$$u_k = -L\hat{x}_{k/_y k}$$

and therefore

$$u = -L(Iz - (I-\bar{K}H)(F-GL)^{-1}\ Kzy)$$

It will be noted that these two control signals provide 2n modes in the closed loop system which are eigenvalues of F−KH (or $(I-\bar{K}H)F$ which are equal to them) and F−GL. The controller has n poles which are different in the two cases, but there is a special difference between the numerators: the "predictor" is strictly proper (as in the continuous case, because of the existence of a white measurement noise), whereas the "controller" includes direct transmission.

## 5.3.2 – SPECIAL FORMULATIONS

### 5.3.2.1 – Correlated noises; combined state/control criterion

If the state and measurement noises are correlated, the formulation should be modified slightly to put the equations into the canonical form. If we introduce

$$E\left[\,w_k\ v_1^T\,\right] = \Omega_k\ \delta_{k1}$$

the state equation is modified by adding and subtracting the observation $z_k$ multiplied by a gain $E_k$ which will be determined in such a way as to make the intercorrelation disappear; it becomes

$$x_{k+1} = F_k x_k + G_k u_k + E_k y_k - E_k(H_k x_k + v_k) + w_k$$

$$= (F_k - E_k H_k)x_k + (G_k u_k + E_k y_k) + \underbrace{w_k - E_k v_k}_{w'_k} \tag{53}$$

and $E_k$ is determined in such a way that

$$E\left[w'_k v_l^T\right] = 0, \quad \text{i.e.} \quad E\left[w_K v_l^T\right] = E_k \times E\left[v_k v_l^T\right]$$

and therefore $\quad E_k = \Omega_k N_k^{-1}$.

Consequently, the new equations without correlation between the state and the measurement are written thus:

$$x_{k+1} = \underbrace{(F_k - \Omega_k N_k^{-1} H_k)}_{F'_k} x_k + G_k u_k + \Omega_k N_k^{-1} y_k + W'_k$$

$$y_k = H_k x_k + v_k$$

with

$$E\left[W'_k W'^T_l\right] = M'_k \delta_{kl}$$

$$= (M_k - \Omega_k N_k^{-1} \Omega_k^T)\delta_{kl}$$

$$E\left[v_k v_l^T\right] = N_k \delta_{kl} \qquad E\left[W'_k v_l^T\right] = 0$$

The equations are then deduced without difficulty: for example, in the case of the "predictor",

$$\hat{x}_{k+1/_y k} = (F_k - \Omega_k N_k^{-1} H_k)\hat{x}_{k/_y k-1} + G_k u_k + \Omega_k N_k^{-1} y_k + K_k(y_k - H_k \hat{x}_{K/_y k-1}) \tag{54}$$

with

$$K_k = (F_k - \Omega_k N_k^{-1} H_k) P_{k/_y k-1} H_k^T (H_k P_{k/_y k-1} H_k^T + N_k)^{-1} \tag{55}$$

and

$$P_{k+1/_y k} = (F_k - \Omega_k N_k^{-1} H_k - K_k H_k) P_{k/_y k-1} (F_k - \Omega_k N_k^{-1} H_k - K_k H_k)^T + K_k N_k K_k^T + M_k - \Omega_k N_k^{-1} \Omega_k^T \tag{56}$$

It is also clear that the system of equations (54) (55) (56) can be reinterpreted by introducing the modified gain:

$$K'_k = K_k + \Omega_k N_k^{-1}$$

In the case of a combined state/control quadratic criterion:

$$E\left[\sum_{k=1}^{N} x_k^T Q_k x_k + \sum_{k=0}^{N-1} u_k^T R_k u_k + 2\sum_{k=0}^{N} u_k^T T_k x_k\right] \qquad (57)$$

The formulation of the control problem and its solution are dualised versions of those above. The following formulae may therefore be obtained by inspection:

$$u_k = -L_k\hat{x}_{k/I_k} - R_k^{-1}T_k\hat{x}_{k/I_k} \qquad (58)$$

where

$$L_k = (R_k+G_k^T S_{k+1}G_k)\quad G_k^{-1} S_{k+1}^T (F_k-G_k R_k^{-1}T_k) \qquad (59)$$

and

$$S_k = (F_k-G_kL_k-G_kR_k^{-1}T_k)^t \quad S_{k+1}(F_k-G_kL_k-G_kR_k^{-1}T_k) + L_k^TR_kL_k + Q_k - L_k^T R_k^{-1}T_k \qquad (60)$$

It is also possible to perform the proof in a different way by introducing a modified control signal u' to eliminate the common terms of the state and control signal.

### 5.3.2.2 – **Coloured measurement noise** (singular measurement noise)

The colouring of the measurement noise is normally treated by increasing the dimension of the state vector by the number of states necessary to represent the noise. This increase presents no problems as regards formulation, but, from the point of view of solution,

- the estimation problem becomes singular, because there is not more measurement noise; and
- uncontrollable states appear in the control problem.

The first difficulty can be solved by modifying the measurement vector to make the state equation appear in the observation. The equations are written thus:

$$\begin{bmatrix} x_{k+1} \\ w'_{k+1} \end{bmatrix} = \begin{bmatrix} F_k & 0 \\ 0 & D_k \end{bmatrix} \begin{bmatrix} x_k \\ w'_k \end{bmatrix} + \begin{bmatrix} G_k \\ 0 \end{bmatrix} u_k + \begin{bmatrix} w_k \\ v_k \end{bmatrix} \qquad (61)$$

$$y_k = \begin{bmatrix} H_k & I \end{bmatrix} \begin{bmatrix} x_k \\ w'_k \end{bmatrix} \qquad (62)$$

introducing the correlated noise:

$$w'_{k+1} = D_k w'_k + v_k$$

To solve the problem of the singularity of the measurement noise, the following new observation is created:

$$z_k = y_{k+1} - H_{k+1} G_k u_k$$

and this brings the equations into the canonical form with intercorrelation between the state noise and the measurement noise:

$$z_k = [H_{k+1} \quad I] \begin{bmatrix} F_k & 0 \\ 0 & D_k \end{bmatrix} \begin{bmatrix} x_k \\ w'_k \end{bmatrix} + [H_{k+1} \quad I] \begin{bmatrix} G_k \\ 0 \end{bmatrix} u_k$$

$$+ [H_{k+1} \quad I] \begin{bmatrix} w_k \\ v_k \end{bmatrix} - H_{k+1} G_k u_k = [H_{k+1} F_k \quad D_k] \begin{bmatrix} x_k \\ w'_k \end{bmatrix} + H_{k+1} w_k + v_k$$

In this form, the problem of estimation has the dimension n(dimension of $x_k$) + q(dimension of $w'_k$), and therefore does not take into account the reduction of dimensions that can be obtained as a result of the absence of measurement noise in the initial problem. With the new observation

$$z'_k = z_k - D_k y_k$$

the system of equations of dimension n

$$x_{k+1} = F_k x_k + G_k u_k + w_k$$

$$z'_k = (H_{k+1} F_k - D_k H_k) x_k + H_{k+1} w_k + v_k$$

is in canonical form with intercorrelation between the state noise and the measurement noise. The filter obtained from these equations operates on the observations

$$z'_k = y_{k+1} - D_k y_k - H_{k+1} G_k u_k$$

It therefore operates by using the information available at the instant k+1 (because of $y_{k+1}$) and will therefore be in the form of an "estimator" but of dimension n only.

The problem of the uncontrollable modes does not affect the solution of the control problem; the state feedback obtained will not change the modes in question. The gains on the corresponding states will be zero, if the quadratic criterion only affects the deterministic states.

### 5.3.3 - EXAMPLE

To illustrate the theoretical results above, we shall return to the example already used in section 5.2.

Consider the system described by the following equations:

$$x_{k+1} = \begin{bmatrix} 0.5 & 0 & 0 \\ 0 & 0 & 1 \\ 0 & 0 & 0 \end{bmatrix} x_k + \begin{bmatrix} 1 \\ 0 \\ 1 \end{bmatrix} u_k + \begin{bmatrix} 1 \\ 0 \\ 1 \end{bmatrix} w_k$$

$$y_k = \begin{bmatrix} 1 & 0 & 0 \\ 0 & 1 & 0 \end{bmatrix} x_k + \begin{bmatrix} v_{k1} \\ v_{k2} \end{bmatrix}$$

with $Ew_k \, . \, w_k^T) = 1$

$$E(v_k \, . \, v_k^T) = \begin{bmatrix} 0 & 0 \\ 0 & 1 \end{bmatrix}$$

The criterion to be minimised is:

$$J = E\{ u_k^2 + 7 \, x_{1k}^2 + 7 \, x_{2k}^2 \}$$

It is simple to check that the sufficient conditions for the existence and unit nature of the matrices P and S are satisfied ((F,G) and (F, $M^{1/2}$) are stabilisable, and (F, $Q^{1/2}$) and (F,H) are detectable) and therefore that the problem can be solved.

The solution of the Ricatti equation for the control part results in

$$S = \begin{bmatrix} 8 & 0 & 0 \\ 0 & 7 & 0 \\ 0 & 0 & 7 \end{bmatrix}$$

from which the following feedback gain vector is deduced:

$$L = [\tfrac{1}{4} \; , \; 0 \; , \; 0 \; ]$$

Since only the first state $x_1$ is present and since it is measured without noise, the solution could be halted at this stage. However, in order to illustrate the theoretical part, the Kalman filter will be formulated.

The prediction covariance matrix P is calculated by solving the equation

$$P = FPF^T + M - F\bar{K}HPF^T$$

The division of P into sub-blocks:

$$P = \begin{bmatrix} P_o & C \\ C^T & p_{33} \end{bmatrix}$$

makes the solution evident:

$$C^T = [1, \, 0] \quad \text{and } p_{33} = 1$$

The comparison of the remaining equalities with those for $\bar{K}$, namely

$$\bar{K}\,(N + HPH^T) = PH^T = \begin{vmatrix} P_o \\ C^T \end{vmatrix}$$

leads to the following semi-positive matrix P:

$$P = \begin{vmatrix} 1 & 0 & 1 \\ 0 & 0 & 0 \\ 1 & 0 & 1 \end{vmatrix}$$

and the gain

$$\bar{K} = \begin{vmatrix} 1 & 0 \\ 0 & 0 \\ 1 & 0 \end{vmatrix}$$

The filter equations are then written thus:

$$\hat{x}_{1k+1/y^{k+1}} = y_{1k+1}$$

$$\hat{x}_{2k+1/y^{k+1}} = \hat{x}_{3k/y^k}$$

$$\hat{x}_{3k+1/y^{k+1}} = -\frac{1}{2}\,\hat{x}_{1k/y^k} + y_{1k+1}$$

In the case of a zero measurement noise, the calculation results in the same value of P, implying that the matrix $R' + HPH^T$ is non-invertible. The value of $\bar{K}$ may then be found by considering a pseudo-inverse of this matrix which, in the present case, leads to

$$\bar{K} = \begin{vmatrix} 1 & \alpha_1 \\ 0 & \alpha_2 \\ 1 & \alpha_3 \end{vmatrix}$$

where the scalar quantities $\alpha_i$ may have any value.

The equations of the resulting Kalman filter remain the same as before except for the addition of the terms $\alpha_i \cdot (y_{2k+1} - \hat{x}_{3k/yk})$, $\{i = 1,2,3\}$. In the case of a zero measurement noise, this quantity is itself zero.

## *5.4 - COMPARISON OF THE TWO APPROACHES*

The feedback difference matrices which play an essential part in current studies of robustness enable a connection to be made between the state space and the transfer function approaches which have been described.

The basis of this comparison is the writing of the stationary Ricatti equation in a form which clearly displays the different transfer matrices of the system and consequently the dynamic of the closed loop system. The two methods which will be investigated for the purpose of comparison are based on rational matrices in one case, and on polynomial matrices in the other case. Both of these transform the solution of the Ricatti equation into a factorisation and polynomial identification.

### 5.4.1 – RATIONAL MATRICES

This method, historically the first to be developed, was initially used for continuous systems. Elementary manipulations of the Ricatti equation lead to the following identities:

* Estimation

$$E(z) \,.\, (N+HPH^T)\; E^T(z^{-1}) = N+H(zI-F)^{-1}M(z^{-1}I-F)^{-1}H^T$$
$$= \Phi(z) \qquad (63)$$

* Control signal

$$D^T(z^{-1}).(R+G^TSG)D(z) = R+G^T(z^{-1}I-F^T)^{-1}Q(zI-F)^{-1}\; G = \Psi(z) \qquad (64)$$

where

$$E(z) = I + H(zI-F)^{-1}\; F\; \bar{K} \qquad (65)$$

$$D(z) = I + L(zI - F)^{-1}\; G \qquad (66)$$

which are simply the feedback difference matrices in $\epsilon$ and u (Figs. 13 and 14).

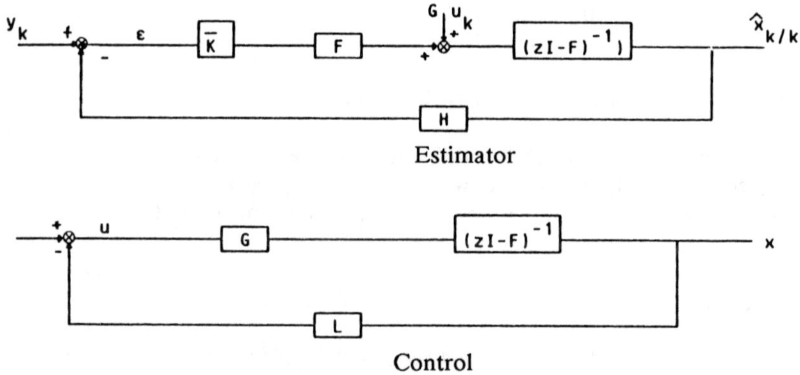


Figs. 13 and 14

It may be noted that the matrix $\Phi(z)$ corresponds to the spectral density of the measurement and state noise related to the output.

By examining the above equalities, it will be seen that:

– the poles of E(z) and D(z) are the poles of the system to be controlled;

– the poles of $E^{-1}(z)$ and $D^{-1}(z)$ are the poles of the optimal closed-loop system and must therefore have a modulus of less than 1 to ensure stability.

It may be deduced from this that the para-Hermitian matrices $\Phi(z)$ and $\Psi(z)$, when factorised in such a way as to satisfy the two preceding conditions, enable the matrices E(z) and D(z) to be found. The practical procedure is as follows:

1) Factorisation of the matrices $\Phi(z)$ and $\Psi(z)$ in the form:

$$\Delta(z^{-1}) \, . \, \Delta^T(z) = \Phi(z)$$
$$\Delta_1^T(z) \, . \, \Delta_1(z^{-1}) = \Psi(z)$$

by choosing the factors $\Delta(z^{-1})$ and $\Delta_1(z^{-1})$ in such a way that

$\Delta(z^{-1})$ has the same poles as $I - z^{-1} F$

$\Delta^{-1}(z^{-1})$ is asymptotically stable.

2) Change to the z transform by letting

$$y(z) = \Delta(z^{-1})$$

$$y_1(z) = \Delta_1(Z^{-1})$$

which leads to the following relationships:

$$E(z) \, (N+HP \, H^T)^{1/2} = y(z)\{ \lim_{z^{-1}\to 0} \Delta(z^{-1})\}$$

$$(R + G^T SG)^{1/2} \, D(z) = \{ \lim_{z^{-1}\to 0} \Delta_1(z^{-1})\} \, . \, y_1(z)$$

3) Calculation of the gains L and $F\bar{K}$ by polynomial identification
The above factorisations always exist and are unique apart from constant unitary matrices if and only if the rank of the matrices $\Phi(z)$ and $\Psi(z)$ is complete. If this is not the case, the problem becomes singular and corresponds to the case in which $R+G^T SG$ and/or $N+HPH^T$ are no longer invertible. These circumstances are indicated by the matrices $\Delta(z^{-1})$ and/or $\Delta_1(z^{-1})$ which are no longer square, and their inverses must be replaced by a left-hand inverse in the case of the estimation and a right-hand inverse for the control signal.

The above example will now be used to detail the different stages of this approach which leads to the optimal controller:

a) the control part requires the generalised factorisation of

$$\Psi(z)=1+\frac{z}{(z-0.5)(z^{-1}-0.5)}\times\frac{7}{z^{2}\,z^{-2}}=\frac{4z^{2}(z-1/4)}{z^{2}(z-0.5)}\times\frac{4z^{-2}(z^{-1}-1/4)}{z^{-2}(z^{-1}-0.5)}$$

In this way the function D(z) is obtained:

$$D(z) = 1 + L(zI-F)^{-1}G = \frac{4z^{2}(z-1/4)}{z^{2}(z - 0.5)} \times \frac{1}{4}$$

In the last expression, the factor 1/4 arises from the equality

$$\{\lim_{z\to\infty} y_1(z)\}^{-1} = (R + G^{T}SG)^{-1/2}$$

By identification, the following optimal controller is finally obtained:

$$L = [1/4, \quad O, \quad 0]$$

b) The prediction part requires the general factorisation of

$$\Phi(z^{-1}) = \begin{bmatrix} \dfrac{1}{(1-0.5z^{-1})(1-0.5z)} & \dfrac{z}{1-0.5z^{-1}} \\ \dfrac{z^{-1}}{1 - 0.5\,z} & 2 \end{bmatrix} = \Delta(z^{-1})\ \Delta^{T}(z)$$

We obtain

$$\Delta(z^{-1}) = \begin{bmatrix} \dfrac{1}{1-0.5z^{-1}} & 0 \\ z^{-1} & 1 \end{bmatrix}$$

i.e.

$$E(z) = I + H(zI-F)^{-1}K' = \begin{bmatrix} \dfrac{z}{z-0.5} & 0 \\ \dfrac{1}{z} & 1 \end{bmatrix}$$

By identification, the matrix $K' = F\overline{K}$ is as follows:

$$K' = \begin{bmatrix} 1/2 & 0 \\ 1 & 0 \\ 0 & 0 \end{bmatrix}$$

## 5.4.2 – POLYNOMIAL MATRICES

The above approach using rational matrices has the merit of having been the first to demonstrate the relationships between the state space approach and the spectrum approach. However, it has the disadvantages of being a transcription from continuous to discrete, of requiring special spectrum factorisations, and of using rational matrices. At present, therefore, an approach using polynomial matrices according to the findings of Kucera (1981) appears to be simpler and to be a better match to the techniques set out in this chapter.

Consider the right-hand and left-hand prime representations of the following transfer matrices:

$$(I-z^{-1}F)^{-1}z^{-1}G = B_1(z^{-1})\,A_1^{-1}(z^{-1}) \tag{67}$$

$$z^{-1}H(I-z^{-1}F)^{-1} = A_2^{-1}(z^{-1})\,B_2(z^{-1}) \tag{68}$$

To simplify the representation, the indeterminate $z^{-1}$ will be omitted in the following text. By carrying these inequalities into expressions (63) and (64), and after appropriate multiplication by the matrices $A_1$ and $A_2$, the following relationships are found:

$$(\bar{A}_1+\bar{B}_1L^T)\,(R+G^TSG)\,(A_1+LB_1) = \bar{A}_1RA_1 + \bar{B}_1QB_1$$

$$(A_2+B_2K')\,(N+HPH^T)(\bar{A}_2+K'^T\bar{B}_2) = A_2N\bar{A}_2 + B_2M\bar{B}_2$$

where

$$K' = F\bar{K} \qquad \bar{B}_i = B_i^T(z) \qquad \bar{A}_i = A_i^T(z) \qquad \text{and } (i = 1,2)$$

The matrices E(z) and D(z) are then expressed by

$$E(z) = A_2^{-1}(A_2 + B_2K')$$

$$D(z) = (A_1 + LB_1)\,A_1^{-1}$$

Since their inverses must be stable, the factorisation consists in finding the polynomial matrices $J(z^{-1})$ and $I(z^{-1})$ having stable inverses and meeting the conditions

$$\bar{J}J = \bar{B}_1QB_1 + \bar{A}_1RA_1 \tag{69}$$

$$I\bar{I} = B_2M\bar{B}_2 + A_2N\bar{A}_2 \tag{70}$$

The matrices J and I are unique apart from constant unitary matrices $v_1$ and $v_2$. We may therefore write

$$J = v_1(R+G^TSG)^{1/2}(A_1+LB_1) \tag{71}$$

$$I = (A_2+B_2K')\,(N+HPH^T)^{1/2}v_2 \tag{72}$$

To summarise, the application of this approach consists in the solution of the following standard problems:

1) Factorisation of (69) and (70) leading to matrices J, $\bar{J}$, I and $\bar{I}$.

2) Finding the constant solutions of the polynomial equations

$$X_1A_1 + Y_1B_1 = J \tag{73}$$

$$A_2X_2 + B_2Y_2 = I \tag{74}$$

where

$$L = X_1^{-1} Y_1 \tag{75}$$

$$K' = Y_2 X_2^{-1} \tag{76}$$

The singularity of the problems is shown here by the non-invertibility of the constant matrices $X_1$ and/or $X_2$ arising from the fact that the matrices J and/or I are not square. As with the state vector approach, one method of solving this problem is to use pseudo-inverses.

Application

The above theoretical results will now be applied to the previous example.

The first step is to find matrices $B_1$, $A_1$, $B_2$, $A_2$ in agreement with equations (76) and (68), as follows:

$$A_1 = 1 - 0.5\, z^{-1} \qquad B_1 = \begin{bmatrix} z^{-1} \\ z^{-2}(1-0.5z^{-1}) \\ z^{-1}(1-0.5z^{-1}) \end{bmatrix}$$

$$A_2 = \begin{bmatrix} 0 & 1 \\ 0.5z^{-1}-1 & 0 \end{bmatrix} \qquad B_2 = \begin{bmatrix} 0 & z^{-1} & z^{-2} \\ -z^{-1} & 0 & 0 \end{bmatrix}$$

The second step is to perform the factorisations indicated in relationships (69) and (70), obtaining

$$7 + 8\,(1-0.5z^{-1})(1-0.5z) = (4 - z^{-1})\,(4 - z) = J\,\bar{J}$$

$$A_2 \begin{bmatrix} 0 & 0 \\ 0 & 1 \end{bmatrix} \bar{A}_2 + B_2 \begin{bmatrix} 1 & 0 & 1 \\ 0 & 0 & 0 \\ 1 & 0 & 1 \end{bmatrix} \bar{B}_2 = \begin{bmatrix} 1 & -z^{-1} \\ 0 & 0 \end{bmatrix} \begin{bmatrix} 1 & 0 \\ -z & 1 \end{bmatrix} = I\,\bar{I}$$

The third step is to find the solutions of the polynomial equations (73) and (74). Relationship (73) for the control signal is written as follows:

$$X_1(1-0.5\, z^{-1}) + Y_1 \begin{bmatrix} z^{-1} \\ z^{-2}(1-0.5z^{-1}) \\ z^{-1}(1-0.5z^{-1}) \end{bmatrix} = 4 - z^{-1}$$

The constant solution $X_1$, $Y_1$ is as follows:

$$X_1 = 4 \quad \text{and} \quad Y_1 = [1 \quad 0 \quad 0]$$

and the gain of the controller, according to (75), becomes

$$L = [1/4 \quad 0 \quad 0]$$

For estimation, relationship (74) is written

$$\begin{bmatrix} 0 & 1 \\ 0.5z^{-1}-1 & 0 \end{bmatrix} X_2 + \begin{bmatrix} 0 & z^{-1} & z^{-2} \\ -z^{-1} & 0 & 0 \end{bmatrix} Y_2 = \begin{bmatrix} 1 & -z^{-1} \\ 0 & 1 \end{bmatrix}$$

and therefore

$$X_2 = \begin{bmatrix} 0 & -1 \\ 1 & 0 \end{bmatrix} \quad \text{and} \quad Y_2 = \begin{bmatrix} 0 & -1/2 \\ 0 & -1 \\ 0 & 0 \end{bmatrix}$$

and the gain of the predictor becomes, according to (76),

$$K' = F\,\bar{K} = \begin{bmatrix} 1/2 & 0 \\ 1 & 0 \\ 0 & 0 \end{bmatrix}$$

In the singular case in which N = 0, the factorisation of equation (70) results in

$$I = \begin{bmatrix} -z^{-1} \\ 1 \end{bmatrix}$$

Solution of the polynomial equation (74) leads to

$$X_2 = \begin{bmatrix} -1 \\ 0 \end{bmatrix} \quad \text{and} \quad Y_2 = \begin{bmatrix} -1/2 \\ -1 \\ 0 \end{bmatrix}$$

The gain of the predictor is then obtained according to relationship (76) by using the pseudo-inverse matrix

$$K' = \begin{bmatrix} 1/2 & \alpha_1 \\ 1 & \alpha_2 \\ 0 & \alpha_3 \end{bmatrix}$$

where the scalar quantities $\alpha_i$ may have any value.

## *5.5 - CONCLUSION*

In these sections a connection has been established between the state space approach and the frequency approach by using the research of Shaked (1979) and Kucera (1979, 1981). It is interesting to note that the solution of the Ricatti equation is split into two problems, namely

- a problem of spectrum factorisation;
- a problem of solving a polynomial matrix equation.

***REFERENCES***

M. Aoki, Optimization of stochastic systems, Academic Press, 1967

P.J. Antsaklis, Some relations satisfied by prime polynomial matrices and their role in linear multivariable system theory, IEEE AC, No. 4, August 1979

K.J. Astrom, Introduction to stochastic control theory, New York Academic, 1970

S. Barnett, Matrix in control theory, VNR Co., London, 1971

C. Castel and C. Reboulet, Algebraic approach to stochastic optimal digital controller design, Int. J. Control, v. 37, No. 5, 1983

S. Fenyo, Modern mathematical methods in technology, v. 2, North Holland Publishing Company, 1975

T. Kailath, Linear systems, Prentice Hall, 1980

V. Kucera, New results in state estimation and regulation, Automatica, No. 5, 1981

V. Kucera, Discrete linear control, the polynomial equation approach, Chichester, Wiley, 1979

Sivan Kwakernaak, Linear optimal control systems, Wiley Intersciences, 1972

M. Labarrère, J.P. Krief, and B. Gimonet, Le filtrage et ses applications, CEPADUES Editions, Toulouse, 1983

V. Peterka, On steady state minimum variance control strategy, Kybernetika, v. 8, No. 3, 1972

C. Reboulet and C. Castel, Algebraic approach to stochastic optimal controller design multivariable case, Int. J. Control, 1984

U. Shaked, A transfer function approach to the linear discrete stationary filtering and the steady state discrete optimal control problems, Int. J. Control, v. 29, No. 2, 1979

W.G. Tuel, Modern mathematical methods in technology, v. 2, IBM Journal, March 1968

W.A. Wolovich, Linear multivariable systems, Springer Verlag, New York, 1974

D.C. Youla, J.J. Bongiorno, and H.A. Jabr, Modern Wiener Hopf design of optimal controllers – Part I : The single input output case, IEEE AC, 21, No. 1, 1976; Part II : The multivariable case, IEEE AC 21, No. 3, 1976.

# *CHAPTER 6*

# *ADAPTIVE CONTROL OF STOCHASTIC SYSTEMS*

## *6.1 - INTRODUCTION*

Algorithms of the proportional, integral and derivative (PID) type are used to control many industrial processes. Their success is due to their simplicity, robustness, and good performance in many cases. The choice of the parameters of these regulators is often simple and is based on a rough knowledge of the nature of the process.

The limitations of these conventional methods rapidly become apparent in many applications. In fact, no regulator with constant parameters can deal with the slow or rapid time variations of the system to be controlled.

In parallel with these requirements, the recent developments in mini- and micro-computing have made it possible to use complex control laws requiring substantial processing. These technological developments, combined with the requirements of specifications, explain the current interest in self-adaptive control. The term self-adaptation is taken to refer to regulators with variable parameters which minimise a performance criterion for variant systems.

This chapter is not concerned with the class of self-adaptive systems using a reference model with a deterministic approach. Our discussion will essentially centre on self-adaptive regulators of stochastic systems which impose the principle of separation (self tuning controllers). Because of their simplicity, robustness, and level of performance tested in numerous industrial processes, these systems appear to be the most promising and the most open to future applications. However, before this synthesis is carried out, the systems will be located in the general class of adaptive control systems with or without the duality characteristic introduced by Feldbaum (1960).

## *6.2 - CLASSIFICATION OF SELF-ADAPTIVE CONTROL SYSTEMS*

### 6.2.1. - SELF-ADAPTIVE STOCHASTIC OPTIMAL CONTROL

For linear systems with known parameters, disturbed by white state and measurement noises, the minimisation of a quadratic criterion leads to the standard linear quadratic Gaussian regulator. The method of dynamic programming provides the analytical solution. A recursive equation for the feedback, derived from the Bellman principle, is thus obtained, and this equation determines the optimal solution.

However, this method requires complete knowledge of the model representing the system. The parameters of the latter are often imprecisely known or unknown, and/or may vary. Furthermore, in most cases the disturbances are only approximately modelled. To overcome these difficulties and apply dynamic programming, the unknown parameters are treated as random variables whose statistical characteristics (mean, variance, etc.) must be identified in real time. This leads to a problem of non-linear self-adaptive stochastic control which may be formulated as follows.

Consider a non-linear stochastic system represented by the following discrete equations:

$$x(k+1) = F(k, x(k), \theta(k), u(k)) + \xi(k) \quad (1)$$

$$y(k) = H(k, x(k), \theta(k), + \eta(k) \quad k = 0,1,\ldots N+1 \quad (2)$$

where

$$x(k) \in R^n \quad u(k) \in R^q \quad y(k) \in R^p$$

It is assumed that $\{x(o), \{\xi(k), \eta(k+1)\}_{k=0}^{N-1}$ are vectors of Gaussian and independent white noises.

It is also assumed that the parameters $\theta(k)$ follow a first-order Markov law of the form

$$\theta(k+1) = \Phi \ \theta(k) + v(k) \quad (3)$$

The aim is to find control signals minimising a general criterion of the form

$$J = E(\Psi(X(N)) + \sum_{k=0}^{N-1} L^*(x(k), u(k), k)) \quad (4)$$

where the expected value E refers to all random values. These acceptable control signals will be in the form

$$u(k) = u(k, Y^k, U^{k-1}) \quad (5)$$

where

$$Y^k = (y(1),\ldots, y(k))$$

and

$$U^{k-1} = \{u(o), u(1),\ldots, u(k-1)\}$$

The aim is therefore to find an optimal control sequence

$$\{u\ (k)\}_{k=0}^{N-1}$$

in the form of (5) which minimises criterion (4) and is subject to dynamic constraints (1) to (3).

To solve this problem, the technique of dynamic programming associated with the Bayes laws is used, supported by the fundamental stochastic control lemma of Aström (1970).

Unfortunately, in most cases the recursive equation resulting from the Bellman principle cannot be derived analytically from the optimal law. Because of this, Feldbaum (1960) introduced the concept of **dual control** which was taken up later by Bar Shalom and Tse (1976). The demonstration of the conflict between the quality of identification of the parameters and the quality of control lies at the heart of this duality.

Another obstacle to the solution of the problem of optimal self-adaptive control is the dimension of the problem (the dimension of the state vector increases very rapidly with time). Moreover, the optimal criterion associated with information on the states is not an explicit function. For all these reasons, sub-optimal dual or non-dual strategies are used. Before making this classification, we shall describe in detail the concepts of duality, neutrality, equivalence and separation which define these strategies.

## 6.2.2 - DEFINITION OF THE CONCEPTS

### 6.2.2.1 - The concept of neutrality

As has been indicated (an example in a subsequent section will make this clearer), an optimal self-adaptive control system cannot be formulated analytically in most cases. On the other hand, the solution of the optimal stochastic control law for known linear systems is well established. This is due to the neutrality of these problems.

By definition, a stochastic control problem is neutral if the calculated optimal control signal has no effect on the quality of the estimate of the state of the system.

### 6.2.2.2 - The concept of duality

On the other hand, if the optimal control signal not only affects the state of the system but also affects the quality of the estimate of the state, it has a **dual effect** in the Feldbaum sense. Such laws form a compromise between the quality of identification and the quality of control.

### 6.2.2.3 - The concept of separation

Another important concept appears in relation to self-adaptive control, namely the notion of **separation.** A stochastic control problem is said to be separable if the feedback control law is a function only of the mean values of the random variables and not of higher-order terms.

On the basis of these definitions, the most commonly used principle in adaptive control can be stated. This is the principle of equivalence, whose theorem is given below.

### 6.2.2.4 - Principle of equivalence

Consider a system with a dynamic disturbed by a white noise:

$$x(k+1) = F(k)\ x(k) + G(k),\ u(k) + \xi(k) \tag{6}$$

whose observation equation is

$$y(k) = h_k\ (x(k),\ \eta(k)) \tag{7}$$

where $\eta(k)$ is a measurement noise whose statistical values are known but arbitrary. The functional cost to be minimised takes the following form:

$$J = E\{x(N)^T Q(N) X(N) + \sum_{i=0}^{N-1} x^T(i) Q(i) x(i) + u^T(i) R(i) u(i)\} \tag{8}$$

<u>Theorem</u>

The optimal stochastic control signal for the system (6,7) minimising the quadratic criterion (8) with $Q(i) \geqslant 0$ and $R(i) \geqslant 0$ has the property of equivalence if and only if the control signal has no second-order dual effect, in other words if the covariance $P_{k/k}$ of the error of the state estimate is not a function of previous control signals $U^{k-1}$.

This theorem was proved for linear quadratic Gaussian systems by Aoki (1967), and was then generalised to linear quadratic systems with non-Gaussian dependent noises by Akashi (1975) and Gessing.

The theorem remains very important, since it relates the dual effect to the property of equivalence for a certain class of systems. For this type of system, therefore, neutrality is a sufficient condition for the application of the principle of equivalence. Moreover, the concept of separation is a necessary condition for the principle of equivalence.

### 6.2.3 - EXAMPLE

Before classifying the various forms of adaptive control, we consider it important to provide a concrete illustration of the various concepts described above in a very simple example.

Consider the following first-order discrete system in which **the state is measurable**:

$$x_{i+1} = a\ x_i + b\ u_i + \xi_i \tag{9}$$

where only **the gain parameter, b, is unknown**. The state noise, $\xi_i$, is assumed to be white and Gaussian with a zero mean and a variance Q:

$$E(\xi_i) = 0 \quad \forall i$$

$$E(\xi_i \xi_j) = 0 \quad \text{if } j \neq i \tag{10}$$

$$E(\xi_i^2) = Q$$

The aim is to find the optimal self-adaptive control signals $u_o^*$, $u_1^*, \ldots, u_{N-1}^*$ which minimise the terminal criterion $E(x_N{}^2)$. It is also assumed that there is a priori knowledge, before any measurement, of the distribution of the parameter b, assumed to be Gaussian, with a mean $b_o$ and variance $\Gamma_o$.

$$N_o(b) = N_o(b_o, \Gamma_o)$$

This hypothesis is not restrictive. If there is no a priori information, a uniform distribution law can easily be added to the preceding hypothesis.

#### 6.2.3.1 - Finding the control signal $u^*_{N-1}$

To solve this problem, the method of dynamic programming will be applied; the first stage is to find the control signal $u_{N-1}$ minimising the criterion

$$E(x_N^2/x^{N-1})$$

The expected value is calculated on the set of random variables, with

$$x^{N-1} = \{x_o, x_1, \ldots, x_{N-1}\}$$

We can write

$$\lambda_N = E(x_N^2/x^{N-1}) = \int x_N^2\ p(x_N/x^{N-1})\ dx_N \tag{11}$$

p denotes the conditional probability density. The above equation can be rewritten to make the parameter b appear and by using the Bayes rule

$$\lambda_N = \int x_N^2 \; p\,(x_N/x_{N-1},\ u_{N-1},\ b)\ p(b/x^{N-1})\ db\ dx_N$$

The first integration on the set of $x_N$ leads to:

$$\lambda_N = \int [(ax_{N-1} + b\ u_{N-1})^2 + Q]\ p(b/x^{N-1})\ db$$

which leads to

$$\lambda_N = (a\ x_{N-1} + \hat{b}_{N-1}u_{N-1})^2 + Q + u_{N-1}^2\ \Gamma_{N-1} \qquad (12)$$

with

$$p(b/x^{N-1}) = N\ [\hat{b}_{N-1},\ \Gamma_{N-1}] \qquad (13)$$

The values $\hat{b}_{N-1}$ and $\Gamma_{N-1}$, the mean and variance of the estimate of b, are obtained by the following Kalman filter equations:

$$\hat{b}_{i+1} = \hat{b}_i + \frac{\Gamma_{i+1}}{Q}\ u_i\ (x_{i+1} - ax_i - \hat{b}_i u_i) \qquad (14)$$

$$\frac{1}{\Gamma_{i+1}} = \frac{1}{\Gamma_i} + \frac{u_i^2}{Q} \qquad (15)$$

Since $\hat{b}_{N-1}$ and $\Gamma_{N-1}$ do not depend on $u_{N-1}$, it is sufficient to derive expression (12) to determine the control

signal $u_{N-1}^*$ and obtain

$$u_{N-1}^* = \frac{-\hat{b}_{N-1}\ a\ x_{N-1}}{\hat{b}_{N-1}^2 + \Gamma_{N-1}} \qquad (16)$$

The corresponding performance index is:

$$\gamma_N^* = \frac{\Gamma_{N-1}}{\hat{b}_{N-1}^2 + \Gamma_{N-1}}\ a^2\ x_{N-1}^2 + Q \qquad (17)$$

### 6.2.3.2 - Finding the control signal $u_{N-2}^*$

Since the criterion function depends only on $x_N$, the second stage of the dynamic programming is a matter of minimising the expression

$$\gamma_{N-1} = \int \gamma_N^*\ p(x_{N-1}/\ x^{N-2})dx_{N-1}$$

or

$$\gamma_{N-1} = \int \gamma_N^*\ p(x_{N-1}/\ x_{N-2}, b, u_{N-2})p(b/x^{N-2})\ dx_{N-1}\ db$$

It is immediately seen that the calculation of the first integration (with respect to $x_{N-1}$) leads to the calculation of an expression of the form

$$I = \int_{-\infty}^{+\infty} \frac{x^2}{(ax+b)^2 + c^2} \exp^{-(x-\mu)^2/2} dx$$

In fact, in $\gamma_N^*$ the value $x_{N-1}$ appears both explicitly in the numerator and implicitly in the denominator in the calculation of $\hat{b}_{N-1}$ resulting from expression (14).

This simple example demonstrates the impossibility of providing analytical relationships for the calculation of $u_j^*$ ($j \leqslant N-2$). These values may be obtained either numerically or by an approximation. This gives **optimal dual control** or **sub-optimal dual control**. In addition to its effect on the state, the control signal affects the quality of the future estimates of the parameters (in other words, there is active learning, as shown by equation (15)).

It is also possible to use an approximation in which each stage is optimised; this leads to

$$u_i = \frac{-\hat{b}_i \; a \; x_i}{\hat{b}_i^2 + \Gamma_i} \tag{18}$$

This form of control will be described as prudent, since any inaccurate identification of $b_i$, characterised by a higher variance $\Gamma_i$, reduces the amplitude of the control signal $u_i$. A subsequent section will demonstrate the effects of this strategy, which may be inconvenient at times.

If the principle of separation is applied, a control signal of the following form must be used:

$$u = - \frac{a \; x_i}{\hat{b}_i} \tag{19}$$

in which only the mean of the estimate appears. This results in a radical separation of the identification and control functions, removing all the dual effect from the latter. If there is any learning, it can only be accidental.

## 6.2.4 - CLASSIFICATION OF CONTROLLERS

This example partially illustrates the classification which will now be described in detail. At a given instant, any regulator uses only present and past measurements, since the control is not anticipatory.

However, at this given instant, it has access to information on the future which it may or may not use, namely the **future observation program** (the observation equations and the statistical values associated with it). The classification described here is based on this essential statement and makes use of the work of Bar Shalom and Tse (1974).

Let the system be described by equations (6) and (7) for which the functional cost to be minimised is

$$J = E\{ L^N(X_o^N, U_o^{N-1}) \} \tag{20}$$

where $L^N$ is a real variable function:

$$U_o^{N-1} = \{u_i\}_{i=0,\ldots,N-1} \qquad X_o^N = \{x_i\}_{i = 0,\ldots,N}$$

The sequence of observations between the times i and j corresponding to a

control sequence $U_i^{j-1}$ applied to the system is denoted thus:

$$Y^j_i = \{y_k\} \quad k=i,\ldots,j$$

The dynamic of the system is known through

$$D = \{f_k(\ )\}_{k=0,\ldots,N-1}$$

The measurement program between two instants i and j is denoted thus:

$$M_i^j = \{h_k(\ )\}_{k=1,\ldots,j}$$

The joint probability density of $x_o$ and the random variables $\xi_k$ and $\eta_k$ will be denoted thus:

$$S^k = p(x_o, \xi, \xi_1,\ldots,\xi_{N-1}, \eta_1,\eta_2,\ldots,\eta_{\beta k})$$

with

$$S^o = p_o(x_o, \xi_o,\ldots,\xi_{N-1})$$

The values D, M and S distinguish the regulators by their presence or absence in the laws and enable an initial classification to be made.

### 6.2.4.1 - Open-loop control

In open-loop control no knowledge of the measurements is used. The control signal is therefore written

$$u_k^{BO} = u_k^{BO}(D, S^o, L^N) \tag{21}$$

### 6.2.4.2 - Feedback control

In this type of control, all the measurements made up to the present instant, k, are used for the production of the control signal, but no knowledge of future events is used:

$$u_k^r = u_k^r (Y^k, U^{k-1}, D, M^k, S^k, L^N) \tag{22}$$

This category includes most of the self-adaptive control systems currently in use (open-loop and closed-loop control, self tuning adaptive regulators and cautious regulators, and essentially Bayesian multi-model self-adaptive control systems). These forms of control have no duality and almost all of them use the principle of separation.

### 6.2.4.3 – **Feedback control** $M_{k+n}$

In this case, the regulator uses the observation program and the associated statistics for the next n stages:

$$u_k^{r_n} = u_k^{r_n}(Y^k, U^{k-1}, D, M_o^{k+n}, S_o^{k+n}, L^N) \tag{23}$$

This group includes practically all sub-optimal dual controllers.

### 6.2.4.4 – **Closed-loop control (dynamic programming)**

This strategy integrates the whole future observation program, in other words the fact that the loop remains closed **until the end**. Therefore,

$$u_k^{BF} = u_k^{BF} (Y^k, U^{k-1}, D, M_o^{N-1}, S_o^{N-1}, L^N) \tag{24}$$

This type of control represents dual optimal control.

Although this type of classification demonstrates the effect of the future observation program on the nature of the control (whether dual or not), it does not take the nature of the system into account. For example, for a neutral system, feedback control will also be optimal closed-loop control.

To conclude, a second classification will be mentioned. According to the preceding concepts, the control laws complying with the theorem of separation or the principle of equivalence result in accidental learning (the control signal is not calculated with a view to improving the quality of the estimates of the parameters). Conversely, dual regulation systems show active learning behaviour. These concepts are expressed in the following synoptic diagram (Fig. 1).

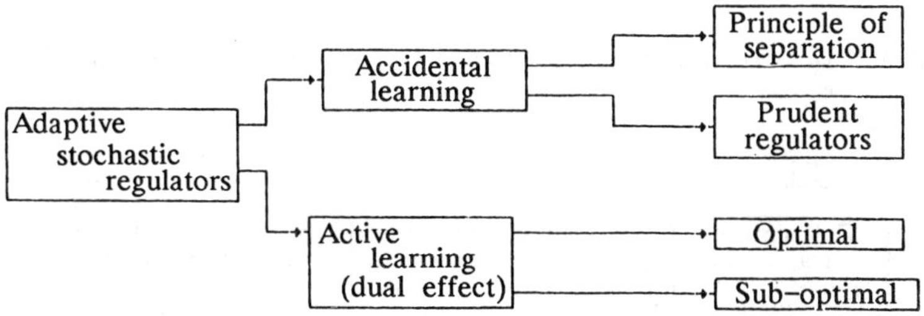


Classification of adaptive controllers

Fig. 1

## *6.3 - INTRODUCTION TO SELF-TUNING CONTROLLERS*

Among the different methods of adaptive control, the model reference adaptive control (Model Reference Adaptive Systems (MRAS)) and self tuning control (Self Tuning Regulators (STR) for the regulation problem, and Self Tuning Controllers (STC) for the servo problem) are the two most commonly used classes of adaptive systems.

Originally, these two types of approach were developed to solve the two following problems:

- closed loop control for continuous time deterministic systems;
- regulation for discrete time stochastic systems.

Although developed separately, the first approach being based on the theory of stability (second method of Lyapunov and Popov's theory of hyperstability) while the second uses the principle of certainty equivalence, these two approaches can be presented in a unified form (Egard (1978-1980), Jung and Landau (1978)).

In Section 6.5, different algorithms of the self tuning adaptive control type will be presented. As indicated before, these algorithms are based on the principle of certainty equivalence. This principle consists in the separation of the problems of identification and control. Since the control strategy is determined by assuming that the parameters of the model are known, at each sampling instant there is a stage of parameter estimation followed by the calculation of the control signal on the basis of the estimated parameters. The resulting regulator is called "self tuning" in the sense that its parameters are adjusted in the course of the process, thus offering the possibility of adaptation to a modification of the operation or the environment of the process. From the practical point of view, this may, for example, make it possible to follow a variation (slow or rapid) of the parameters of the model. In contrast to "cautious regulators" (Wieslander and Wittenmark (1970), Aström and Wittenmark (1971), Wittenmark (1975)), STRs do not take uncertainties (covariances of estimation error relative to the estimated parameters) into account.

The basic structure of an STR or STC is represented in Fig. 2:

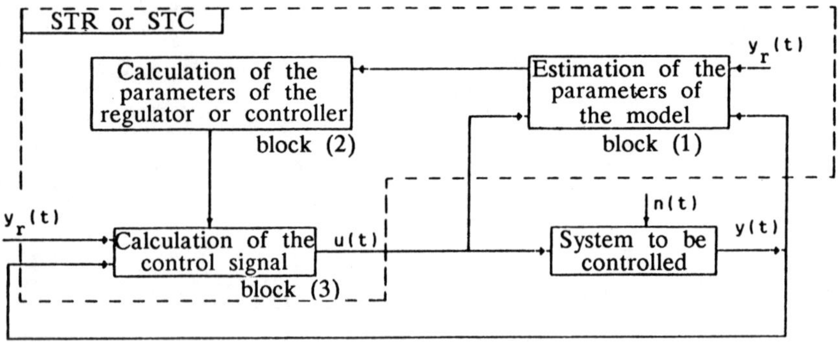


Basic diagram of an STR or STC

Fig. 2

The signals y(t), $y_r(t)$, u(t), and n(t) appearing in Fig. 2 represent the output of the system to be controlled, the reference (or reference output) to be followed, the control signal to be applied to the system, and a stochastic disturbance, assumed to be a zero mean white noise with a standard deviation $\sigma$. This block diagram shows the three basic blocks forming an STR or an STC, namely

- a block (1) for estimating the parameters of the process model from the input-output data;
- a block (2) for calculating the parameters of the regulator or controller in relation to a fixed control strategy;
- a block (3) for the calculation of the control signal.

An algorithm to produce these three blocks is called a "self tuning control algorithm". Two classes of adaptive control algorithms of the self tuning type may be distinguished (Aström 1980)):

- "Explicit algorithms" based on the estimation of the parameters of the process model and its environment, i.e. using an "explicit identification" of the process model. In the terminology of MRAS, algorithms of this type are also called "indirect algorithms" because the parameters of the regulator are adjusted indirectly in accordance with the estimation of the process parameters.

- "Implicit algorithms" or "direct algorithms" in the MRAS terminology, using an "implicit identification" of a model of the process. These algorithms are obtained by considering a modified model of the process, called "implicit model", which is expressed directly as a function of the parameters of the regulator. In this case, block (2) of Fig. 2 is omitted, thus reducing the amount of calculation. Of course, the possibility of using an implicit algorithm for a given control strategy is dependent on the existence of an implicit model for this strategy.

Different STRs or STCs may be obtained by combining a recursive method of parameter estimation with a control strategy.

With regard to the control system, the principal strategies used are as follows (see chapters 3, 4, and 5):

- the minimum output variance control law (Aström and Wittenmark (1973));
- the minimum generalised output variance control law (Clarke and Gawthrop (1975, 1979));
- the linear quadratic Gaussian control law (Peterka and Aström (1973));
- the pole zero placement strategy (Wellstead, Prager, and Zanker (1979), Aström and Wittenmark (1980)).

With regard to the estimation, the methods normally used are as follows:

- the method of recursive least squares (standard method - Aström and Eykhoff (1971)) or factorised and adaptive versions (Favier (1982, 1984, 1987));

- the extended least squares method, also called the approximate maximum likelihood method or RML1 method (Young (1968), Panuska (1968 and 1969));
- stochastic approximation (Saridis and Stein (1968)).

In the following section, the least squares estimation algorithm is briefly described.

## *6.4 - WEIGHTED LEAST SQUARES ESTIMATION ALGORITHM*

Consider model (25) which is linear with respect to the parameter vector $\theta$, the parameters being assumed to be deterministic and constant:

$$y(t) = \varphi^T(t)\ \theta + \epsilon(t,\theta) \tag{25}$$

$\epsilon(t,\theta)$ represents the modelling error, also called the equation error; $\varphi^T(t)$ is called the observation vector, and contains the previous values of the input-output signals of the system.

The problem of identification consists in finding the best possible estimate (in the sense of a certain criterion) of $\theta$ from the knowledge of input-output observations over a certain time interval.

The weighted least squares method corresponds to the minimisation of the following quadratic criterion:

$$J(K,\theta) = \sum_{t=1}^{K} w(t)\, \epsilon^2(t,\theta) \qquad (26)$$

where w(t) is a non-negative scalar weighting. Criterion (26) may be rewritten in matrix form:

$$J(K,\theta) = E^T(K,\theta)\ W(K)\ E(K,\theta) = \|E(K,\theta)\|^2_{W(k)} \qquad (27)$$

where $\| \cdot \|_{W(K)}$ represents the Euclidean norm weighted by W(K), and the quantities W(K) and $E(K,\theta)$ are defined by:

$$W(K) \triangleq \begin{bmatrix} w(1) & \cdots & 0 \\ 0 & \cdots & w(K) \end{bmatrix} \qquad (28)$$

$$E(K,\theta) \triangleq [\epsilon(1,\theta) \ldots \epsilon(K,\theta)]^T \qquad (29)$$

$$= Y(K) - \Phi(K)\ \theta \qquad (30)$$

where

$$Y(K) \triangleq [y(1) \ldots y(K)]^T \qquad (31)$$

$$\Phi(K) \triangleq [\varphi(1) \ldots \varphi(K)]^T \qquad (32)$$

The estimate $\hat{\theta}(K)$ of $\theta$ in the sense of the minimisation of the quadratic criterion (26) is obtained by writing that the partial derivative of $J(K,\theta)$ with respect to $\theta$ is zero for $\theta = \hat{\theta}(K)$. This produces the following equation:

$$[\Phi^T(K)\ W(K)\ \Phi(K)]\ \hat{\theta}(K) = \Phi^T(K)\ W(K)\ Y(K) \qquad (33)$$

These equations are called normal equations.
Assuming that the matrix $\Phi^T(K)\ W(K)\ \Phi(K)$ is not singular,

$$\boxed{\hat{\theta}(K) = [\Phi^T(K)\ W(K)\ \Phi(K)]^{-1}\ \Phi^T(K)\ W(K)\ Y(K)} \qquad (34)$$

Replacing Y(k) by its expression taken from (30) gives

$$E_s[\hat{\theta}(K)] = \theta + E_s\{[\Phi^T(K)\ W(K)\ \Phi(K)]^{-1}\ \Phi^T(K)\ W(K)\ E(K,\theta)\} \qquad (35)$$

where $E_s$ represents the mathematical expectation.

It may therefore be concluded that, in a dynamic system, i.e. when the observation vector $\varphi^T(t)$ depends on the past values $y(t-\tau)$ of the output, $\tau \epsilon [1,n]$, the weighted least squares estimator will generally be biased, since the second term of the right-hand side of (35) is not zero for all values of K. This term represents the bias of the estimator.

### 6.4.1 - ORDINARY LEAST SQUARES ESTIMATION ALGORITHM

This estimator is obtained from the weighted least squares estimator by specifying that

$$w(t) = 1 \qquad \forall t \in [1,K] \tag{36}$$

and therefore

$$\hat{\theta}(K) = [\Phi^T(K)\ \Phi(K)]^{-1}\ \Phi^T(K)\ Y(K) \tag{37}$$

The asymptotic properties of this estimator will now be examined ($K \to \infty$). By incorporating (30) into (37) we get:

$$\hat{\theta}(K) = \theta + [\Phi^T(K)\ \Phi(K)]^{-1}\ \Phi^T(K)\ E(K,\theta) \tag{38}$$

According to the definitions of E and $\Phi$ [(29), (32)],

$$\frac{1}{K}\ \Phi^T(K)\ \Phi(K) = \frac{1}{K} \sum_{t=1}^{K} \varphi(t)\ \varphi^T(t) \tag{39}$$

$$\frac{1}{K}\ \Phi^T(K)\ E(K,\theta) = \frac{1}{K} \sum_{t=1}^{K} \varphi(t)\ \epsilon(t,\theta) \tag{40}$$

Therefore, assuming that the signals are ergodic,

$$\frac{1}{K}\ \Phi^T(K)\ \Phi(K) \xrightarrow{\text{proba}} E_s[\varphi(t)\ \varphi^T(t)] \tag{41}$$

$$\frac{1}{K}\ \Phi^T(K)\ E(K,\theta) \xrightarrow{\text{proba}} E_s[\varphi(t)\ \epsilon(t,\theta) \tag{42}$$

which indicates that the time means are consistent estimators of the autocorrelation and intercorrelation functions. Application of Slutsky's theorem (Elgerd (1967)) to expression (38) gives

$$\hat{\theta}(K) \xrightarrow{\text{proba}} \theta + \{E_s[\varphi(t)\ \varphi^T(t)]\}^{-1} E_s[\varphi(t)\ \epsilon(t,\theta)] \tag{43}$$

It may therefore be concluded that, if the intercorrelation between $\varphi(t)$ and $\epsilon(t,\theta)$ is zero, i.e.

$$E_s[\varphi(t)\ \epsilon(t,\theta)] = 0 \tag{44}$$

then

$$\hat{\theta}(K) \xrightarrow{\text{proba}} \theta \tag{45}$$

in other words, $\hat{\theta}$ is a consistent estimator of $\theta$; conversely, if the intercorrelation between $\varphi(t)$ and $\epsilon(t,\theta)$ is not zero, $\hat{\theta}$ does not generally converge to $\theta$.

It should be noted that, in the case of a closed loop dynamic system, condition (44) will always be satisfied if the error $\epsilon(t,\theta)$ is a sequence of uncorrelated random variables.

### 6.4.2 – RECURSIVE WEIGHTED LEAST SQUARES ESTIMATION ALGORITHM

In order to present in a unified way the following different versions of recursive weighted least squares:

- algorithm with multiplicative weighting,
- algorithm with constant forgetting factor,
- algorithm with variable forgetting factor,

it will be assumed that the weighting w(t) appearing in criterion (26) has the following general form:

$$w(t) = \mu(t) \prod_{i=t}^{K} \lambda(i) \tag{46}$$

According to expression (34), it may be concluded that the calculation of the weighted least squares estimator requires the inversion of the matrix $\Phi^T(K)\ W(K)\ \Phi(K)$ of dimension $(n_\theta,\ n_\theta)$, where $n_\theta$ represents the dimension of $\theta$. The purpose of the recursive formulation of the weighted least squares is to enable $\hat{\theta}(K)$ to be calculated from $\hat{\theta}(k-1)$ in such a way as to avoid the inversion of this matrix.

The equations for the recursive algorithm can be derived by partitioning the matrices Y(K), Φ(K) and W(K) in the following form:

$$Y(K) = \begin{bmatrix} Y(K-1) \\ \cdots\cdots \\ y(K) \end{bmatrix},\quad \Phi(K) = \begin{bmatrix} \Phi(K-1) \\ \cdots\cdots \\ \varphi^T(K) \end{bmatrix},\quad W(K) = \begin{bmatrix} \lambda(K)W(K-1) & \vdots & 0 \\ \cdots\cdots & \vdots & \cdots\cdots \\ 0 & \vdots & \lambda(K)\mu(K) \end{bmatrix} \tag{47}$$

By defining

$$P(K) \triangleq [\Phi^T(K)\ W(K)\ \Phi(K)]^{-1} \tag{48}$$

we obtain

$$P^{-1}(K) = \lambda(K)[P^{-1}(K-1)+\mu(K)\varphi(K)\varphi^T(K)] \tag{49}$$

or by applying the matrix inversion lemma

$$P(K) = \frac{1}{\lambda(K)} \left[ P(K-1) - \frac{P(K-1)\varphi(K)\varphi^T(K)P(K-1)}{\frac{1}{\mu(K)} + \varphi^T(K)\ P(K-1)\varphi(K)} \right] \tag{50}$$

Defining the gain G(K) such that

$$G(K) = \frac{P(K-1)\varphi(K)}{\frac{1}{\mu(K)} + \varphi^T(K)\ P(K-1)\varphi(K)} \tag{51}$$

we obtain

$$P(K) = \frac{1}{\lambda(K)} [I - G(K)\ \varphi^T(K)]\ P(K-1) \tag{52}$$

The recursive calculation of $\hat{\theta}(K)$ may be carried out by using the following formula:

$$\boxed{\hat{\theta}(K) = \hat{\theta}(K-1) + G(K)\ [y(K) - \varphi^T(K)\ \hat{\theta}(K-1)]} \tag{53}$$

Another expression of the gain G(K) as a function of P(K) may be obtained from equation (50). Multiplying the two members of (50) by $\varphi(K)$ and simplifying, we obtain

$$G(K) = \lambda(K)\ \mu(K)\ P(K)\ \varphi(K) \tag{54}$$

This expression clearly demonstrates the fact that G(K) is proportional to P(K).

Notes

1. The recursive weighted least squares algorithm requires the initialisation of the estimated parameters vector $\hat{\theta}$ as well as the matrix P.

- If only a few measurements are available (low values of K) or the parameters may vary fairly rapidly with time, it is preferable to perform the initialisation with the aid of a block of data ($t \epsilon [1, K_i]$ with $K_i \geqslant n_\theta$), i.e.

$$P(K_i) = [\Phi^T(K_i)\ W(K_i)\ \Phi(K_i)]^{-1} \tag{55}$$

$$\hat{\theta}(K_i) = P(K_i)\ \Phi^T(K_i)\ W(K_i)\ Y(K_i) \tag{56}$$

- If a fairly large number of measurements is available and there is no a priori information on $\theta$, the initialisation values will be taken to be

$$\hat{\theta}(o) = 0,\ P(o) = p_o I \tag{57}$$

where $p_o$ is a positive scalar quantity and I is the identity matrix.

2. It should be noted that the inversion of the matrix $\Phi^T W \Phi$ of dimension ($n_\theta$, $n_\theta$) required by the non-recursive formulation is replaced in the recursive version by the inversion of a scalar quantity appearing in the calculation of the gain G.

3. Case of a simple multiplicative weighting.
In this case, it is assumed that $\lambda(K) = 1$, $\forall K \geqslant 1$, so that

$$w(K) = \mu(K) \geqslant 0 \tag{58}$$

From a practical point of view, this type of weighting has the purpose of giving a greater or lesser weight to the observation y(K) as a function of the quality of the measurement made at the instant K. If the modelling error $\epsilon(K,\theta)$ is a white noise with variance $\sigma^2(K)$, the Markov estimator is found by specifying

$$\mu(K) = 1/\sigma^2(K) \tag{59}$$

Equations (51) – (53) are then strictly identical to the Kalman filter equations built on the following state space model:

$$\theta(K) = \theta(K-1) \tag{60}$$

$$y(K) = \varphi^T(K)\ \theta(K) + \epsilon(K,\theta) \tag{61}$$

The matrix P(K) may then be interpreted as the covariance matrix of the parameter estimation error, i.e.

$$P(K) = E_s[\tilde{\theta}(K)\ \tilde{\theta}^T(K)] \tag{62}$$

where

$$\tilde{\theta}(K) \triangleq \theta(K) - \hat{\theta}(K) \tag{63}$$

It should be noted that if the variance $\sigma^2(K)$ is constant and equal to $\sigma^2$, the specification $\mu = 1/\sigma^2 = 1$ in the Kalman filter equations may be reduced to the division of P(k) by $\sigma^2$, without introducing any modification for G(K).

Finally, it should be noted that, in the case of simple multiplicative weighting, according to definition (48),

$$P^{-1}(K) = \sum_{t=1}^{K} w(t)\ \varphi(t)\ \varphi^T(t) \tag{64}$$

so that $\|P^{-1}\|$ tends towards infinity when K becomes very large, provided that $\|\varphi\|$ does not tend towards 0.

Therefore, according to expression (54) of G, it may be deduced that $\|G\|\to 0$ and therefore that $\hat{\theta}$ tends towards a vector of constant parameters. This implies that after a finite period the algorithm entirely loses its capacity of adaptation in response to any variations in the parameters $\theta$. To remedy this, one of the three following methods may be used:

a) – **Modification of the dynamic equation relative to the parameters to be estimated:**

A dynamic noise $\eta(K-1)$ is introduced into the difference equation (60):

$$\theta(K) = \theta(K-1) + \eta(K-1) \tag{65}$$

with

$$E_s[\eta(K)\ \eta^T(K)] = Q(K) \tag{66}$$

This dynamic noise can be added to the model either permanently or temporarily after detecting a variation in the model.

Assuming that the noises $\eta(K)$ and $\epsilon(K,\theta)$ are uncorrelated, only equation (50) is modified, becoming

$$P(K) = P(K-1) - \frac{P(K-1)\ \varphi(K)\ \varphi^T(K)\ P(K-1)}{\sigma^2(K) + \varphi^T(K)\ P(K-1)\varphi(K)} + Q(K-1) \tag{67}$$

The introduction of the term Q(K–1) into equation (67) has the effect of increasing the matrix P(K) and consequently the adaptation gain G(K).

b) – **Reinitialisation of P(K)**

In order to maintain an adaptation gain which is not negligible, the matrix P(K) is reinitialised either at regular time intervals or after a variation has been detected in the model. As in the previous case, this has the effect of increasing the gain G(K).

c) – **Introduction of a forgetting factor**

This method is described below for each of the two following cases:

- the case of a constant forgetting factor;
- the case of a variable forgetting factor.

* Algorithm with a constant forgetting factor

The standard formulation of the algorithm with a constant forgetting factor is obtained by specifying

$$\mu(t) = \frac{1}{\lambda} \qquad \forall t \in [1,\ K] \tag{68}$$

and

$$\lambda(t) = \lambda \qquad \forall t \in [1,\ K] \tag{69}$$

where

$$\lambda_{min} \leqslant \lambda \leqslant 1 \qquad \text{in practice} \quad \lambda \in [0.95-0.99] \tag{70}$$

in expression (46) of w(t).

The effect of such a forgetting factor will now be analysed. The criterion to be minimised is written

$$J(K,\theta) = \sum_{t=1}^{K} \lambda^{K-t} \epsilon^2(t,\theta) \tag{71}$$

$J(K,\theta)$ can be interpreted as the output of the system having the transfer function $1/(1-\lambda Z^{-1})$, with the sequence $\epsilon^2(t,\theta)$ as its input. Equations (49) – (51) then become

$$P^{-1}(K) = \lambda P^{-1}(K-1) + \varphi(K)\,\varphi^T(K) \tag{72}$$

$$P(K) = \frac{1}{\lambda}\left[P(K-1) - \frac{P(K-1)\varphi(K)\varphi^T(K)P(K-1)}{\lambda + \varphi^T(K)\,P(K-1)\varphi(K)}\right] \tag{73}$$

$$G(K) = \frac{P(K-1)\varphi(K)}{\lambda + \varphi^T(K)\,P(K-1)\varphi(K)} \tag{74}$$

The use of a constant forgetting factor $\lambda$ is equivalent to introducing an exponential window of length N centred on observation y(K). The length N represents the number of measurements actually taken into account in the minimisation of (71), in other words those for which the weighting $\lambda^{K-t}$ is higher than a certain threshold S.

Table 1 shows the value of $\lambda^N$ for different values of $\lambda$ and N.

| $\lambda$ \ N | 10 | 20 | 50 | 100 |
|---|---|---|---|---|
| 0.99 | 0.904 | 0.818 | 0.605 | 0.366 |
| 0.95 | 0.599 | 0.358 | 0.077 | 0.006 |
| 0.9 | 0.349 | 0.122 | 0.005 | 0.000 |
| 0.8 | 0.107 | 0.012 | 0.000 | 0.000 |

Table 1: Values of $\lambda^N$

Thus N is equal to 10 for $\lambda = 0.8$ and a threshold $S = 10^{-1}$. According to equation (72), it can easily be deduced that the introduction of a forgetting factor $\lambda<1$ entails a diminution of $P^{-1}(K)$ and therefore an increase in P(K) and consequently in the correction gain G(K). Thus the forgetting factor $\lambda$ can be used to improve the adaptation capacity of the estimation algorithm, and this is particularly useful both during the initialisation period and after a model variation.

However, the use of a constant forgetting factor with a value strictly less than 1 may have the disadvantage of causing the algorithm to "blow up" in one of the two following situations:

$\varphi(K) \to 0$, which may occur when the system is in a state close to equilibrium;

$\varphi(K) \perp P(K-1)$, i.e. $P(K-1)\varphi(K) = 0$.

In this case, equation (73) which updates P is then simplified to

$$P(K) = \frac{1}{\lambda} P(K-1) \tag{75}$$

and P(K) increases exponentially.

One method of avoiding this risk of blow-up is to freeze the identification algorithm when the tracking error becomes negligible. This is equivalent to defining a dead zone relative to the identification algorithm. Another solution is proposed by Dexter (1983). It consists of using two different values of the forgetting factor $\lambda$ ($\lambda_1 = 1$ and $\lambda_2 < 1$) and choosing between these two values at each sampling instant on the basis of comparison of a short term estimate with a long term estimate of the variance of the prediction error $\hat{\varepsilon}(K) = y(K) - \varphi^T(K) \hat{\theta}(K-1)$.

* Algorithm with variable forgetting factor

There are various methods of updating the forgetting factor:

- updating of the exponential type (Söderstrom, Ljung, and Gustavsson (1974));

- updating enabling a constant trace to be provided for the matrix P(K) or a limited trace for the matrix $P^{-1}(K)$ and leading to the constant trace (Irving (1979)) or limited trace (Lozano (1982)) algorithms;

- updating by using the prediction error (Wellstead and Sanoff (1981), Fortescue, Kershenbaum, and Ydstie (1981), Sanoff and Wellstead (1983), Saelid and Foss (1983)).

The first two methods of updating are described below.

* Exponential forgetting factor

Since the weighting $\mu(t)$ is determined by (68), the law of variation of the forgetting factor $\lambda$ is then such that

$$\lambda(K) = \lambda_o \lambda(K-1) + (1-\lambda_o) \tag{76}$$

with

$$\lambda_{min} \leqslant \lambda_o, \lambda(o) \leqslant 1 \tag{77}$$

From the initial value $\lambda(o)$, $\lambda(K)$ tends exponentially towards 1 when $K \to \infty$. This enables the phenomenon of blow-up to be avoided.

* Constant trace algorithm

In this case, the weighting $\lambda(K)$ is calculated in such a way as to maintain the trace of the matrix $P(K)$ at a constant value $D_o$. The value of $\lambda(K)$ may easily be deduced from equation (50).

$$\lambda(K) = \frac{1}{D_o} \, tr \left[ P(K-1) - \frac{P(K-1)\varphi(K)\varphi^T(K)P(K-1)}{\frac{1}{\mu(K)} + \varphi^T(K)\, P(K-1)\varphi(K)} \right] \tag{78}$$

* Limited trace algorithm

This algorithm is obtained by specifying $\lambda(K)$ such that

$$tr\ [P^{-1}(K)] \leqslant D_1 < \infty \tag{79}$$

According to (49) and allowing for the fact that $\varphi(K)$ is a column vector,

$$tr\ [P^{-1}(K)] = \lambda(K)\ \{tr[P^{-1}(K-1)] + \mu(K)\varphi^T(K)\varphi(K)\} \tag{80}$$

Condition (79) therefore implies that

$$\lambda(K) \leqslant \frac{D_1}{tr[P^{-1}(K-1)] + \mu(K)\varphi^T(K)\varphi(K)} \tag{81}$$

One solution is to specify that

$$\lambda(K) = \frac{D_1(1-\eta)}{D_1 + \mu(K)\ \varphi^T(K)\ \varphi(K)}, \qquad \eta > 0 \tag{82}$$

The last two algorithms are designed to establish a minimum adaptation gain in steady–state conditions.

## *6.5 – DESCRIPTION OF SELF TUNING CONTROLLERS*

### 6.5.1 – MINIMUM OUTPUT VARIANCE CONTROL STRATEGY

It was seen in Chapter 3 that for a minimum phase system the control signal u(t) which minimised the output variance, i.e.

$$\underset{u(t)}{Min}\ \{E_s[y^2(t+d)/M_t^*]\} \tag{83}$$

is obtained by writing that the d–step–ahead prediction of the output is equal to zero, i.e.

$$\hat{y}(t+d/t) = \frac{Y_d}{C}\, y(t) + \frac{\beta_d}{C}\, u(t) = 0 \tag{84}$$

where

$$\beta_d = B_1 X_d, \quad \delta(\beta_d) = \delta(B_1) + d-1 \tag{85}$$

where $X_d$ and $Y_d$ are the solution of the polynomial equation

$$C = A X_d + z^{-d} Y_d \tag{86}$$

with

$$\delta(X_d) = d-1, \quad \delta(Y_d = \text{Sup}\ [\delta(A)-1,\ \delta(C)-d] \tag{87}$$

A direct adaptive control algorithm may therefore be obtained from the construction of the self tuning d-step-ahead predictor of the output y(t), based on the use of the following implicit model:

$$y(t) = \hat{y}(t/t-d) + \tilde{y}(t/t-d) \tag{88}$$

where $\hat{y}(t/t-d)$ is obtained from expression (95) in Chapter 3, letting T = 1 and k = d, i.e.

$$y(t) = \varphi^T(t)\ \theta + \tilde{y}(t/t-d) \tag{89}$$

with

$$\Phi^T(t) \triangleq [u(t-d)..u(t-d-\delta(\beta_d))y(t-d)...y(t-d-\delta(Y_d)) \\ -\hat{y}(t-1/t-d-1)...-\hat{y}(t-\delta(C)/t-d-\delta(C))] \tag{90}$$

$$\theta \triangleq [\beta_o^{(d)} .. \beta_{\delta(\beta_d)}^{(d)}\ y_o^{(d)} .. y_{\delta(Y_d)}^{(d)}\ c_1 ...\ c_{\delta(C)}]^T \tag{91}$$

The corresponding adaptive control algorithm is summarised in Table 2:

| |
|---|
| 1. Estimate the parameters vector $\theta$ of the d-step predictive model of the output, defined by equations (89-91) with the aid of a factorised version of the weighted least squares algorithm shown in paragraph 6.4 (see Favier (1987)): $\longrightarrow \hat{\theta}(t)$ |
| 2. Calculate the control signal u(t) from the equation $\hat{y}(t+d/t) = \varphi^T(t+d)\ \hat{\theta}(t) = 0$ (92) |

Table 2: Self tuning regulator based on the strategy of minimum output variance control - Direct adaptive control algorithm

Notes

1. Equation (92) can be explicited for the calculation of u(t):

$$u(t) = \frac{1}{\hat{\beta}_o^{(d)}(t)} \left[ - \sum_{i=1}^{\delta(\beta_d)} \hat{\beta}_i^{(d)}(t)\, u(t-i) - \sum_{i=0}^{\delta(Y_d)} \hat{y}_i^{(d)}(t)\, y(t-i) \right.$$

$$\left. + \sum_{i=1}^{\delta(C)} \hat{c}_i(t)\, \hat{y}(t+d-i/t-i) \right] \qquad (94)$$

Since this calculation requires division by $\hat{\beta}_o^{(d)}(t)$, this estimated value must not be zero.

It should be noted that, according to (85) and (86),

$$\beta_o^{(d)} = b_{1_o} x_o^{(d)} = b_{1_o} \qquad (95)$$

In practice, this coefficient $\beta_o^{(d)}$ is specified a priori, in order to avoid problems of identifiability which may arise when an output feedback control is applied, if the system is not subject to excitation by an external signal (such as a reference) (Söderstrom, Gustavsson, and Ljung (1975)).

2. Taking into account the fact that the control signal u(t) is calculated at each instant in such a way as to cancel the d–step–ahead prediction of the output y(t), Aström and Wittenmark (1973) have proposed a simplified version of the algorithm described in Table 2.

This simplification is obtained by eliminating the terms $\hat{y}(t-i/t-d-i)$ in the vector $\varphi^T(t)$. Thus, by taking into account the two previous notes, the simplified direct algorithm in Table 3 is obtained:

1. Estimation of the parameter vector $\theta$ of the predictive model

$$y(t) - \hat{\beta}_o^{(d)} u(t-d) = \varphi^T(t)\ \theta + \epsilon(t,\theta) \qquad (96)$$

with

$\hat{\beta}_o^{(d)}$ set a priori

$$\varphi^T(t) \triangleq [u(t-d-1)..u(t-d-\delta(\beta_d)),y(t-d)..y(t-d-\delta(Y_d))] \qquad (97)$$

$$\theta \triangleq [\beta_1^{(d)}...\beta_{\delta(\beta_d)}^{(d)},y_o^{(d)}...y_{\delta(Y_d)}^{(d)}]^T \qquad (98)$$

$\delta(\beta_d)$ and $\delta(Y_d)$ defined in (85) and (87)

$\longrightarrow \hat{\theta}(t)$

2. Calculation of the control signal u(t)

$$u(t) = \frac{1}{\hat{\beta}_o^{(d)}} \varphi^T(t+d)\ \hat{\theta}(t) \qquad (99)$$

Table 3: Self tuning regulator based on the strategy of minimum output variance control. Simplified direct adaptive control algorithm.

It should be noted that the simplification which consists in eliminating the terms $\hat{y}$ in the vector $\varphi$ is equivalent to assuming a simplified model of the process, for which the stochastic part is characterised by $C(z^{-1}) = 1$. Aström and Wittenmark (1973) have shown that if the estimates of the model parameters (96) converge (not necessarily to exact values) the self tuning regulator described in Table 3 converges towards the optimal regulator in the sense of the minimisation of the criterion (83).

### 6.5.2 - MINIMUM GENERALISED OUTPUT VARIANCE CONTROL STRATEGY

This section presents two direct adaptive control algorithms which are associated with the control law corresponding to the minimisation of the following quadratic criterion:

$$\underset{u(t)}{\mathrm{Min}}\ \{E_s\{\{[\Gamma\ y(t+d)-Z\ y_r(t+d)]^2+[\Lambda\ u(t)]^2\}/\ M_t^*\}\} \qquad (100)$$

The minimisation of this criterion can be seen as a compromise between:

- the tracking of the output of a reference model with transfer function $Z/\Gamma$ (see equation (133) of Chapter 3, corresponding to the case where

$\Lambda = 0$); and

– a minimisation of the energy type corresponding to the presence of the term $[\Lambda\ u(t)]^2$ in criterion (100).

Direct algorithm No. 1

This algorithm is based on the use of equation (137) in Chapter 3 for the calculation of the control signal u(t):

$$\hat{\Phi}(t+d/t) = \frac{Y_d}{C}\ y(t) + \left[\frac{B_1 X_d + C\Lambda^M}{C}\right] u(t) - Z\ y_r(t+d) = 0 \tag{101}$$

where $\hat{\Phi}(t+d/t)$ denotes the d–step–ahead prediction of the auxiliary output $\Phi$ defined as

$$\Phi(t) \stackrel{\Delta}{=} \Gamma\ y(t) - Z\ y_r(t) + \Lambda^M\ u(t-d) \tag{102}$$

where

$$\Lambda^M \stackrel{\Delta}{=} \frac{\lambda_o \Lambda}{b_{1_o}} \tag{103}$$

$X_d$ and $Y_d$ are solutions of the polynomial equation

$$AX_d + z^{-d} Y_d = C\ \Gamma \tag{104}$$

with

$$\delta(X_d) = d-1;\ \delta(Y_d) = \sup\ [\delta(A)-1, \delta(C) + \delta(\Gamma)-d]$$

Consequently, by defining the polynomials

$$X \stackrel{\Delta}{=} B_1\ X_d + C\Lambda^M = \sum_{i=0}^{\delta(X)} x_i\ z^{-i}$$

$$\delta(X) = \mathrm{Sup}[\delta(B_1) + d-1,\ \delta(C) + \delta(\Lambda)] \tag{105}$$

$$Q \stackrel{\Delta}{=} -CZ = \sum_{i=0}^{\delta(Q)} q_i\ z^{-i},\ q_o = -z_o,\ \delta(Q) = \delta(C) + \delta(Z) \tag{106}$$

and the vectors

$$\theta \stackrel{\Delta}{=} \left[y_o^{(d)} \ldots y_{\delta(Y_d)}^{(d)}, x_o \ldots x_{\delta(X)}, q_1 \ldots q_{\delta(Q)}, c_1 \ldots c_{\delta(c)}\right]^T \tag{107}$$

$$\varphi^T(t) \stackrel{\Delta}{=} \Big[y(t-d)..y(t-d-\delta(Y_d)), u(t-d)..u(t-d-\delta(X)), y_r(t-1)..$$

$$y_r(t-\delta(Q)), -\hat{\Phi}(t-1/t-d-1)..-\hat{\Phi}(t-\delta(C)/t-d-\delta(c))\Big] \tag{108}$$

an implicit model of the process y(t) is obtained from the d–step–ahead predictive model of $\Phi$:

$$\Phi(t) = \hat{\Phi}(t/t-d) + \tilde{\Phi}(t/t-d) \tag{109}$$

and this is written

$$\Phi(t) + z_o\ y_r(t) = \varphi^T(t)\theta + \tilde{\Phi}(t/t-d) \tag{110}$$

where $\tilde{\Phi}$ is given by expression (139) in Chapter 3.

It has been seen that, in order to cancel the steady state servo error for a reference in step form, the polynomial Z must be chosen such that

$$Z = \eta Z_1 \tag{111}$$

with

$$\eta = \frac{\Gamma(1)}{Z_1(1)} + \frac{A(1)\ \Lambda^M(1)}{B_1(1)\ Z_1(1)} \tag{112}$$

or, using (104) and (105),

$$\eta = \frac{\Gamma(1)}{Z_1(1)} + \frac{\Lambda^M(1)}{Z_1(1)}\ \frac{C(1)\ \Gamma(1) - Y_d(1)}{X(1) - C(1)\Lambda^M(1)} \tag{113}$$

The second expression of $\eta$ has the advantage of being a function of the coefficients of the parameter vector $\theta$ relative to the implicit model (110).

Direct algorithm No. 1 is based on the calculation of a self-tuning d-step-ahead predictor of $\Phi(t)$. This algorithm is described in Table 4.

1. Choice of weightings $\Gamma$, Z and $\Lambda^M$
2. Calculation of the auxiliary output $\Phi(t)$

$$\Phi(t) = \Gamma\ y(t) - \hat{\eta}(t-1)\ Z_1\ y_r(t) + \Lambda^M\ u(t-d) \tag{114}$$

3. Recursive estimation of the parameter vector $\theta$ of the implicit model

$$\Phi(t) + \hat{\eta}(t-1)\ z_{1o} y_r(t) = \varphi^I(t)\ \theta + \tilde{\Phi}(t/t-d) \tag{115}$$

using a factorised version of the recursive least squares with the observation vector $\varphi(t)$ defined in (108), see Favier (1987)

$$\rightarrow \hat{\theta}(t) \longleftrightarrow \hat{Y}_{dt},\ \hat{X}_t,\ \hat{Q}_t,\ \hat{C}_t$$

4. Calculation of the normalisation gain $\eta$ by using (113)

$$\hat{\eta}(t) = \frac{\Gamma(1)}{Z_1(1)} + \frac{\Lambda^M(1)}{Z_1(1)}\ \frac{\hat{C}_t(1)\ \Gamma(1) - \hat{Y}_{d_t}(1)}{\hat{X}_t(1) - \hat{C}_t(1)\ \Lambda^M(1)} \tag{116}$$

5. Calculation of the control signal u(t) from:

$$\hat{\Phi}(t+d/t) = \varphi^T(t+d)\ \hat{\theta}(t) - \hat{\eta}(t)\ z_{1o}\ y_r(t+d) = 0 \tag{117}$$

Table 4: Self tuning controller based on the minimum generalised output variance control strategy. Direct algorithm No. 1.

Notes

1. Simplified version of direct algorithm No. 1.
This simplified version, proposed by Clarke and Gawthrop (1975), is a result of the remark made concerning the minimum output variance control strategy, namely that the control signal u(t) is determined in such a way as to cancel the d–step–ahead prediction of the auxiliary output $\Phi$.

The simplification of the algorithm is therefore achieved by eliminating the terms $\hat{\Phi}$ in the observation vector $\varphi(t)$.
The vectors $\theta$ and $\varphi(t)$ then become

$$\theta \triangleq \left[y_o^{(d)} \ldots y_{\delta(Y_d)}^{(d)},\ x_o \ldots x_{\delta(X)}, q_1 \ldots q_{\delta(Q)}\right]^T \tag{118}$$

$$\varphi^T(t) = \left[y(t-d)\,..\,y(t-d-\delta(Y_d)), u(t-d)\,..\,u(t-d-\delta(X)), y_r(t-1)\,.. \; y_r(t-\delta(Q))\right] \tag{119}$$

Consequently, since the polynomial C is no longer estimated directly, it is necessary to modify the means of updating $\eta$ (116).

It may be deduced from (106) and (113) that

$$\eta = \frac{\Gamma(1)\ X(1) - \Lambda^M(1)\ Y_d(1)}{Z_1(1)X(1) + \Lambda^M(1)\ Q(1)/\eta} \tag{120}$$

This new expression of $\eta$ is now a function of the polynomials X, $Y_d$ and Q, whose coefficients constitute the parameter vector $\theta$ defined in (118).
The formula for updating $\eta$ therefore becomes

$$\hat{\eta}(t) = \frac{\Gamma(1)\ \hat{X}_t(1) - \Lambda^M(1)\ \hat{Y}_{d_t}(1)}{Z_1(1)\hat{X}_t(1) + \Lambda^M(1)\ \hat{Q}_t(1)/\hat{\eta}(t-1)} \tag{121}$$

This formulation was proposed by Allidina and Hughes (1979) in a slightly different form.
As indicated in Chapter 3, the incorporation of a term $(1-z^{-1})$ in the weighting $\Lambda^M$ makes it possible to eliminate the updating of the normalisation gain $\eta$ which is then equal to $\Gamma(1)/Z_1(1)$.

2. Direct algorithm No. 1 and its simplified form have the disadvantage of depending on the estimation model defined in (115) with respect to the weighting $\Lambda^M$ through the coefficients $x_i$ of the vector $\theta$. This implies that the estimated model varies following a modification of $\Lambda^M$.

   The following section will show how this interaction between the choice of the control strategy through the polynomial $\Lambda^M$ and the estimation model can be eliminated.

Direct algorithm No. 2

This second algorithm is based on formulation (144) (Chapter 3) of the control law which is expressed in terms of a d-step-ahead predictor of the auxiliary output $y^f$ defined as:

$$y^f(t) \triangleq \Gamma\, y(t) \tag{122}$$

The control signal u(t) is then obtained by writing

$$\hat{y}^f\,(t+d/t) = Z\, y_r(t+d) - \Lambda^M\, u(t) \tag{123}$$

where $\hat{y}^f$ is given by formula (142) in Chapter 3

$$\hat{y}^f(t+d/t) = \frac{Y_d}{C}\, y(t) + \frac{B_1\, X_d}{C}\, u(t) \tag{124}$$

Therefore, by defining the polynomial

$$\beta_d \triangleq B_1\, X_d = \sum_{i=0}^{\delta(\beta_d)} \beta_i^{(d)}\, z^{-i}, \qquad \delta(\beta_d) = \delta(B_1) + d-1 \tag{125}$$

and the vectors

$$\theta \triangleq \left[y_o^{(d)} \ldots y_{\delta(Y_d)}^{(d)},\ \beta_o^{(d)} \ldots \beta_{\delta(\beta_d)}^{(d)}, c_1 \ldots c_{\delta(c)}\right]^T \tag{126}$$

$$\varphi^T(t) \triangleq \big[y(t-d)\ldots y(t-d-\delta(Y_d)), u(t-d)\ldots u(t-d-\delta(\beta_d)),$$
$$-\hat{y}^f(t-1/t-d-1)\ldots -\hat{y}^f(t-\delta(C)/t-d-\delta(C))\big] \tag{127}$$

an implicit model of the process y(t) is obtained from the d-step-ahead predictive model of $y^f(t)$

$$y^f(t) = \hat{y}^f(t/t-d) + \tilde{y}^f(t/t-d) \tag{128}$$

where

$$\tilde{y}^f(t/t-d) = \tilde{\Phi}(t/t-d) = X_d\ n(t) \tag{129}$$

and

$$\hat{y}^f(t/t-d) = \varphi^T(t)\ \theta \tag{130}$$

and therefore

$$y^f(t) = \varphi^T(t)\ \theta + \tilde{y}^f(t/t-d) \tag{131}$$

Additionally, according to (105), (113) and (125), the normalisation gain $\eta$ can be calculated from the following expression:

$$\eta = \frac{\Gamma(1)}{Z_1(1)} + \frac{\Lambda^M(1)}{Z_1(1)} \frac{C(1)\ \Gamma(1) - Y_d(1)}{\beta_d(1)} \tag{132}$$

This leads to direct algorithm No. 2 shown in Table 5.

1. Choice of weightings $\Gamma$, Z and $\Lambda^M$
2. Calculation of the auxiliary output $y^f$ (t)

$$y^f(t) = \Gamma\ y(t) \tag{133}$$

3. Recursive estimation of the parameter vector $\theta$ of the implicit model

$$y^f(t) = \varphi^T(t)\ \theta + \tilde{y}^f(t/t-d) \tag{134}$$

using a factorised version of the recursive least squares algorithm with the observation vector $\varphi(t)$ being defined in (127)

$$\rightarrow \theta(t) \longleftrightarrow Y_{d_t}, \beta_{d_t}, C_t \tag{135}$$

4. Updating of the normalisation gain $\eta$ by using formula (132):

$$\hat{\eta}(t) = \frac{\Gamma(1)}{Z_1(1)} + \frac{\Lambda^M(1)}{Z_1(1)} \frac{\hat{C}_t(1)\ \Gamma(1) - \hat{Y}_{d_t}(1)}{\hat{\beta}_{d_t}(1)} \tag{136}$$

5. Calculation of the control signal u(t) from:

$$\hat{y}^f(t+d/t) = \varphi^T(t+d)\ \hat{\theta}(t) = Z\ y_r(t+d) - \Lambda^M\ u(t) \tag{137}$$

Table 5: Self tuning controller based on the minimum generalised output variance control strategy. Direct algorithm No. 2.

Notes

1. This second formulation based on the calculation of a self tuning predictor of the auxiliary output $y^f$ was developed by Clarke and Gawthrop (1979 a and b, 1981).
   As indicated in the previous section, it is possible to eliminate the updating of $\eta$ by adding a factor $(1-z^{-1})$ to the weighting $\Lambda^M$.

2. We may compare the numbers of parameters to be estimated in the three following cases:

   - Indirect algorithm

$$\theta \triangleq [a_1 \ldots a_{\delta(A)}, b_{1o} \ldots b_{1_{\delta(B_1)}}, c_1 \ldots c_{\delta(c)}]^T \tag{138}$$

and therefore

$$\delta(\theta) = \delta(A) + \delta(B_1) + \delta(C) + 1 \tag{139}$$

- Simplified direct algorithm No. 1

$$\theta \triangleq \left[y_o^{(d)} \ldots y_{\delta(Y_d)}^{(d)}, x_o \ldots x_{\delta(X)}, q_1 \ldots q_{\delta(Q)}\right]^T \tag{140}$$

with

$$\delta(Y_d) = \text{Sup}\ [\delta(A)-1,\ \delta(C) + \delta(\Gamma) - d] \tag{141}$$

$$\delta(X) = \text{Sup}\ [\delta(B_1)+d-1,\ \delta(C) + \delta(\Lambda)] \tag{142}$$

$$\delta(Q) = \delta(C) + \delta(Z) \tag{143}$$

and therefore

$$\delta(\theta) = \text{Sup}[\delta(A),\ \delta(C)+\delta(\Gamma)-d+1] + \text{sup}[\delta(B_1)+d, \delta(C)+\delta(\Lambda)+1] + \delta(C) + \delta(Z) \tag{144}$$

- Direct algorithm No. 2

$$\theta \triangleq \left[y_o^{(d)} \ldots y_{\delta(Y_d)}^{(d)},\ \beta_o^{(d)} \ldots \beta_{\delta(\beta_d)}^{(d)}, c_1 \ldots c_{\delta(c)}\right]^T \tag{145}$$

where

$$\delta(\beta_d) = \delta(B_1) + d - 1 \tag{146}$$

and therefore

$$\delta(\theta) = \text{Sup}[\delta(A),\ \delta(C)+\delta(\Gamma)-d+1] + \delta(B_1)+d+\delta(C) \tag{147}$$

Thus, for example, in the case where

$$\delta(A) = \delta(B_1) = \delta(C) = n,\ \delta(\Lambda) \leqslant d-1, \text{ and } \delta(\Gamma) \leqslant d-1 \tag{148}$$

the following is true:

| | | |
|---|---|---|
| Indirect algorithm | : $\delta(\theta) = 3n + 1$ | (149) |
| Direct algorithm No. 1 | : $\delta(\theta) = 3n + \delta(Z)+d$ | (150) |
| Direct algorithm No. 2 | : $\delta(\theta) = 3n + d$ | (151) |

We may therefore conclude that, from the point of view of numerical complexity, direct algorithm No. 2 is preferable to direct algorithm No. 1 which requires an excessive number of parameters to be estimated, due to the estimation of the polynomial Q. In fact, this polynomial Q contains, as a factor, the polynomial Z which is set by the user and which, consequently, does not have to be estimated, with the exception of the multiplier $\eta$ which is updated elsewhere.

Finally, it should be noted that the indirect algorithm appears to be the one with fewest parameters to be estimated. On the other hand, this algorithm requires the solution, at each sampling instant, of the polynomial equation (104) with respect to the polynomials $X_d$ and $Y_d$. This calculation, which corresponds to block (2) of Fig. 2, is equivalent to the least squares estimation of an additional number of parameters equal to:

$$\delta(\theta_e) = \delta(X_d) + \delta(Y_d) + 1 \qquad (152)$$

i.e.

$$\delta(\theta_e) = \text{Sup}\ [\delta(A)-1,\ \delta(C)+\delta(\Gamma)-d] + d \qquad (153)$$

Thus for $\delta(A) = \delta(B_1) = \delta(C) = n$ and $\delta(\Gamma) \leqslant d-1$ this is equivalent to the

estimation of a total number of parameters equal to

$$\delta(\theta) + \delta(\theta_e) = 4n + d \qquad (155)$$

Thus it may be stated that the volume of additional calculation as compared with direct algorithm No. 2 is equivalent to the estimation of n parameters.
The use of the direct algorithm No. 1 described in Table 4 will now be illustrated by means of the numerical example considered in section 3.4.4 of Chapter 3.

### 6.5.3 - NUMERICAL EXAMPLE

The system to be controlled is assumed to be described by the equation

$$A\ y(t) = z^{-d}\ B_1\ u(t) + C\ n(t) \qquad (156)$$

with

$$A(z^{-1}) = 1-0.5\ z^{-1},\ B_1(z^{-1}) = C(z^{-1}) = 1,\ d = 2 \qquad (157)$$

The delay d and the order of the polynomials A, $B_1$ and C are assumed to be known, while the coefficients of these polynomials are unknown. Since the coefficients $a_o$ and $c_o$ can always be assumed to be normalised to 1, there are two unknown parameters, $a_1$ and $b_{1o}$.

The self tuning controller described below is associated with the direct algorithm No. 1 and makes it possible to realise the minimum generalised output variance control law resulting from the minimisation of the following criterion:

$$\underset{u(t)}{\text{Min}}\ \{E_s\{\{[y(t+d)-\eta\ y_r(t+d)]^2 + \lambda\ u^2(t)\ \}/\ M_t^*\}\} \qquad (158)$$

in other words, $\Gamma = Z_1 = 1$, $\Lambda = \surd\lambda$, $\Lambda^M = \lambda/b_{1o}$ (159)

The output $\Phi(t)$ defined in (102) is as follows:

$$\Phi(t) = y(t) - \eta \; y_r(t) + \Lambda^M \; u(t-2) \tag{160}$$

and the two-step-ahead prediction of $\Phi(t)$ is given by

$$\hat{\Phi}(t/t-2) = \varphi^T(t) \; \theta - \eta \; y_r(t) \tag{161}$$

with

$$\theta \triangleq [y_o^{(2)} x_o \; x_1] \tag{162}$$

where

$$\varphi^T(t) \triangleq [y(t-2) \; u(t-2) \; u(t-3)] \tag{163}$$

$$X \triangleq B_1 \; X_2 + C \; \Lambda^M = b_{1o}(x_o^{(2)} + x_1^{(2)} z^{-1}) + \Lambda^M = x_o + x_1 z^{-1} \tag{164}$$

Noting that $x_o^{(2)} = \gamma_o = 1$, we find that

$$x_o = b_{1o} + \Lambda^M, \quad x_1 = b_{1o} \; x_1^{(2)} \tag{165}$$

Additionally, from expression (113) of the normalisation gain,

$$\eta = 1 + \Lambda^M \; \frac{1 - y_o^{(2)}}{x_o + x_1 - \Lambda^M} \tag{166}$$

The corresponding equations of direct algorithm No. 1 are shown in Table 6.

1. Choice of the weighting $\Lambda^M$
2. Calculation of the auxiliary signal

$$\Phi_m(t) \triangleq \Phi(t) + \hat{\eta}(t-1)\, y_r(t) = y(t) + \Lambda^M\, u(t-2) \qquad (167)$$

3. Estimation of the parameter vector $\theta$ of the implicit model

$$\Phi_m(t) = \varphi^T(t)\, \theta + \tilde{\Phi}\, (t/t-2) \qquad (168)$$

using the recursive least squares algorithm, the vector $\varphi(t)$ being defined in (163)

$$\rightarrow \hat{y}_o^{(2)}(t),\ \hat{x}_o(t),\ \hat{x}_1(t)$$

4. Calculation of the normalisation gain

$$\hat{\eta}(t) = 1 + \Lambda\, \frac{1 - \hat{y}_o^{(2)}(t)}{\hat{x}_o(t) + \hat{x}_1(t) - \Lambda^M} \qquad (169)$$

5. Calculation of the control signal u(t) from (117) or by an equivalent formula

$$u(t) = \frac{1}{\hat{x}_o(t)}\, [\hat{\eta}(t)\, y_r(t+2) - \hat{y}_o^{(2)}(t)\, y(t) - \hat{x}_1(t)\, u(t-1)] \qquad (170)$$

Table 6: Direct algorithm No. 1

The block diagram for the self tuning controller associated with the direct algorithm No. 1 is shown in Fig. 3.

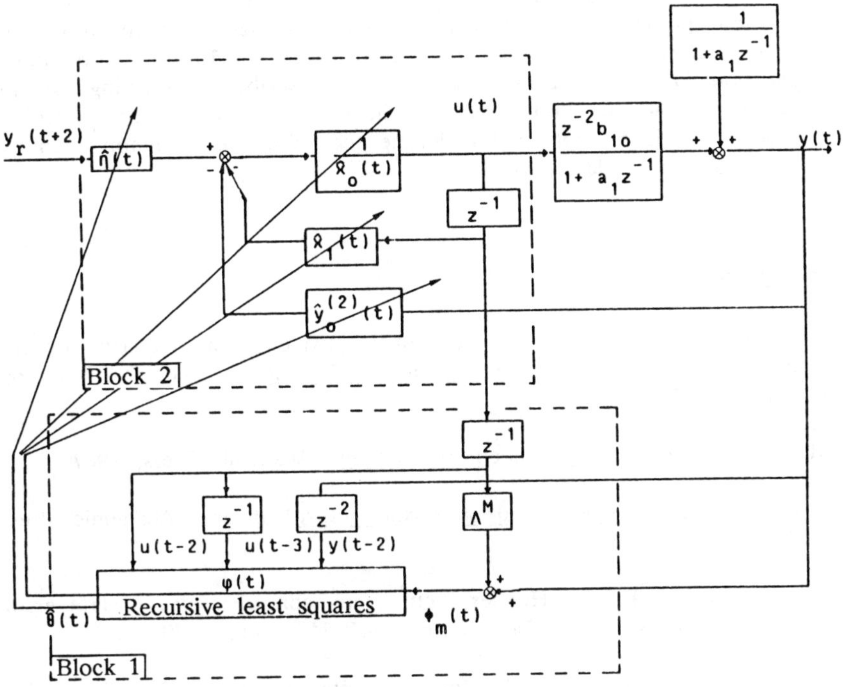


Self tuning controller using direct algorithm No. 1

Fig. 3

This block diagram shows the two calculation blocks characterising a direct adaptive control algorithm, namely:

- block 1: identification block
- block 2: control calculation block.

The arrows passing through blocks $\hat{\eta}(t)$, $1/\hat{x}_0(t)$, $\hat{x}_1(t)$ and $\hat{y}_0^{(2)}(t)$ indicate that the corresponding coefficients are updated at each sampling instant on the basis of the vector of estimated parameters $\hat{\theta}(t)$.

## *6.6 – CONCLUSION*

In this chapter, the adaptive control methods enforcing the separation principle and dissociating the parameter identification and control phases have been discussed. In Chapter 13, an aeronautical application of these techniques will be described, consisting of the reduction of the vibration level of a helicopter by multicyclic control of the blades. This example will underline the relevance of adaptive control systems to present-day automation.

## *REFERENCES*

H. Akashi and K. Nose, On certainty equivalence of stochastic optimal control problem, International Journal Control, v. 21, No. 5, pp. 857–863 (1975)

M. Aoki, Optimisation of stochastic systems, Academic Press, 1967

K.J. Aström, Introduction to stochastic control theory, Academic Press, 1970

A.Y. Allidina and F.M. Hughes, Self tuning controller steady state error, Electronics letters, v. 15, No. 12, pp. 346–347, June 1979

K.J. Aström, Design principles for self tuning regulators, Int. Symposium on Adaptive Systems, Bochum, FRG, 1980

K.J. Aström and P. Eykhoff, System identification. A survey, Automatica, v. 7, pp. 123–162, 1971

K.J. Aström and B. Wittenmark, Problems of identification and control, Journal of Math. Analysis and Applications, v. 34, pp. 90–113, 1971

K.J. Aström and B. Wittenmark, On self tuning regulators, Automatica, v. 9, pp. 185–199, 1973

K.J. Aström and B. Wittenmark, Self tuning controllers based on pole-zero placement, IEE Proc., v. 127, pt. D, No. 3, pp. 120–130, May 1980.

Y. Bar Shalom and E. Tse, Dual effect, certainty equivalence and separation in stochastic control, IEEE AC 19, No. 5, pp. 494–500, Oct. 1974

D.W. Clarke and P.J. Gawthrop, Self tuning controller, Proc. IEE, v. 122, No. 9, pp. 929–934, Sept. 1975

D.W. Clarke and P.J. Gawthrop, Self tuning control, Proc. IEE, v. 126, No. 6, pp. 633–640, June 1979

D.W. Clarke and P.J. Gawthrop, Implementation and application of microprocessors based self-tuners, Proc. 5th IFAC Symposium on Identification and System Parameter Estimation, Darmstadt, pp. 197–208, Sept. 1979. Published in Automatica, v. 17, No. 1, pp. 233–244, 1981

A.L. Dexter, Self tuning control algorithm for single chip microcomputer implementation, IEE Proc., v. 130, Part D, No. 5, pp. 255–260, Sept. 1983

B. Egardt, A unified approach to model reference adaptive systems and self tuning regulators, Report TFRT 7134, Dept of Automatic control, Lund Inst. of Techn., Sweden, Jan. 1978

O.I. Elgerd, Control systems theory, McGraw Hill, New York, 1967

G. Favier, Filtrage, modélisation et identification de systèmes linéaires stochastiques à temps discret, Editions du CNRS, 15 quai A. France, Paris, 1982

G. Favier, Regulateurs numériques auto-ajustables, Rapport final LASSY, Contrat DRET No. 81/548, Jan. 1984

G. Favier, Computationally efficient adaptive identification algorithms, Proc. IEEE Int. Conf. on "Acoustics, Speech & signal Processing", Dallas, April 1987

A.A. Feldbaum, Dual control theory – Part 1: v. 21, No. 9, pp. 1240–1249 (1960) – Part 2: v. 21, No. 11, pp. 1453–1464 (1960) – Part 3: v. 22, No. 1, pp. 3–16 (1961) – Part 4: v. 22, No. 2, pp. 129–142 (1961)

T.R. Fortescue, L.S. Kershenbaum, and B.E. Ydstie, Implementation of self tuning regulators with variable forgetting factors, Automatica, v. 17, No. 6, pp. 831–835, 1981

R. Gessing, The generalized certainty equivalence principle, Preprints 7th IFAC Congress, Helsinki, v. 3, pp. 2175–2180

E. Irving, New developments in improving power network stability with adaptive control, Proc. Workshop on Applications of Adaptive Control, Yale Univ., New Haven, 1979

L. Ljung and I.D. Landau, Model reference adaptive systems and self tuning regulators. Some connections, Proc. 7th IFAC Congress, Helsinki, June 1978

R. Lozano, Independent tracking and regulation adaptive control with forgetting factor, Automatica, v. 18, No. 4, pp. 455–459, 1982

V. Panuska, A stochastic approximation method for identification of linear systems using adaptive filtering, JACC, University of Michigan, pp. 1014–1021, 1968

V. Panuska, An adaptive recursive least squares identification algorithm, Proc. IEEE Symp. on Adaptive Processes, Pennsylvania State University, 1969

V. Peterka and K.J. Aström, Control of multivariable systems with unknown but constant parameters, Proc. 3rd IFAC Symposium on Identification & System Parameter Estimation, The Hague, 1973

S. Saelid and B. Foss, Adaptive controllers with a vector variable forgetting factor, Proc. IEEE Conf. on Decision and Control, 1983

S.P. Sanoff and P.E. Wellstead, Comments on: Implementation of self tuning regulators with variable forgetting factors, Automatica, v. 19, No. 3, pp. 345–346, 1983

G.N. Saridis and G. Stein, Stochastic approximation algorithms for discrete time system identification, IEEE AC, v. 13, pp. 515–523, 1968

T. Soderstrom, L. Ljung, and I. Gustavsson, A comparative study of recursive identification methods, Report 7427, Dept. Automatic control, Lund Inst. of Techn., Sweden, Jan. 1974

T. Soderstrom, I. Gustavsson, and L. Ljung, Identifiability conditions for linear systems operating in closed loop, Int. Journal Control, v. 21, No. 2, pp. 243–255, 1975

P.E. Wellstead, D. Prager, and P. Zanker, Pole assignment self tuning regulator, Proc. IEE, v. 126, No. 8, pp. 781–787, Aug. 1979

P.E. Wellstead and S.P. Sanoff, Extended self tuning algorithm, Int. J. Control, v. 34, No. 3, pp. 433–455, 1981

J. Wieslander and B. Wittenmark, An approach to adaptive control using real time identification, Proc. IFAC Symp. on Identification & Process Parameter Estimation, Prague, 1970

B. Wittenmark, Stochastic adaptive control methods: a survey, Int. J. of Control, v. 21, No. 5, pp. 705–730, 1975

P.C. Young, The use of linear regression and related procedures for the identification of dynamic processes, Proc. 7th IEEE Symp. on Adaptive Process, UCLA, 1968

# CHAPTER 7

# BRIEF DESCRIPTION OF ALGEBRAIC AND GEOMETRICAL METHODS FOR NON-LINEAR CONTROL

## 7.1 - INTRODUCTION

Non-linear control has made rapid advances over the last fifteen years. Although some successful attempts have been made to develop industrial applications, the engineering community has not yet given general acknowledgement to this new state of affairs. One reason often quoted for this is the nature of the mathematics used. This is claimed to be difficult to deal with, but is the differential geometry currently used really more difficult than the Fourier or Laplace transform techniques? The fact is that this form of geometry is not taught on engineering courses, which prefer to give priority to conventional functional analysis, a domain whose application to non-linear control has remained very modest.

This digression is not without importance, since it provides the justification for the form of this chapter. Whereas the state space of a linear system is a vector space, the same is not true of a non-linear system, whose state space is a differentiable manifold, in other words something generalising the concept of surface. A simple practical example of this is given. Following the introduction of this concept of state space, the controllability is studied, using the work of R.E. Kalman. The study of this problem led to the introduction of current non-linear methods at the end of the 1960's, following the work of R. Hermann and C. Lobry. It is then necessary to define as concisely as possible the concept of Lie brackets of vector fields, which enables the conventional criterion of controllability to be generalised.

In addition to state space, a key concept of linear control is that of the transfer function. It is based on the concept of the Laplace transformation, valid only for linear differential equations with constant coefficients, and is therefore not generalisable as such. It is the concept of generating series with several non-commutative variables, resulting from work done twenty years ago in automata and language theory, which provides the better analogy. An illustration of this will be given for the systems known as "bilinear".

## 7.2 - *STATE SPACE IN CONTINUOUS TIME*

### 7.2.1 - ON THE CONCEPT OF VECTOR FIELDS

Consider the following differential vector equation:

$$\begin{cases} \dot{q}^1(t) = a_1(q) \\ \dots\dots\dots\dots\dots\dots \\ \dot{q}^N(t) = a_N(q) \\ y(t) = h(q) \end{cases}$$

The state $q = (q^1,\dots,q^N)$ belongs to $R^N$. The output h is a function with values in R. The calculation of the total derivative dy/dt is elementary:

$$\frac{dy}{dt} = \sum_{k=1}^{N} a_k(q) \frac{\partial h}{\partial q^k}$$

The linear differential operator of the first order then appears

$$\sum_k a_k(q) \frac{\partial}{\partial q^k}$$

This expresses the directional derivative of the function h when there is an infinitesimal displacement $\Delta q_1,\dots,\Delta q_N$ parallel to the vector $(a_1(q),\dots,a_N(q))$.

The function of this operator is such that, in the current language of differential geometry, $(a_1(q),\dots,a_N(q))$ is often deliberately equated with $\Sigma\, a_k(q)\ \partial/\partial qk$ under the name of vector field.

### 7.2.2 - LIE BRACKETS

Given two vector fields

$$A = \sum_k a_k(q) \frac{\partial}{\partial q^k}, \qquad B = \sum_k b_k(q) \frac{\partial}{\partial q^k}$$

an essential operation is the **Lie bracket**, denoted [A,B], defined by

$$[A,B] = AB - BA$$

Elementary calculation shows that

$$[A,B] = \sum_{k=1}^{N} \left( \sum_{1=1}^{N} a_1 \frac{\partial b_k}{\partial q_1} - b_1 \frac{\partial a_k}{\partial q_1}\right) \frac{\partial}{\partial q_k}$$

This is also a vector field since the second order derivations disappear.

An illustration, instead of an axiomatic statement, will now be provided. Consider the system

$$\dot{q}(t) = u(t)\ A(q) + v(t)\ B(q)$$

where u and v are two control signals. Let $\Delta t > o$ denote a very short time interval. Let the control signals have the following values:

$$u(t) = \begin{cases} 1, & \text{if } 0 \leqslant t < \Delta t/4 \\ 0, & \text{if } \Delta t/4 \leqslant t < \Delta t/2 \text{ or } 3\Delta t/4 \leqslant t \leqslant \Delta t \\ -1, & \text{if } \Delta t/2 \leqslant t < 3\Delta t/4 \end{cases}$$

$$v(t) = \begin{cases} 1, & \text{if } \Delta t/4 \leqslant t < \Delta t/2 \\ 0, & \text{if } 0 \leqslant t < \Delta t/4 \text{ or } \Delta t/2 \leqslant t < 3\Delta t/4 \\ -1, & \text{if } 3\Delta t/4 \leqslant t \leqslant \Delta t \end{cases}$$

A calculation of limited expansions, which will not be reproduced here but which the reader is strongly advised to perform for himself, gives the result

$$q(\Delta t) = q(0) + [A,B]\ (q(0))\ \left(\frac{\Delta t}{4}\right)^2 + 0((\Delta t)^2)$$

Note. Although the natural state space of a linear system is a vector space, the same is not true for the non-linear case. To demonstrate this, it is only necessary to think of a rotating body such as a gyroscope. This gives curved surfaces such as a circle or sphere. Therefore the state space used must be the abstract generalisation of such surfaces, in other words differentiable manifolds.

### 7.2.3 – CONTROLLABILITY

Consider a system

$$\dot{q}(t) = \sum_{i=1}^{m} u_i(t)\ A_i(q),$$

for which the state $q = (q^1,\ldots,q^N)$ belongs to $\mathbf{R}^N$, the vector fields $A_1,\ldots,A_m$ are defined as above, and $u_1,\ldots,u_m$ represent the control signals.

The system is said to be controllable in the neighbourhood of a point $q_o$ if all the points in the neighbourhood can be attained. To provide the solution, the **Lie algebra** generated by $A_1,\ldots,A_m$ must be defined. This is the real vector space generated by $A_1,\ldots,A_m$, the Lie brackets $[A_i, A_j]$ and the iterated brackets

$$[A_{i_\nu}, \ldots, [A_{i_2}, A_{i_1}] \ldots ]$$

A necessary and sufficient condition for controllability is that the dimension of this space evaluated at the point q(o) is equal to N.

A vector field $A_0$ will be introduced here:

$$\dot{q}(t) = A_o(q) + \sum_{i=1}^{m} u_i(t)\, A_i(q)$$

The above criterion, calculated by the Lie algebra generated by $A_0, A_1, \ldots, A_m$, also provides a controllability where going backward in time is authorised.

Example: For the linear system

$$\dot{q}(t) = F\ a(t) + u_1(t)\ G,$$

the corresponding vector fields are

$$A_o = (q^1, \ldots, q^N)\,{}^tF \begin{bmatrix} \partial/\partial q^1 \\ \vdots \\ \partial/\partial q^N \end{bmatrix}$$

$$A_1 = {}^tG \begin{bmatrix} \partial/\partial q^1 \\ \vdots \\ \partial/\partial q^N \end{bmatrix}$$

where M denotes the transpose of the matrix M. It becomes

$$[A_o, A_1] \Big|_{q(o)=0} = {}^t(FG) \begin{bmatrix} \partial/\partial q^1 \\ \vdots \\ \partial/\partial q^N \end{bmatrix}$$

This brings us back to the Kalman criterion of controllability, namely that the rank of the matrix $[G, FG, \ldots F^{N-1}\, G]$ must be equal to N.

## *7.3 – GENERATING SERIES*

### 7.3.1 – BILINEAR SYSTEMS IN DISCRETE TIME

A **bilinear** system in discrete time takes the following form:

(B)
$$\left[\begin{array}{l} q(t+1) = (M_o + \sum_{i=1}^{m} u_i(t)\, M_i)\, q(t) \\ y(t) = \lambda\, q(t) \end{array}\right.$$

The state q belongs to a real vector space Q, of finite dimension N. The applications $M_o, M_1, \ldots, M_m$: $Q \longrightarrow Q$ and $\lambda : Q \longrightarrow \underline{R}$ are linear (a scalar output is chosen for simplicity). If Q denotes the space $R^N$ of the column matrices of dimension N, then $M_o, M_1, \ldots, M_m$ and $\lambda$ are square and row matrices of order N respectively. The control signals $u_1, \ldots, u_m$ : $[0, 1, 2, \ldots[ \rightarrow \underline{R}$ have real values.

After choosing the initial state q(0), it is a simple matter to calculate the output at successive instants. For the sake of uniformity,

the imaginary control $u_o \equiv 1$ is introduced, and the first line of (B) is replaced with

$$q(t+1) = (\sum_{j=0}^{m} u_j(t)\, M_j)\, q(t)$$

This gives:

$$y(o) = \lambda\, q(o)$$

$$y(1) = \sum_{j_o=0}^{m} \lambda\, M_j\, q(o)\, u_j(o)$$

$$y(2) = \sum_{j_o, j_1=0}^{m} \lambda\, M_{j_1} M_{j_o}\, q(o)\, u_{j_1}(1)\, u_{j_o}(0),$$

---

$$y(t+1) = \sum_{j_o, j_1, \ldots, j_t=0}^{m} \lambda\, M_{j_t} \ldots M_{j_o}\, q(o)\, u_{j_t}(t) \ldots u_{j_o}(o)$$

Let a finite set of symbols $X = \{x_0, x_1, \ldots, x_m\}$ be introduced. Let the following formal series correspond to (B):

$$g = \lambda\, q(o) + \sum_{\nu \geqslant 0} \sum_{j_o, \ldots, j_\nu=0}^{m} \lambda\, M_{j_\nu} \ldots M_{j_0}\, q(0)\, x_{j_\nu} \ldots x_{j_0}$$

Attention should be paid to the non-commutativity of the symbols: $x_0\, x_1 \neq x_1\, x_0$, which indicates that in general $u_0(1)u_1(0) \neq u_1(1)u_0(0)$. The above formulae demonstrate the correspondence between g and the input-output behaviour. The series g is called the **generating series** of (B) for which it acts in an analogous way to the transfer function.

To illustrate this point, consider the following linear system with scalar control signal and output.

$$(\Lambda) \quad \left[ \begin{array}{l} \eta(t+1) = F\,\eta(t) + u_1(t)\,G \\ y(t) = H\,\eta(t) \end{array} \right.$$

It has the same input–output behaviour as a bilinear system (B) where

$$M_o = \left[ \begin{array}{ccc|c} & F & & \begin{matrix} 0 \\ \vdots \\ 0 \end{matrix} \\ \hline 0 & \cdots & 0 & 1 \end{array} \right] \qquad M_1 = \left[ \begin{array}{c|c} (0) & G \\ \hline 0 \quad \cdots & 0 \end{array} \right]$$

$$\lambda = (H, 0), \qquad \eta(0) = \left[ \begin{array}{c} q(0) \\ 1 \end{array} \right]$$

The transfer function of $(\Lambda)$ is

$$H(z-F)^{-1}\,G = \sum_{\nu \geqslant 0} \frac{H\,F^{\nu}\,G}{z^{\nu+1}},$$

while the generating series is

$$\sum_{\nu \geqslant 0} H\,F^{\nu}\,G\,x_o^{\nu}\,x_1$$

The two objects are identical, except for an obvious change in variables. It is also clear that the multiplication by 1/z, like multiplication on the left by $x_o$, corresponds to a delay.

### 7.3.2 – BILINEAR SYSTEMS IN CONTINUOUS TIME

The system is as follows:

$$\left[ \begin{array}{l} \dot{q}(t) = (M_o + \sum_{i=1}^{m} u_i(t)\,M_i)\,q(t) \\ y(t) = \lambda\,q(t) \end{array} \right.$$

The definition is identical to that of the previous section, with the exception of the control signals $u_1,\ldots,u_m = [0, \infty[ \longrightarrow \underline{\mathrm{R}}$ which are functions.

It will be useful to recall some features of the linear differential equations:

$$\dot{q}(t) = M^*(t)\,q(t),$$

where M* is a square matrix of order N, whose coefficients may vary with time. It can be shown that the solution is expressed in the form of an infinite series according to the Peano–Baker formula:

$$q(t) = (1 + \int_0^t M^*(\tau)\, d\tau + \int_0^t M^*(\tau_2)\, d\tau_2 \int_0^{\tau_2} M^*(\tau_1)\, d\tau_1 + \ldots$$

$$+ \int_0^t M^*(\tau_n) d\tau_n \int_0^{\tau_n} \ldots\ldots \int_0^{\tau_2} M^*(\tau_1)\, d\tau_1 + \ldots) q(0)$$

The proof is obtained by term–by–term derivation.

Translation by the bilinear system results in:

$$y(t) = \lambda\, q(o) + \sum_{\nu \geqslant 0} \sum_{j_\nu, \ldots, j_o = 0}^{m} \lambda\, M_{j_\nu} \ldots M_{j_o}\, q(o) \int_0^t d\xi_{j_\nu} \ldots d\xi_{j_o}$$

The iterated integral $\int_0^t d\,\xi_{j_\nu} \ldots d\,\xi_{j_o}$ is defined by induction on the length:

$$\xi_o(\tau) = \tau,\ \xi_i(\tau) = \int_0^\tau u_i(\sigma)\, d\sigma \quad (i = 1, \ldots, m),$$

$$\int_0^t d\xi_j = \xi_j(t) \qquad (j=0, 1, \ldots, m),$$

$$\int_0^t d\xi_{j_\nu} \ldots d\xi_{j_\xi} = \begin{cases} \int_0^t (\int_0^\tau d\xi_{j_{\nu-1}} \ldots d\xi_{j_o})\, d\tau, & \text{if } j_\nu = 0 \\ \int_o^t (\int_0^\tau d\xi_{j_{\nu-1}} \ldots d\xi_{j_o})\, u_{j_\nu}(\tau)\, d\tau, & \text{if } j_\nu = 1, \ldots, m \end{cases}$$

As in the discrete time case, the result is the generating series

$$\underline{\underline{g}} = \lambda\, q(o) + \sum_{\nu \geqslant 0} \sum_{j_o \ldots j_\nu = 0}^{m} \lambda M_{j_\nu} \ldots M_{j_o}\, q(o)\, x_{j_\nu} \ldots x_{j_o},$$

which also characterises the input–output behaviour.

### 7.3.3 – IMPLEMENTATION

Among the numerous applications of transfer function matrices, the Hankel matrix implementation should be mentioned (see the so–called Ho–Kalman algorithm).

There is an analogous case for bilinear systems. For a generating series g an infinite table is constructed, whose rows and columns are indexed by the words $x_{j_v} \ldots x_{j_o}$ without forgetting the "empty" word 1. If $w_1$ and $w_2$ denote such words, the coefficient of index $(w_1, w_2)$ is the concatenation $w_1 w_2$. The rank is defined as for finite matrices by calculation of the sub-determinants. It can be shown that it characterises the minimal dimension of the state space. This state space is unique up to an isomorphism.

## *7.4 - FURTHER INFORMATION*

Applications arising from the non-linear theory will be briefly reviewed here.

- **Non-linear identification**: a remarkable property of approximation by bilinear systems has resulted in a technique of identification by state space in non-linear systems, which has been applied in industry.

- **Decoupling and rejection of disturbances**: on the basis of geometric techniques appearing in controllability studies, it has been possible to decouple non-linear systems and provide rejection of disturbances by a similar method to that used in the linear case. Many applications are currently under development.

- **Linearisation**: in spite of recent discoveries, the theory of linear systems remains much more tractable than that of non-linear systems. This is why it would be useful to find the feedback laws which transform a non-linear system into a linear one. This has been done and, as expected, has immediately led to practical applications.

## *7.5 - BRIEF REFERENCES*

This is not the place to give a list of the articles describing the development of the discipline over the last fifteen years. Further information and bibliographical references will be found in the two following volumes:

I.D. Landau (ed.), Outils et modèles mathématiques pour l'automatique, l'analyse de systèmes et le traitement du signal, vv. 1 and 3, C.N.R.S., Paris, 1981 and 1983.

# PART TWO

# APPLICATIONS OF DIGITAL CONTROL TO MACHINES AND ROBOTS

## *CHAPTER 8*

## *DIGITAL CONTROL OF SYSTEMS AT THE LIMIT OF STABILITY: APPLICATION TO THE STABILISATION OF SATELLITES*

### *8.1 - INTRODUCTION*

This first chapter on applications, concerned with attitude control of satellites, illustrates the usefulness of the digitisation techniques and the methods of synthesis of digital controllers described in Chapters 2 and 3.

The purpose of attitude control of satellites is to maintain a desired orientation of a body in orbit around the earth (the stabilisation problem) or to make this orientation undergo a predetermined variation (the tracking problem). It is therefore concerned with the dynamics of the vehicle in its movement about its centre of mass, whereas the movement of the centre of mass itself is a matter for orbital control.

The problem of stabilisation is particularly relevant because, except in the case of astronomical missions which need to cover the whole sky (such as IRAS, Hipparcos, the Space Telescope) it determines the performance of the mission; the problems of tracking are most common in the course of positioning, at which time the control system generally forms a reference of short open-loop phases.

The stabilisation of satellites was initially achieved by a purely passive solution: the rapid rotation of the satellite about its own axis provided a "gyroscopic stability" which was sufficient to meet the current requirements. In successive projects, there was a gradual development through rotation stabilisation complemented by an active system (in other words, using means of attitude measurement and activating devices enabling torque to be applied to the vehicle), to the rotation of one part of the satellite only ("dual spin", on-board angular momentum), and finally to the current system of three-axis stabilisation.

With the development of the concept of stabilisation, the more complex equations governing satellite dynamics are now encountered not so much in problems concerning gyroscopic coupling as in those concerning the presence of flexible attachments such as solar panels and antennae. Consequently the application of control techniques to the field of satellite stabilisation will be demonstrated with the help of examples illustrating each of these two cases.

## *8.2 – STABILISATION OF SATELLITES IN ROTATION*

### 8.2.1 – OUTLINE OF THE PROBLEM

The equations governing the dynamic behaviour of the attitude of a spinning satellite are relatively complex and are non–linear if no approximation is made for the sake of simplification. Consequently, the following assumptions are generally made:

– the deviations concerned are small;

– the satellite is of revolution in the dynamic sense; in other words, its inertia tensor is diagonal in the satellite reference: since the y axis is conventionally the axis of spin, the inertias about the axes x and z are equal:

$$I_x = I_z = I$$

– the dynamic equations (relating the angular acceleration to the control torque) are linearised around the nominal spin value $\Omega$:

$$\omega_x = \omega_z = 0; \quad \omega_y = \Omega$$

– the kinetic equations (governing the velocity) are linearised around the orbital pulsation $\omega_o$ (i.e. the angular velocity of the satellite with respect to the inertial frame).

It is important to note that the two above dynamics are very different. The coupling between the roll and yaw equations resulting from the spinning motion leads to the appearance of a "nutation frequency"

$$\omega = \frac{I_y}{I}\,\Omega$$

which depends on the inertia ratio $I_y/I$ (greater than 1 for reasons of stability of movement, but close to 1); this frequency is much higher than the orbital frequency (which is of the order of $10^{-3}$ rad/s in low orbit). The problem of stabilisation may therefore be divided into two subsidiary problems: the short–term control of the angles of deviation of the roll ($\varphi$) and yaw ($\psi$) which uses the dynamic coupling due to the rotation of the satellite about its own axis, and the long–term control which uses the coupling due to the rotation of the satellite around the earth. The attitude sensors normally only provide the roll angle: the yaw angle is not immediately available, since the earth remains in a constant position for a rotation whose axis passes through the yaw axis. Furthermore, the roll sensor is a sweep sensor (the rotation of the satellite about the y axis providing the necessary movement), and therefore the measurement is, by its very nature, sampled at a frequency which is not determined by the designer of the control system. If the torques are also intrinsically sampled (as in the case of gas jet actuators), the control problem is inevitably discrete in nature.

## 8.2.2 - SHORT-TERM CONTROL

The equations governing the dynamics are written in a reference frame associated with the satellite but without its own rotation, which simply describes the orientation of the spin axis:

$$\left.\begin{aligned} \dot{\omega}_x &= \omega\omega_z + u_x \\ \dot{\omega}_z &= -\omega\omega_x + u_z \end{aligned}\right] \quad \text{Linearised dynamic equations}$$

$$\left.\begin{aligned} \dot{\varphi} &= \omega_x \\ \dot{\psi} &= \omega_x \end{aligned}\right] \quad \text{Simplified kinetic equations}$$

These correspond to the following block diagram (Fig. 1):

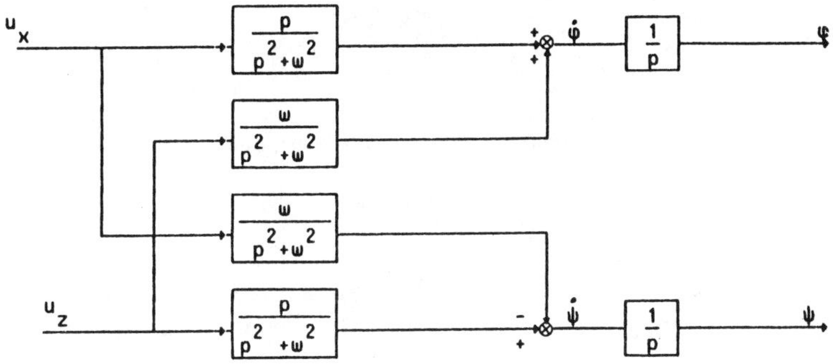


Block diagram of the dynamic model

Fig. 1

The control signals $u_x$ and $u_z$ are homogeneous to torques divided by the transverse inertia I.

This essentially multivariable problem is simplified by assuming that the yaw control signal is always proportional to the roll control signal, i.e.

$$u_z = k\ u_x$$

Only the angle $\varphi$ is measured at the rate $\Omega$; the block diagram is therefore changed to a monovariable type between $u_x$ and $\varphi$ and the problem becomes one of the control of a system including a pure oscillating mode $\omega = I_y\Omega/I$ sampled at frequency $\Omega$.

There are various possible modes of action; two of these will be considered here, depending on whether the control is of the impulse type or through a zero-order hold. Impulse control is encountered in the case of gas jet actuators with width modulation: this mode of action is non-linear, but it is possible to consider a linear approximation of it by making the energy of an impulse similar to that of the interval generated by the width modulation (Fig. 2).

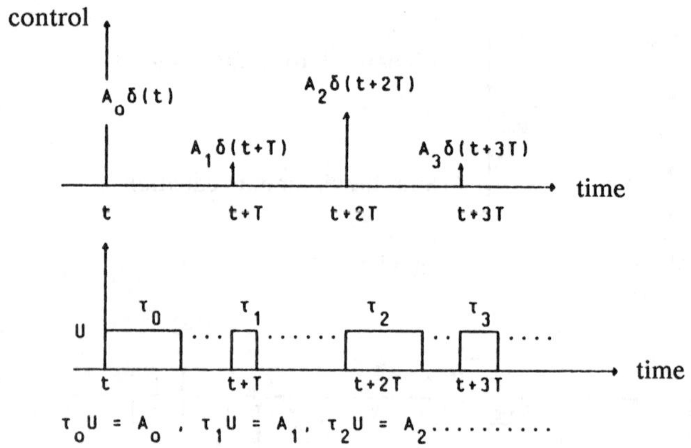


Approximation by width modulation

Fig. 2

This approximation is effective as long as the calculated $\tau_i$ remain reasonably small by comparison with the sampling period T.

### 8.2.2.1 – Impulse control

The roll torque is controlled by a sampled network of the first order:

$$D(Z) = \frac{K(1-a\,Z^{-1})}{1-b\,Z^{-1}}$$

and the gains K, a, b and the proportionality gain between the yaw control and the roll control k are determined by the analysis of the characteristic closed loop equation which is written:

$$\Delta = 1 + K\,\frac{Z-a}{Z-b}\cdot\frac{Z\sin\omega T}{\omega D} + \frac{kK(Z-a)}{(Z-b)\,\omega}\left[\frac{Z}{Z-1} - \frac{Z(Z-\cos\omega T)}{D}\right]$$

where $D = Z^2 - 2Z\cos\omega T + 1$.

The search for a response in minimal time leads to the placement of all the poles of Δ at the origin, which is possible by solving a system of simultaneous equations. For example, for

$\frac{I_y}{I} = 1.4$, $T = 10$ s, and $\omega = 0.89$ rad/s, we obtain

$$\begin{cases} U_x(nT) = 0.136\,[\varphi(nT) - 1.78\,\varphi((n-1)T)] \\ U_z(nT) = 2.22\,U_x(nT) \end{cases}$$

The associated root locus has the following pattern (Fig. 3) with respect to the gain

$$K' = \frac{K(\sin \omega T + k(1-\cos \omega T))}{\omega}$$

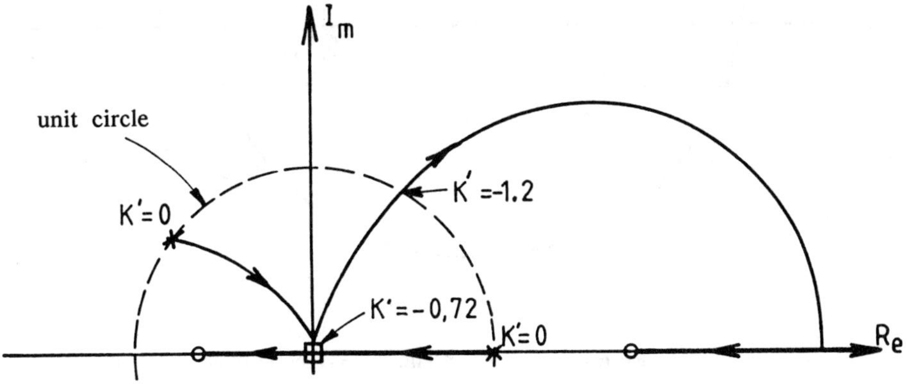


Root locus (K' varying from 0 to −∞)

Fig. 3

Stability can be said to be assured as long as the gain K' remains greater than −1.2. An example of reset of roll and yaw is shown in Fig. 4.

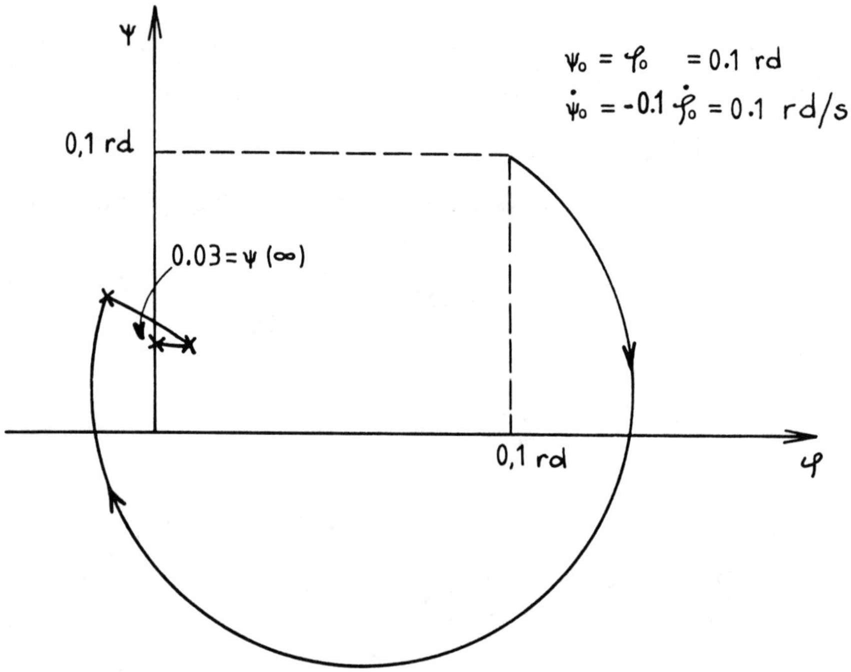


Resetting in the plane (:,=) by impulse control

Fig. 4

It will be noted that the final value $\psi(\infty)$ is not cancelled by this control law: the next section will show how the function of the long-term controller is precisely to cancel $\varphi$ and $\psi$ simultaneously by kinetic coupling (which "does not have time" to act in the short term).

### 8.2.2.2 – Controller with zero-order hold

It will now be assumed that k=1 (identical control for roll and yaw). A controller similar to the previous one must be found, to place the poles of the closed loop in the following way:

- two at the origin;
- two "dominant" poles satisfying $1-SZ^{-1}+PZ^{-2}=0$.

The solution of this system enables a, b, and K to be determined for a given relation between S and P which only depends on T. The parameters are then chosen for maximum damping. For example, using the same conditions as before, with

$$D(Z) = -0.045 \frac{1 - 2.05\ Z^{-1}}{1 + 0.77\ Z^{-1}}$$

the characteristic equation has two poles at the origin and two poles at $-0.4 \pm 0.1j$. It is interesting to examine the sensitivity of this tuning to a variation of ±5% in the nominal spin velocity N (Fig. 5):

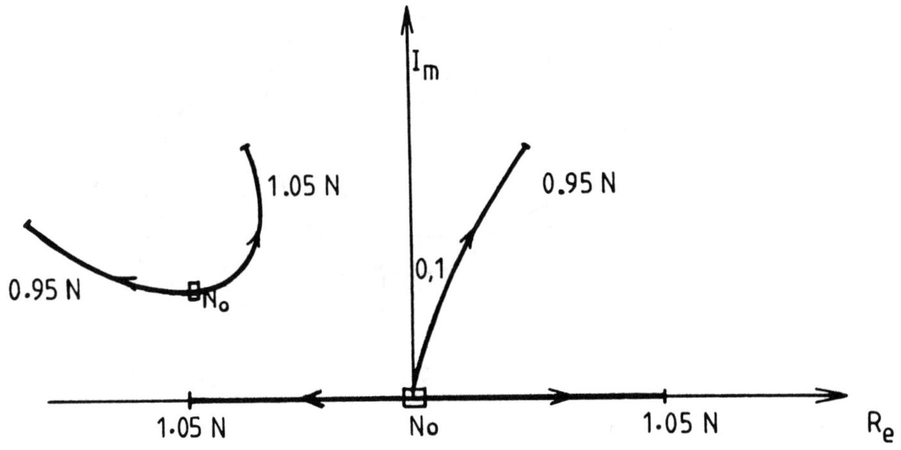


Sensitivity of the poles to a variation of N

Fig. 5

An example of the development of the roll and yaw is given in Fig. 6 for the following initial conditions:

$$(\varphi_o = \psi_o = \dot{\varphi}_o = \dot{\psi}_o = 0.1)$$

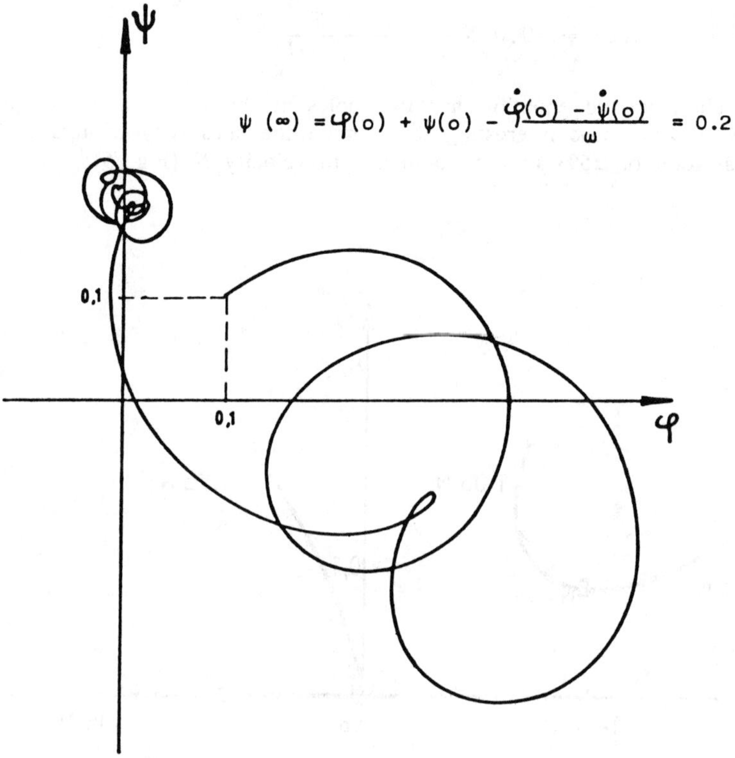


Resetting in the plane (:,=) by control with hold

Fig. 6

### 8.2.3 - LONG-TERM CONTROL

Assuming zero nutation, the equations governing the long-term control are written thus:

$$\dot{\varphi} = -\omega_o \psi + \omega_x \qquad \omega_o = \text{orbital frequency}$$

$$\dot{\psi} = \omega_o \varphi + \omega_z$$

$$\omega_x = \frac{U_z}{H}$$

$$\omega_z = -\frac{U_x}{H}$$

The equations are similar to the dynamic equations, with the exception that one integration has disappeared, since the order of the equations has decreased. The controller with the zero-order hold can then be used to freely place the three roots of the characteristic equation without constraint on the dominant poles. For example, minimum time control may be chosen (with the poles at the origin) and the reset is performed in three stages as in the example below (Fig. 7).

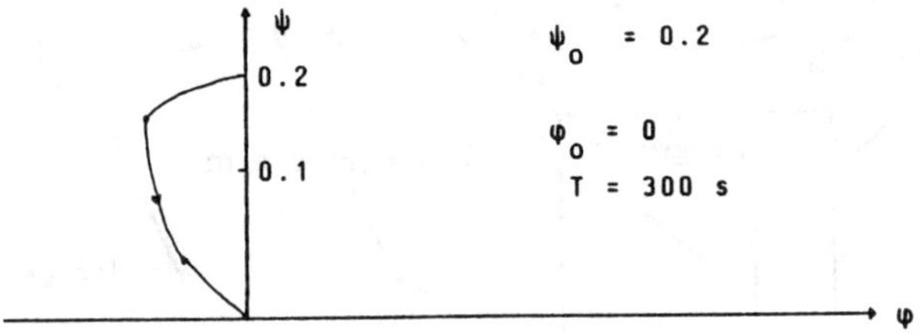


Long-term control

Fig. 7

## *8.3 - THREE-AXIS STABILISATION OF A SATELLITE WITH FLEXIBLE APPENDAGES*

### 8.3.1 - OUTLINE OF THE PROBLEM

As emphasised above, the presence of flexible appendages of increasing size has for some time caused concern to designers of stabilisation sub-systems because of their effect on performance and stability. From the examples described in the literature, we shall briefly describe here the SPOT satellite, intended to be the first application of the "Plate Forme Multimissions" (PFM) (Multiple-mission platform). More detailed information will be found in Chrétien (1980, 1982).

The Plate Forme Multimissions consists of two sub-assemblies, namely a vehicle of cubic form with an edge measurement of approximately 2 m, and a solar generator carried by an articulated arm to permit storage and deployment. The vehicle carries all the measurement and actuation devices required for the satisfaction of a mission of the earth observation type; the orbit is low and

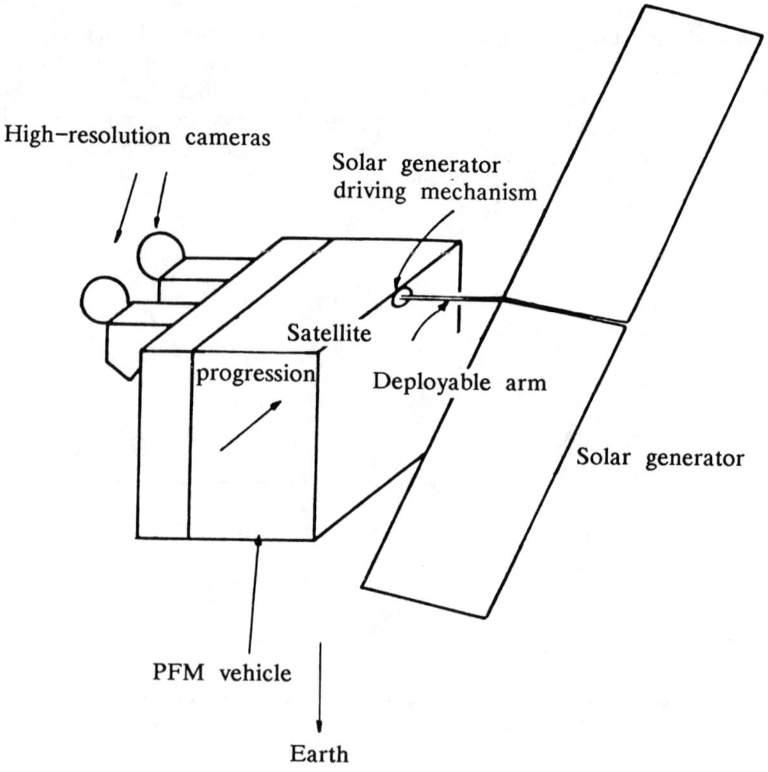


SPOT payload of the multi–mission platform

Fig. 8

heliosynchronous (in other words, the point overflown always has the same local time, chosen once for all at the beginning of the mission). The side opposite that carrying the solar generator is designed to carry the payload: in the case of SPOT, this consists of two high-resolution visible-spectrum cameras.

The general layout is shown in Fig. 8. It will be noted that the solar generator is articulated with respect to the vehicle by means of an actuating mechanism called "MEGS" designed to keep the generator oriented towards the sun while the satellite travels in its orbit.

Earth observation missions differ from telecommunications missions in the high stability required for the tracking; as with a camera, it is necessary to avoid "shake", and this is reflected in the very strict specifications for angular velocity (typically $10^{-3}$·/s for the three axes). The most troublesome disturbances are then those of internal origin: the start of rotation of wheels, such as the reels of the magnetic recorder, the pivoting of return mirrors, and the vibrations of the solar generator. The effect of these factors will now be examined more closely, for the case of the normal operating mode of the satellite (called the fine pointing mode).

## 8.3.2 - DYNAMIC MODEL OF THE SATELLITE

The appropriate representation of the vibrations of the generator is a complex problem; if the primary purpose is to study the effect of these vibrations on the vehicle and its payload, there is a powerful method available which enables the representation to be limited to the dynamic forces applied at the point of attachment of the generator to the vehicle. This method, known as the method of effective masses, expresses numerically the relationships between the forces (torques) at the attachment point and the accelerations (linear and angular) of the carrying vehicle:

$$\begin{bmatrix} F \\ C \end{bmatrix} = M(s) \begin{bmatrix} \gamma \\ \dot{\omega} \end{bmatrix}$$

in the form of a development on modal frequencies:

$$M(s) = M_r + \sum_{k=1}^{N} \frac{M_k \; s^2}{s^2 + \omega_k^2}$$

The symmetrical constant matrices $M_k$ (which are of rank 1) are called effective mass matrices; they are associated with the frequencies $\omega_k$, and it is also necessary to take into account $M_r$, known as the residual mass matrix, which represents the "rigid" part of the generator.

The coupling by MEGS between this dynamic model of the

generator and the dynamic model of the payload + vehicle combination then makes it possible to establish the overall dynamic model of the satellite; this model is characterised by its secular evolution along the orbit, since the satellite rotates around the generator. A reasonable approximation of the model may be obtained, allowing for the symmetry properties of the vehicle and given the fact that the three first flexible modes in increasing order of frequency are in practice associated with each of the axes of symmetry of the generator. This approximation leads to a de facto decoupling of the three axes: the model for the roll or yaw axis is a conventional model with one flexible mode: the case of the pitch is more complex, since it is possible to use the generator driving motor as a secondary actuator, thus making the problem one of the multiple-input multiple-output type.

### 8.3.3 - ROLL AND YAW AXES

The block diagram for the problem of control on the roll and yaw axes is of the following standard type (Fig. 9):

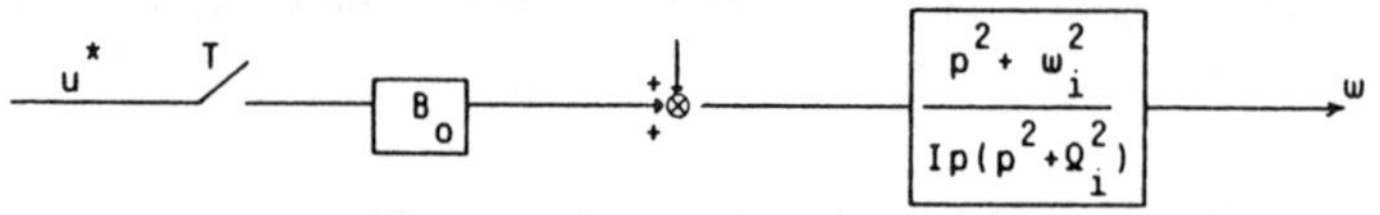


Block diagram of the controller

Fig. 9

and relates the sampled control signal u* to the angular velocity $\omega$ on the corresponding axis. The model comprises one pole at the origin, two imaginary poles at the free frequency $\Omega_i$ and two imaginary zeros at the frequency $\omega_i$. The discrete transfer function between u* and $\omega^*$ is written approximately (up to the third power in $\omega_i T$)

$$\frac{\omega^*}{u^*} = \frac{T}{I}\left[1 - \frac{(\Omega_i^2 - \omega_i^2)\,T^2}{6}\right]\frac{Z^{-1}}{1-Z^{-1}}\;\frac{1 - 2\cos\omega_i T\,Z^{-1} + Z^{-2}}{1 - 2\cos\Omega_i T\,Z^{-1} + Z^{-2}}$$

showing that, when T is sufficiently low with respect to the periods in question, the continuous poles and zeros are transformed into discrete poles and zeros by the transformation $Z = e^{Ts}$ . Thus the vicinity of the point (+1,0) of the plane Z closely resembles the vicinity of the origin of the continuous locus and the synthesis of the digital controller by the technique of the root locus may be performed in either the digital or the digitised continuous mode (Fig. 10).

At the level of the control law synthesis itself, the system in question has the following basic property:

$$\omega_i < \Omega_i$$

Consequently, the formation of a loop by a pure gain (torque generated by the reaction wheel proportional to the measurement of angular velocity by a gyrometer) leads to a closed loop which is unconditionally stable in continuous form and conditionally stable in discrete form (Fig. 10).

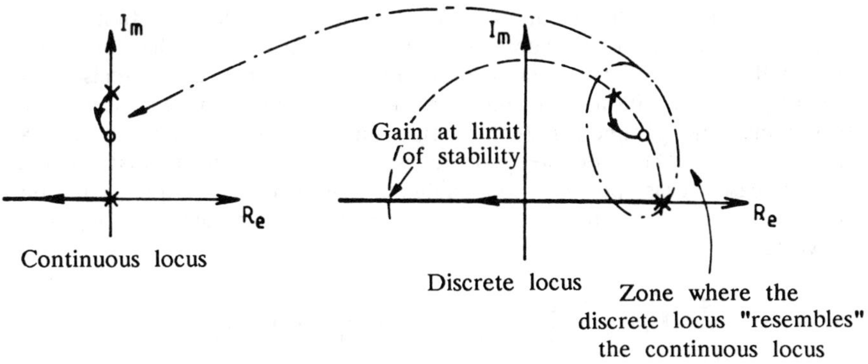


Continuous and discrete loci (control of a flexible mode by proportional feedback)

Fig. 10

This control law, however, does not enable the vibrations of the mode to be damped at will. A more complex solution is provided by the placement of poles with the aid of a second order dynamic controller; it is possible to freely place the five roots in the closed loop, and consequently to regulate the damping of the modes as required, as for example in the locus in Fig. 11.

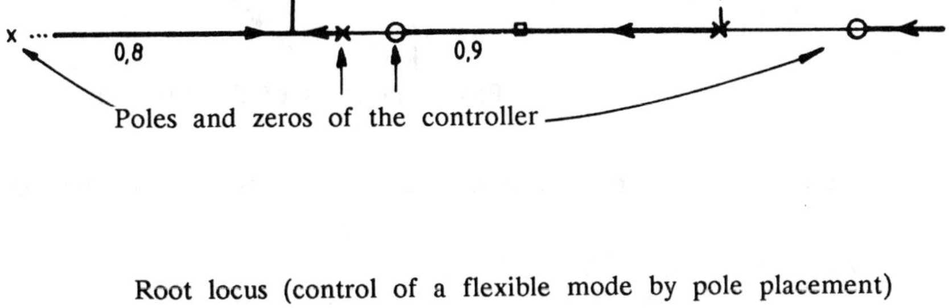

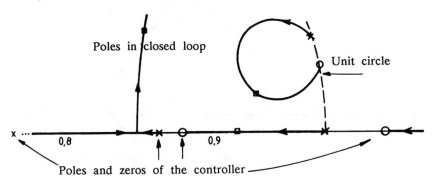

Root locus (control of a flexible mode by pole placement)

Fig. 11

However, the presence of an unstable zero leads to a very low static gain, incompatible with the required precision. Moreover, this solution is very sensitive to the value of the frequencies $\omega_i$ and $\Omega_i$.

An intermediate solution can be provided by a first order controller; among all the possible poles for this, the choice has been restricted to controllers containing an integration. The static precision of the loop is then infinite in the presence of constant disturbances. The two remaining parameters (controller gain and zero position) can be tuned with the help of the root locus. It may be shown that in this special case it is possible to maximise the overall damping. In the continuous case, with the controller $k_1/s + k_2$, the characteristic equation is written:

$$s^4 + k_2 s^3 + (\Omega_i^2 + k_1)\, s^2 + k_2 \omega_i^2\, s + k_1\, \omega_i^2 = 0$$

Identification with a product of two trinomials of the second degree then leads to the root locus shown in Fig. 12:

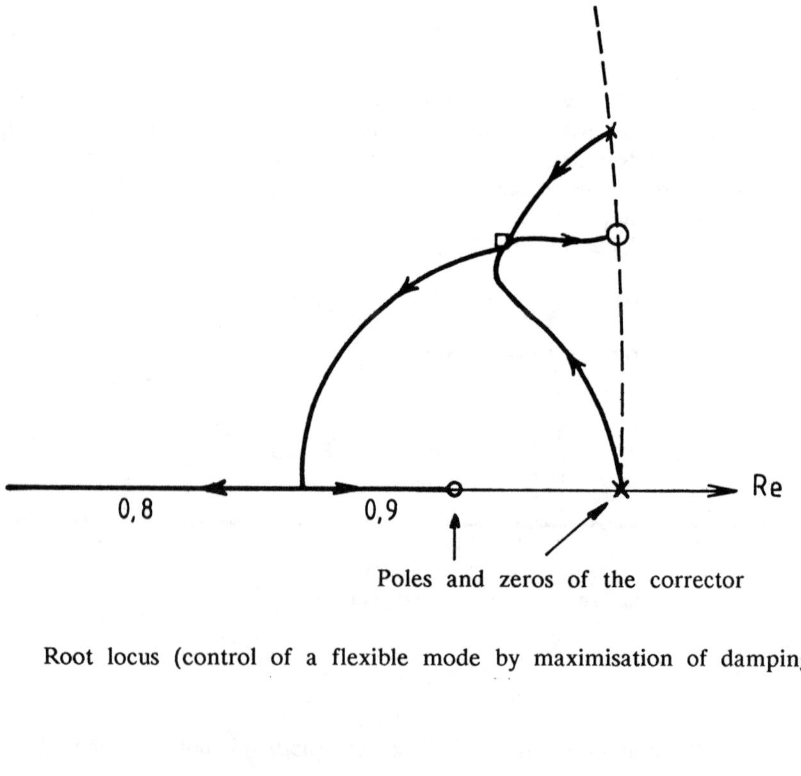


Root locus (control of a flexible mode by maximisation of damping)

Fig. 12

where the characteristic equation has twice the two conjugate complex roots. This solution, which is equivalent to a proportional derivative regulation on the position loop, is found to be much less sensitive to the frequency

values than the previous solution. The discrete controller is then obtained by bilinear transformation (see Chapter 2).

### 8.3.4 - PITCH AXIS

The analysis and synthesis of the control problem for the pitch axis are complicated for two reasons:

- a second complementary actuator may be introduced, consisting of the generator driving motor itself, as well as a secondary sensor which measures that relative torque between the vehicle and the generator;

- the MEGS system does not form a rigid rotoid link; the motor has a certain magnetic compliance which leads to the appearance of a second mode of vibration associated with the movement of the whole "loaded" generator on the corresponding stiffness $K_m$.

The continuous block diagram then takes the following form (Fig. 13):

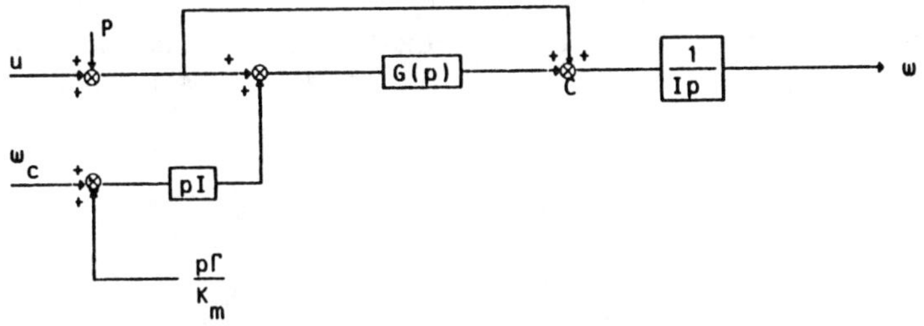


Block diagram of the pitch axis

Fig. 13

The block G(s) is written:

$$G(s) = -\sum_{i=1}^{N} \frac{K_i}{s^2 + \Omega_i^2}$$

where N is the number of modes taken into account in the pitch axis.

The lower part of the diagram shows the contribution of the motor through the possible control signal $\omega_c$ (rotation speed) and the disturbance due to the imperfections of the motor $\Gamma$, which is periodic in nature.

The upper part of the diagram forms a simple rewriting of Fig. 9 in which a direct transmission of the torque to the angular acceleration has been isolated.

The problem is then formulated in the most general way as a problem with two measurements ($\omega$ and C, angular velocity and relative torque between the generator and the vehicle), two actuators (u and $\omega_c$ are the

torque of the reaction wheel and the angular velocity of the motor) and two disturbances P and $\Gamma$ (disturbance torque on the satellite body and disturbance torque due to the driving motor).

It is initially assumed that the torque measurement is not available. The most simple solution then consists of the application of a torque u and a velocity $\omega_c$ proportional to the single measure $\omega$ with the coefficients a and c. Thus the control law (which must reduce to a minimum the effect of the periodic disturbances $\Gamma$) can be synthesised with the aid of a family of root loci. In the continuous form, the starting points of the loci are determined on the imaginary axis by the value of c (Fig. 14), with the end points always being the same, and the zeros in this case corresponding to the frequencies $\Omega_i$. The family shown for one mode may be generalised to a number of modes and the control laws determined in this way are intrinsically stable. It can be shown that the effect of the periodic disturbances $\Gamma$ is reduced as the chosen closed loop poles become "further" from the imaginary axis. The beneficial effect of the introduction of the gain c is apparent in Fig. 14.

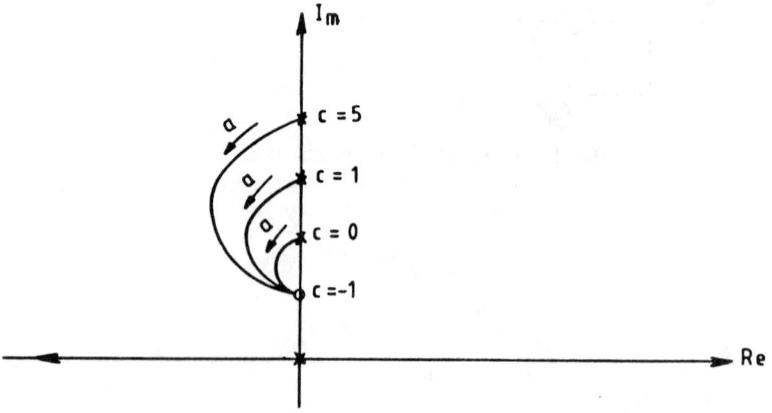


Family of root loci in the continuous case

Fig. 14

The situation is very different in the discrete case (Fig. 15), since the starting points of the locus do not remain on the unit circle, but move away from it into the unstable zone; the vicinity of the point (1,0) closely resembles the vicinity of the origin of the plane s above, but this does not apply to modes with a higher frequency.The margin of stability is much smaller in the discrete domain.

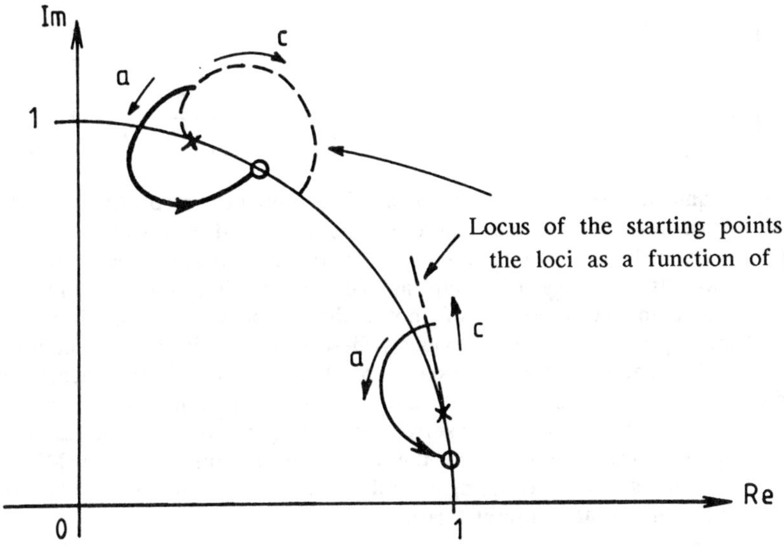


Family of root loci in the discrete case

Fig. 15

It should be noted that, in the continuous case, the solution c=–1 leads to a perfect simplification of the flexible zeros by poles, the transfer function between u and ω being written:

$$u = \frac{\omega}{Is}$$

for any G(s). This solution, which consists in the perfect decoupling of the solar generator from the satellite disturbances, may be useful on occasions when these disturbances are large. On the other hand, it does not damp the motor disturbances.

If the measurement of the torque C is available, the multivariable problem is square and it can be shown that it is possible to determine a decoupling control signal in the sense that each disturbance only affects one output. It is assumed that the control law is a square matrix:

$$\begin{bmatrix} u \\ \omega_c \end{bmatrix} = \begin{bmatrix} k_{11}(s) & k_{12}(s) \\ k_{21}(s) & k_{22}(s) \end{bmatrix} \begin{bmatrix} \omega \\ C \end{bmatrix}$$

The conditions $k_{12}(s) = k_{21}(s) = -1$ then lead to a matrix for the transfer function between the outputs and the disturbance torques

$$\begin{bmatrix} \omega \\ C \end{bmatrix} = \begin{bmatrix} \dfrac{1}{I_s - k_{11}(s)} & 0 \\ 0 & \dfrac{s^2 N I}{K_m(D - N + k_{22}(s) INs)} \end{bmatrix} \begin{bmatrix} P \\ \Gamma \end{bmatrix}$$

where N and D are the numerator and denominator respectively of the transfer function G(s). It may therefore be seen that $k_{11}$ can be chosen to tune as well as possible the control of the satellite body by the reaction wheels and $k_{22}$ to damp as well as possible the vibrations of the generator in the presence of motor disturbances. The condition $k_{22} = -1$ (i.e. $\omega_c = -\omega$) is the previously described condition of decoupling of the solar generator with respect to the satellite disturbances; the condition $k_{12} = -1$ (i.e. $u = -C$) corresponds to a compensation by the reaction wheel of the torques exerted by the generator on the satellite, thus decoupling the satellite from the disturbances caused by the MEGS. These conditions of decoupling applied together can be transferred without difficulty to the discrete case.

## *REFERENCES*

J.P. Chretien, C. Reboulet, and P. Rodrigo, Rapports techniques DERA No. 1 et 2/7345 – Marché CNES No. 800607, 1980

J.P. Chretien, C. Reboulet, P. Rodrigo, and M. Maurette, Attitude control of satellite with a rotating solar array, Journal of Guidance Control and Dynamic, v. 5, No. 6, 1982

J.F. Le Maitre, C. Reboulet, and M. Gauvrit, Rapport technique CNES/DERA, avril 1974.

## *CHAPTER 9*

# *DIGITAL CONTROL SYSTEM OF A LAUNCH VEHICLE*

## *9.1 - INTRODUCTION*

This chapter is concerned with the application of optimal control algorithms for the control of a launch vehicle in its atmospheric phase. It is the result of the study of digital control for the ARIANE IV launch vehicle, the main difficulty of which is associated with the high flexibility of the vehicle, due to its length. The problem is to establish an engine control law which will keep the launch vehicle in its nominal trajectory while minimising the forces on the structure during the atmospheric flight stage, despite disturbances caused by the wind.

The control of such a system has two principal characteristics. On one hand, the control application time is finite and known, since it corresponds to the operating life of the first-stage motors, while on the other hand the system to be guided is non-stationary, because of the change in mass due to the consumption of the propellants. These two factors explain the use of optimal control techniques with a finite horizon for the design of the control system of a launch vehicle (these techniques are described in Chapters 2 and 5).

The application of the linear quadratic Gaussian method consists of the following three stages:

- the writing of a state representation of the system;

- the choice of a criterion and the definition of the state and measurement "noises";

- the calculation of the control and filtering gains.

## *9.2 - MODELLING THE SYSTEM*

### 9.2.1 - NOTATION

| | |
|---|---|
| $A = (a_{ij})$ | dynamic matrix of the state representation |
| $B = (b_{ij})$ | control vector of the state representation |
| $C = (c_{ij})$ | observation matrix of the state representation |
| F | aerodynamic centre |

| | |
|---|---|
| $G$ | overall centre of gravity of the launch vehicle |
| $G_E$ | centre of gravity of the dismantled rigid launch vehicle (without the nozzle) |
| $G_T$ | centre of gravity of the nozzle |
| $G_{Tr}$ | centre of gravity of the nozzle for the rigid launch vehicle |
| $i$ | incidence |
| $J$ | total inertia of the launch vehicle about G |
| $I_E$ | inertia of the dismantled launch vehicle about $G_E$ |
| $I_T$ | inertia of the nozzle about $G_T$ |
| $l_E$ | distance from the centre of gravity $G_E$ to the gimbal point |
| $l_F$ | distance from the aerodynamic centre to the gimbal point |
| $l_T$ | distance from the centre of gravity $G_T$ to the gimbal point O |
| $M$ | total mass of the launch vehicle |
| $M_E$ | mass of the dismantled launch vehicle |
| $M_T$ | mass of the nozzle |
| $M_i$ | generalised mass associated with the 1st bending mode |
| $N_\alpha$ | lift coefficient |
| $O$ | gimbal point |
| $P$ | total thrust |
| $P_o$ | pivoting thrust |
| $T$ | kinetic energy |
| $V$ | velocity of the launch vehicle in its nominal trajectory |
| $V_{x/y}$ | relative velocity of point X with respect to point Y |
| $W$ | lateral wind velocity |
| $X_A$ | drag |
| $Y_{x_i}$ | normalised amplitude of the i-th bending mode at the abscissa x |
| $Y'_{x_i}$ | slope of the i-th bending mode at the abscissa x, $Y'_{x_i} = \delta/\delta x(Y_{x_i})$ |
| $Z$ | lateral displacement of the overall centre of gravity G |
| $\beta$ | steering angle of the nozzle |
| $\eta_i$ | i-th bending mode |
| $\theta$ | angle between the vertical and the longitudinal axis of the rigid launch vehicle |
| $\omega_T$ | natural frequency of the servo motor |
| $\xi_T$ | damping of the servo motor |
| $\omega_i$ | frequency of the i-th bending mode |
| $\xi_i$ | damping of the i-th bending mode |

State variables of the model

$$X^T = (\dot{Z},\ \theta,\ \dot{\theta},\ \beta,\ \dot{\beta},\ W,\ W_1,\ \eta_1,\ \dot{\eta}_1,\ \eta_2,\ \dot{\eta}_2)$$

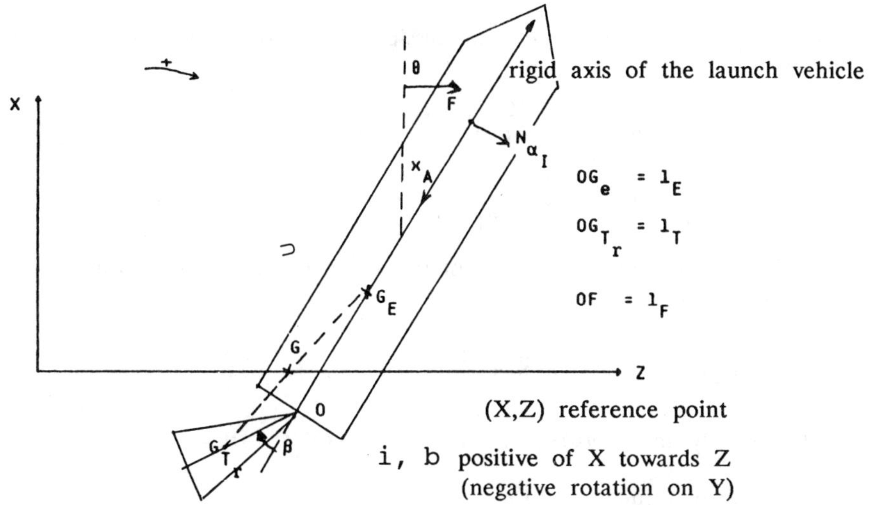


Schematic representation of the rigid launch vehicle

Fig. 1

## 9.2.2 – HYPOTHESES

The launch vehicle model is obtained by writing Lagrange equations for the following hypotheses:

i) there is plane movement about the reference trajectory which is assumed to be in one plane and rectilinear;
ii) there are small movements about the centre of gravity;
iii) the launch vehicle is flexible;
iv) the propellant is solid (or the movements of liquid fuel are disregarded).

## 9.2.3 – EQUATIONS

The total kinetic energy of the system is expressed by the following relationship:

$$2T = M\,\dot{Z}^2 + \left(I_E + I_T + \frac{M_E M_T}{M}(l_E + l_T)^2\right)\dot{\theta}^2 + \sum_i M_i \dot{\eta}_i^2 + \left(I_T + \frac{M_E M_T}{M}\, l_T^2\right)\dot{\beta}^2$$

$$+2(I_T + \frac{M_E M_T}{M} l_T(l_E + l_T))\dot{\beta}\,\dot{\theta} + \sum_i (I_T Y'_{o_i} - M_T l_T (Y_{o_i} - l_T Y'_{o_i}))\dot{\eta}_i\,\dot{\beta}$$

The potential energy and the dissipative energy are formulated as follows:

$$E_p = \frac{1}{2} \sum_i M_i\, \omega_i^2\, \eta_i^2$$

$$E_d = \frac{1}{2} \sum_i 2M_i\, \omega_i\, \xi_i\, \eta_i^2$$

The forces acting on the system, allowing for the approximation of the small angles, are as follows:

- the thrust $\vec{P} = P\,\vec{x} + (P(\theta + \Sigma Y'_{o_i}\eta_i) + P_o\beta)\vec{z}$

- the aerodynamic forces $\vec{A} = -X_A\vec{x} + (X_A\theta + N_\alpha i)\,\vec{z}$

- the control torque $\vec{\Gamma}$

with
- $P$ total thrust
- $P_o$ pivoting thrust
- $X_A$ drag
- $N_\alpha$ lift coefficient
- $i$ incidence

The generalised coordinates chosen to describe the movement of the launch vehicle are as follows:

$$(Z,\ \theta,\ \beta,\ \eta_i)$$

The generalised forces associated with each coordinate are expressed thus:

$$Q_z = P(\theta + \sum_i Y'_{o_i}\eta_i) + P_o\beta - X_A\theta + N_\alpha(\theta - \frac{\dot{Z}-W}{V})$$

$$Q_\theta = [-(P-P_o-X_A)\beta + X_A \sum_i Y'_{o_i}\eta_i]\frac{M_T l_T}{M} - P\Sigma Y_{o_i}\eta_i$$

$$-(P\sum_i Y'_{o_i}\eta_i + P_o\beta)\frac{M_E l_E}{M} + N_\alpha(\frac{M_T l_T}{M} - \frac{M_E l_E}{M} + l_F)(\theta - \frac{\dot{Z}-W}{V})$$

$$Q_\beta = (-(P-P_o-X_A)\beta + X_A\Sigma Y'_{o_i}\eta_i)\frac{M_T l_T}{M} + N_\alpha\frac{M_T l_T}{M}(\theta - \frac{\dot{z}-w}{v}) + \Gamma$$

$$Q_{\eta_i} = (P(\sum_i Y'_{o_i}\eta_i)+P_o\beta)\,Y_{o_i}+[-(P-P_o-X_A)\beta+X_A\sum_i Y'_{o_i}\eta_i+(N_\alpha(\theta-\frac{\dot{z}-w}{V})]\,\frac{M_T l_T}{M}\,Y'_{o_i}$$

The Lagrange equations may then be written thus:

$$M_\alpha = N_d(\frac{M_T l_T}{M} - \frac{M_E l_E}{M} + l_F)$$

$$\frac{d}{dt}(\frac{\partial T}{\partial \dot{q}_i}) - \frac{\partial T}{\partial q_i} + \frac{\partial D}{\partial \dot{q}_i} + \frac{\partial V}{\partial q_i} = Q_i \qquad (1)$$

i.e.

$$M\ddot{z} = (P-X_A+N_\alpha)\theta+P\Sigma Y'_o\eta_i+P_o\beta-N_\alpha\frac{\dot{z}}{V}+N_\alpha\frac{w}{V}$$

$$J\ddot{\theta}+(I_T+\frac{M_T M_E}{M}l_T(l_E+l_T))\ddot{\beta} = M_\alpha\theta+(X_A\frac{M_T l_T}{M} - P\frac{M_E l_E}{M})\Sigma Y'_{o_i}\eta_i$$

$$-P\sum_i Y_{o_i}\eta_i-(\frac{M_T l_T}{M}(P-P_o-X_A)+P_o\frac{M_E l_E}{M})\beta-M_\alpha\frac{\dot{z}}{V}+M_\alpha\frac{w}{V} \qquad (2)$$

$$M_i\ddot{\eta}_i+[I_T Y'_{o_i}-M_T l_T(Y_{o_i}-l_T Y'_{o_i})]\ddot{\beta}+M_i 2\xi_i\omega_i\dot{\eta}_i+M_i\omega_i^2\eta_i =$$

$$(N_\alpha\frac{M_T l_T}{M}Y'_{o_i})\theta+(PY_{o_i}+X_A\frac{M_T l_T}{M}Y'_{o_i})\sum_i Y'_{o_i}\eta_i$$

$$+(P_o Y_{o_i}-(P-P_o X_A)\frac{M_T l_T}{M}Y'_{o_i})\beta-N_\alpha\frac{M_T l_T}{M}Y'_{o_i}\frac{\dot{z}}{V}+N_\alpha\frac{M_T l_T}{M}Y'_{o_i}\frac{w}{V} \qquad (3)$$

$$(I_T+\frac{M_T M_E}{M}l_T^2)\ddot{\beta}+(I_T+\frac{M_T M_E}{M}l_T(l_E+l_T))\ddot{\theta}+\sum_i(I_T Y'_{o_i}-Y'_{o_i}))\ddot{\eta}_i$$

$$= N_\alpha\frac{M_T l_T}{M}\theta+X_A\frac{M_T l_T}{M}\sum_i Y'_{o_i}\eta_i-(P-P_o-X_A)\frac{M_T l_T}{M}\beta-N_\alpha\frac{M_T l_T}{M}\frac{\dot{z}}{V}$$

$$+N_\alpha\frac{M_T l_T}{M}\frac{w}{V}+\Gamma \qquad (4)$$

The last equation may be written formally as follows:

$$I_o\ddot{\beta} = \Gamma_r + \Gamma$$

where $\Gamma_r$ represents all the terms of the inertial coupling between the launch vehicle and the nozzle, and $\Gamma$ denotes the torque applied by the actuators to the nozzle. Since the hydraulic actuators used in the launch vehicles have low compressibility, the following simplified hypothesis may be stated:

$$\Gamma = -\Gamma_r + \Gamma_s$$

where $\Gamma_s$ is the torque determined by the servocontrol.

On this hypothesis, equation (4) is replaced by the equation representing the dynamic of the servo motor which is assumed to be of the second order, i.e.

$$\ddot{\beta} + 2\xi_T\,\omega_T\,\dot{\beta} + \omega_T^2\,\beta = \omega_T^2\,\beta_c$$

### 9.2.4 – STATE REPRESENTATION

The above equations are complemented by a second-order stochastic wind model of the form

$$\begin{vmatrix} \dot{W} \\ W_1 \end{vmatrix} = \begin{vmatrix} 0 & \alpha_o \\ -\alpha_o & \alpha_1 \end{vmatrix} \begin{vmatrix} W \\ W_1 \end{vmatrix} + \begin{vmatrix} 0 \\ b \end{vmatrix} W_b$$

where $W_b$ is a Gaussian variable with unit variance.

The gyroscopic measurements, $\theta_m$, and gyrometric measurements, $\dot{\theta}_m$, are represented by the following equations:

$$\theta_m = \theta + \sum_i Y'_{ci}\,\eta_i + w_1$$

$$\dot{\theta}_m = \dot{\theta} + \sum_i Y'_{Ci}\,\dot{\eta}_i + w_2$$

The above system of equations can easily be expressed in the state form:

$$\dot{X} = A(t)\,X + B(t)\,U + V_b \qquad \text{where} \qquad E|V_b\,V_b^T| = M_b$$

$$Y = C(t)\,X + W_m \qquad\qquad E|W_m\,W_m^T| = N$$

and

$$X = (\dot{Z},\ \theta,\ \dot{\theta},\ \beta,\ \dot{\beta},\ W,\ W_1,\ \eta_1,\ \dot{\eta}_1,\ \eta_2,\ \dot{\eta}_2)^T$$

$$U = \beta_c\ ; \qquad Y = (\theta \quad \dot{\theta})^T$$

The digitisation of this model for the period T, which is chosen according to the desired dynamic in closed loop, results in the required recursive equations:

$$X_{k+1} = F_k X_k + G_k U_k + V_k \qquad E|V_k V_k^T| = M_k$$

$$Y_k = H_k X_k + W_k \qquad E|W_k W_k^T| = N_k$$

where

$$F_k = e^{A(t_k)t_k}; \quad G_k = \int_{t_k}^{t_{k+1}} e^{A(t).t} B(t)dt; \quad H_k = C(t_k)$$

$$M_k = \int_{t_k}^{t_{k+1}} e^{A(t^k).T} M e^{A^T(t)t} dt; \quad N_k = N$$

## *9.3 - CHOICE OF THE CRITERION AND OF THE COVARIANCE MATRICES*

### 9.3.1 - OPTIMISATION CRITERION

The function of the launch vehicle control system is to keep the vehicle on its reference trajectory, while ensuring that the flight conditions are acceptable for the structures by minimising the aerodynamic load factor. This implies that two terms must be taken into account in the criterion, one in $\theta^2$ and the other in $i^2$, where i represents the incidence of the launch vehicle:

$$(i = \theta - \frac{\dot{Z}-W}{V})$$

To increase the control possibilities, the criterion to be minimised in the course of the flight can be written in a more general way:

$$C = E \sum_{k=i}^{N} \left[ k_i (\frac{i}{i_m})^2 + k_\theta (\frac{\theta}{\theta_m})^2 + k_{\dot\theta} (\frac{\dot\theta}{\dot\theta_m})^2 + k_{\dot Z} (\frac{\dot Z}{\dot Z_m})^2 + k_\beta (\frac{\beta}{\beta_m})^2 + k_{\eta_1} (\frac{\eta_1}{\eta_{1_m}})^2 + (\frac{\beta_c}{\beta_{c_m}})^2 \right]_k$$

The values $(.)_m$ represent an estimate of the maximum value of the variable (.) during the flight. By using these values, it is possible to work with a dimensionless criterion, making the values of the weighting coefficients $k_j$ more representative.By expressing the incidence i as a function of the chosen state variables, the criterion C can be written in the standard matrix form:

$$C = E \sum_{k=1}^{N} (X_k^T Q X_k + U_k^T R U_k)$$

### 9.3.2 - COVARIANCE MATRIX

These matrices, which are required for the determination of the Kalman filter gains, are calculated from the definition of the state and measurement noises.

In theory, the state noises affecting the system are those introduced in the modelling of the wind, the corresponding covariance matrix being the matrix $M_k$.

In practice, however, it is often advisable to introduce additional imaginary state noises in order to allow for any inaccuracies in the evaluation of one or more of the model parameters. Thus the matrix used for the calculation of the filter will be in the form

$$M''_k = M_k + M'_k$$

where $M'_k$ is a matrix whose elements are a function of the confidence in the equation governing the variable in question. In fact, these coefficients are the tuning parameters of the Kalman filter.

### 9.3.3 - CHOICE OF PARAMETERS

At the present time, there is no systematic method of choosing weighting coefficients for the performance criterion and the parameters of the covariance matrix of the state noises. An approach by successive iterations enables the required performance of the control system to be obtained. In addition to the variables taken into account in the criterion, the deviation from the reference trajectory and the load factor, the performance in the case of a launch vehicle is judged in terms of the margin of stability and the robustness in relation to modelling errors. The last two considerations are difficult to translate into the form of a quadratic criterion for the states, and this explains the absence of a methodology for this problem.

When seeking suitable parameters, it must be borne in mind that the effect of the criterion and that of the state noise on the resulting performance are closely related, and that, consequently, they cannot be defined independently of one another. This is why both the choice of the performance criterion and the choice of the state noise covariance matrix are included in the same paragraph, so that their interdependence in practical cases can be underlined, although theoretically these two quantities are completely independent. This difference between theory and practice is explained by the fact that the theory ensures the optimality of the control for the vehicle in its nominal configuration, whereas in practice a high degree of sub-optimality, if not optimality, is required for any vehicle approximating the nominal vehicle.

## *9.4 - DETERMINATION OF THE CONTROLLER - APPLICATION*

With the system, the criterion, and the noises defined, it only remains to apply the theoretical equations to determine the control and filtering gains, using the results obtained in Chapter 5.

* Equations of the system

$$X_{k+1} = F_k X_k + G_k U_k + V_k$$

$$Y_k = H_k X_k + W_k$$

with

$$E[V_k V_l^T] = M_k \delta_{kl} \quad ; \quad E[W_k W_l^T] = N_k \delta_{kl}$$

* Criterion to be minimised

$$C = E\left[ \sum_{k=1}^{N} (X_k^T Q_k X_k + U_k^T R_k U_k) \right]$$

According to the hypothesis, since the system is linear, the state and measurement noises are Gaussian, and the criterion to be minimised is quadratic in X and U, the solution of this optimisation problem is provided by a linear quadratic Gaussian controller with the following structure (Fig. 2):

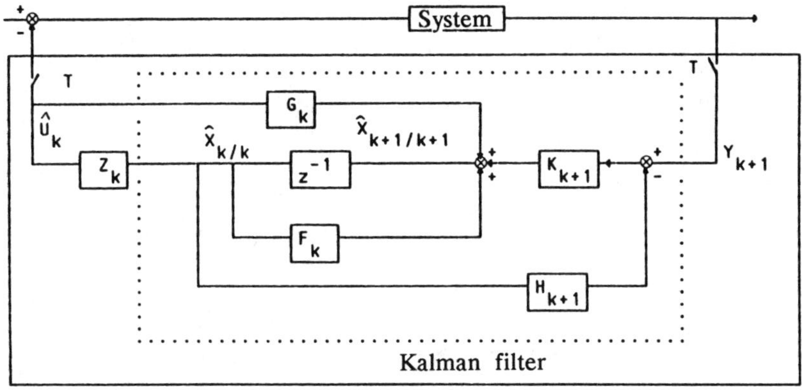


Structure of a linear quadratic Gaussian controller

Fig. 2

### 9.4.1 - CONTROL GAINS

$$L_k = (R_k + G_k^T S_{k+1} G_k)^{-1} G_k^T S_{k+1} F_k ;$$

$$S_k = Q_k + F_k^T S_{k+1} F_k - F_k^T S_{k+1} G_k L_k;$$

$$S_N = Q_N$$

This system of equations is solved in a recursive way by decreasing the time. It provides the sequence of control gains $L_k$ to be applied on the horizon in question.

### 9.4.2 – KALMAN GAINS

$$K_{k+1} = P_{k+1/k} H_{k+1}^T (H_{k+1} P_{k+1/k} H_{k+1}^T + N_{k+1})^{-1}$$

$$P_{k+1/k+1} = (I - K_{k+1} H_{k+1}) P_{k+1/k} (I - K_{k+1}^T H_{k+1})^T + K_{k+1} N_{k+1} K_{k+1}^T$$

$$P_{k+1/k} = F_k P_{k/k} F_k^T + M_k$$

$$P_{0/0} = E[X_0 \, X_0^T]$$

This system of equations is solved by increasing the time, and provides the sequence of Kalman gains $K_k$ to be applied on the horizon in question.

### 9.4.3 – APPLICATION

The application of the linear quadratic Gaussian control consists in the performance of the following calculations in real time:

* Estimator of the system states:

$$\hat{X}_{k+1/k} = F_k \hat{X}_{k/k} + G_k U_k;$$

$$\hat{X}_{k+1/k+1} = \hat{X}_{k+1} + K_{k+1} (Y_{k+1} - H_{k+1} \hat{X}_{k+1/k})$$

* Determination of the controller

$$U_{k+1} = -L_{k+1} \, \hat{X}_{k+1/k+1}$$

It will be seen that it is necessary to include not only the gains $K_k$ and $L_k$, determined "off-line", but also the matrices $F_k$, $G_k$, and $H_k$ of the discrete modelling of the system. Because of the limited space available in the on-board computer, these data are generally compressed with the aid of a polynomial approximation, enabling their development to be represented over the whole horizon with a smaller number of parameters. In addition, the on-board model may be diagonalised by taking into account the development of the diagonalising base (eigenvectors) or may be reduced if this reduction has been taken into account at the time of determination of the gains.

This technique has been applied to the ARIANE IV launch vehicle. The results presented correspond to the digital controller determined in the first stage of ARIANE 44L, using a form of modelling including three modes of vibration of the structures. The sampling frequency of 10 Hz is chosen as a function of the disturbances to be counteracted and the desired performances.

Fig. 3, which shows the Black locus of the complete channel control system at the critical instant of the flight, demonstrates the lack of robustness of the linear quadratic Gaussian controller determined with the nominal characteristics of the state noise $M_k$. The margins of gain and phase obtained do not exist, and the locus passes through the critical point (0dB, $-180^o$). The modification of the noise covariance matrix in the calculation of the Kalman filter by addition of $M_k^{\prime}$ associated with an imaginary noise at the input of the system [Doyle and Stein (1981)] enables the margins to be improved as shown in Fig. 4.

Figs. 5a and 5b show a simulation of the first stage of the launch vehicle controlled with a reduced controller. This reduction, obtained by eliminating the two bending modes of highest frequency in the modelling of the system, provides a performance comparable to the optimum, on condition that a continuous preliminary filtering of the measurements is introduced to eliminate the effect of the disregarded modes on the information from the sensors. This technique can be used to obtain a good knowledge of the structural modes of the system, since, because of the preliminary filtering, the resulting digital control system is very robust in relation to errors of modelling of the vibrations of the launch vehicle.

## *9.5 - COMMENTS*

The application of a linear quadratic Gaussian technique to the particular case of the control of a launch vehicle leads us to make the following comments:

i) The use of an optimal controller on a finite horizon does not ensure local stability of the system. Depending on the criterion in use, it may be more advantageous to let the system deviate for a few moments rather than to try to stabilise it. In the case of the launch vehicle, this phenomenon appears when the aerodynamic conditions are maximal and the incidence term in the criterion becomes preponderant, in which case the stability of the lateral derivative $\dot{Z}$ is no longer provided for a few seconds. Although these local instabilities are theoretically acceptable, in practice the weightings in the criterion are modified in an attempt to eliminate them for reasons of security.

ii) If a criterion of the type with a margin of stability and insensitivity to parameters is taken into consideration, there must be simultaneous regulation of the filter control. At the present time, there is no systematic method available for obtaining a satisfactory compromise between robustness and performance. However, certain approaches, such as those proposed by Doyle, provide useful information for finding a solution.

iii) The problems of computer size or robustness may lead to the use of a controller of reduced dimensions based on the procedures of reduction of linear quadratic Gaussian controllers. The use of these techniques gives satisfactory results if they are combined with appropriate processing of the measurements, so that the hypotheses made concerning the nature of the noises may be checked as thoroughly as possible.

## *9.6 - CONCLUSION*

The linear quadratic Gaussian technique appears highly suitable for the control of a launch vehicle, a system whose behaviour is generally well known. In fact, the optimality of the linear quadratic Gaussian controller is essentially determined by the knowledge of the internal dynamics of the system to be controlled, and its efficiency is closely related to the precision of the model used.

The use of such a technique has the advantage of being highly adaptable to different launch vehicle configurations. The solution of the optimality equations, following the definition of the criteria, provides a systematic determination of a satisfactory control law having an identical structure for all versions of the launch vehicle envisaged.

Its application does not present any major difficulties, the only drawback being the problem of the choice of the criterion including the concepts of robustness. This problem, together with that of the reduction of the controller, has formed the subject of many studies in recent years, and, given the results obtained so far, it may be expected that it will be solved in the near future.

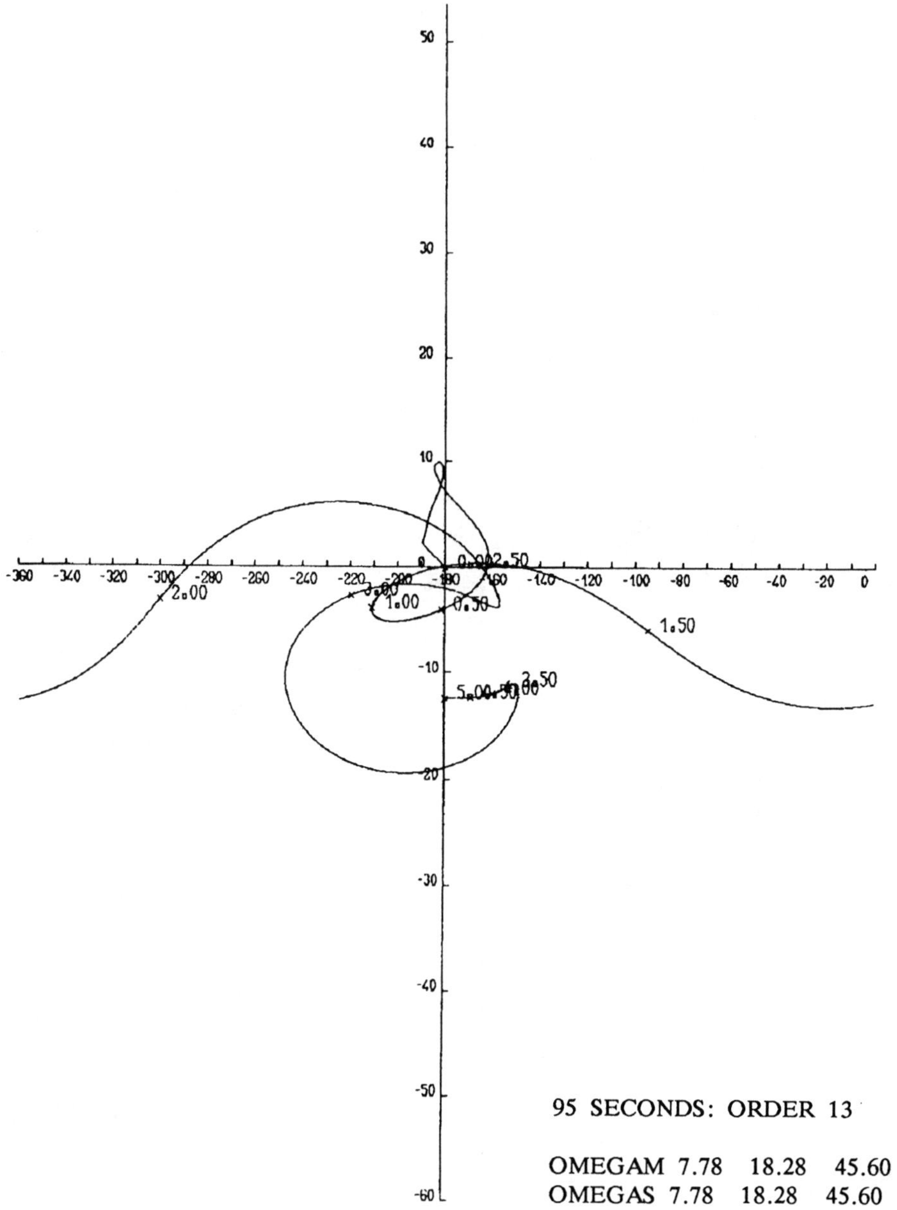


Black locus with nominal Kalman filter

Fig. 3

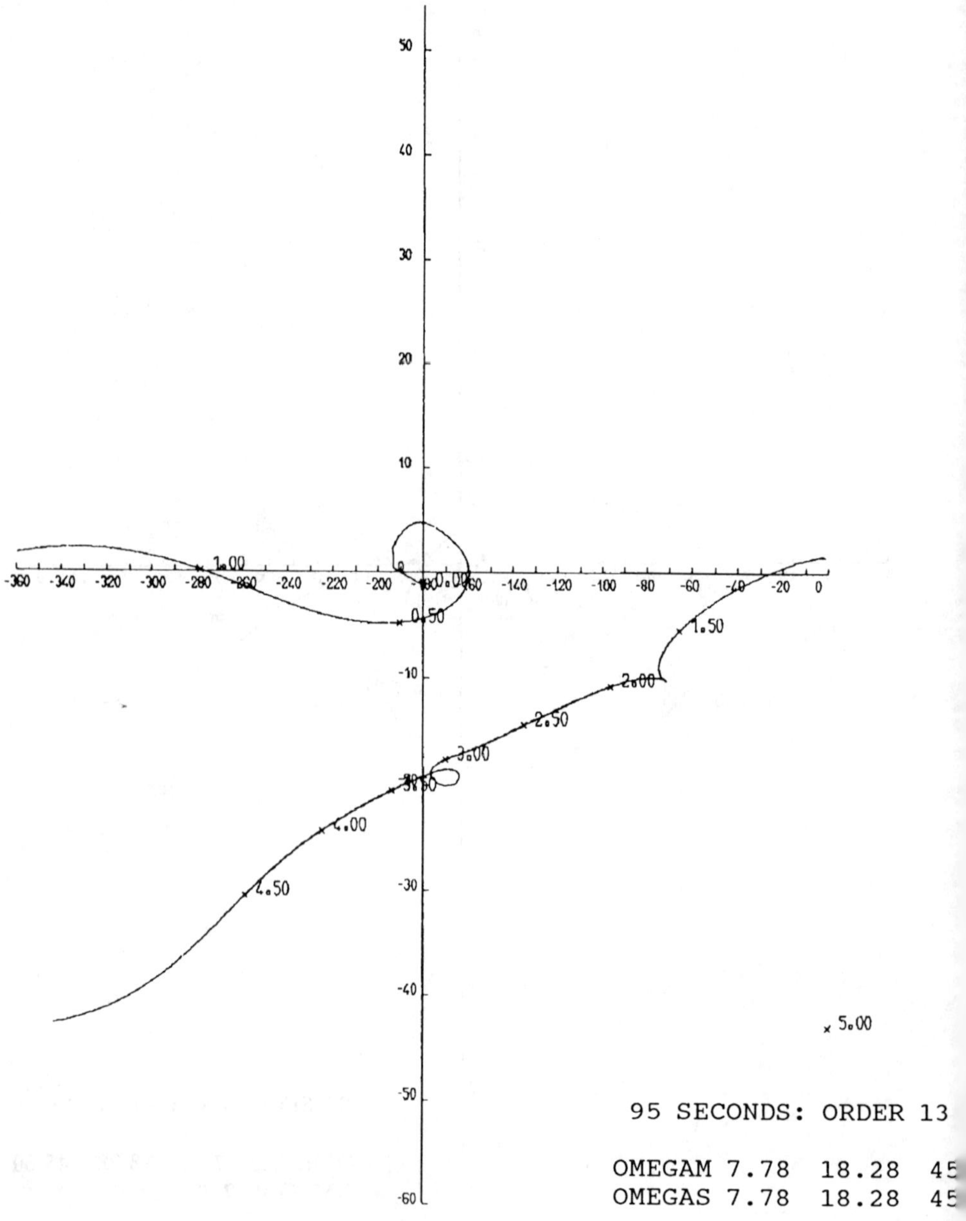


Black locus with modified Kalman filter

Fig. 4

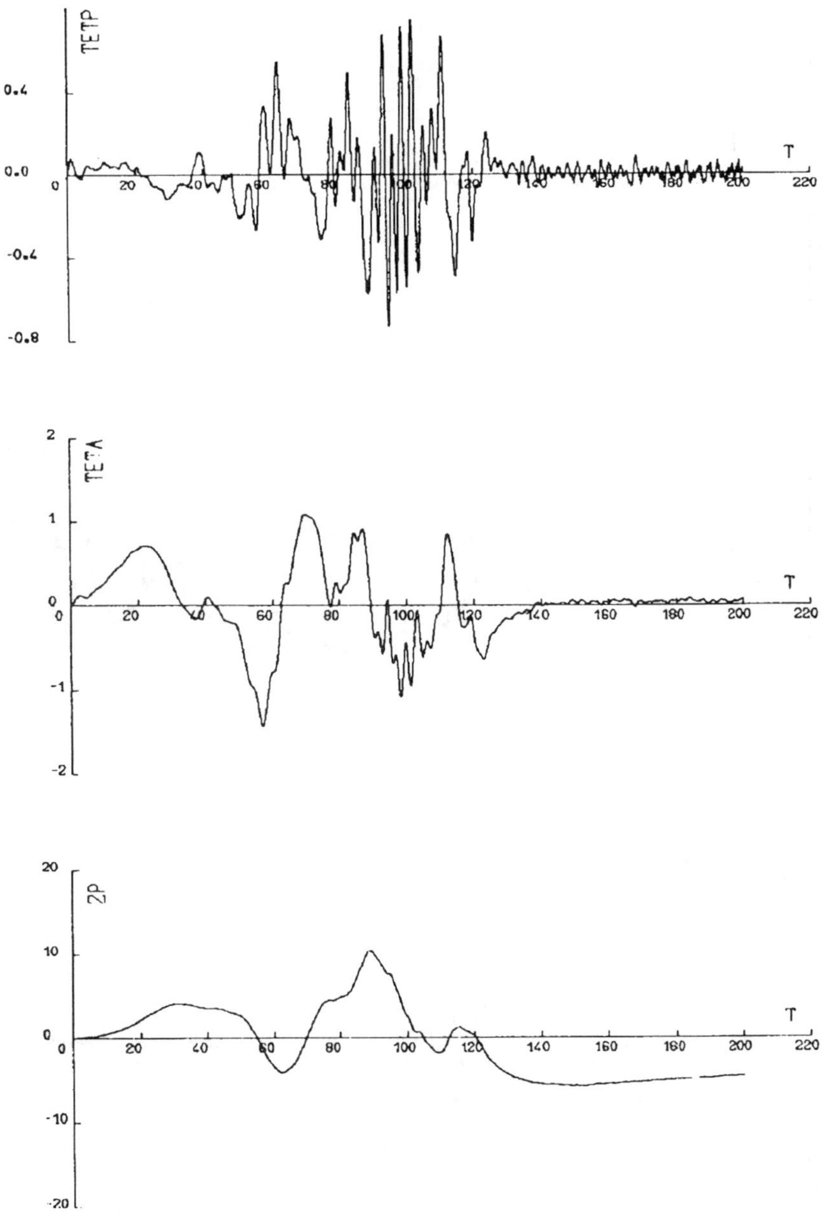


Simulation with prefiltering of the measurement (Cauer at 2.5 Hz) and controller of order 9

Fig. 5.a

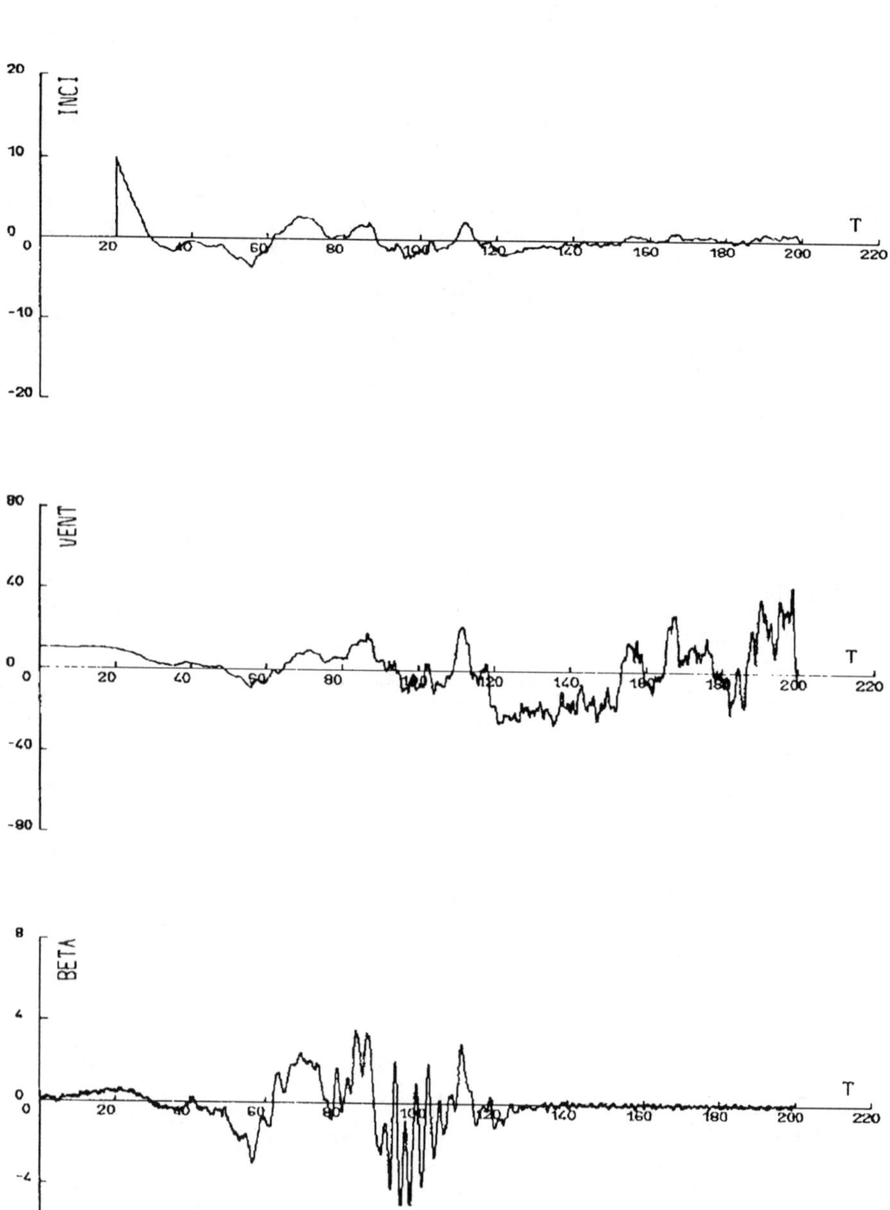
INCI
20
10
0
-10
-20
T
0
20
40
60
80
100
120
140
160
180
200
220
VENT
80
40
0
-40
-80
BETA
8
4
0
-4
-8

Fig. 5.b

***REFERENCES***

C. Champetier, N. Imbert, and A. Piquereau, Etude du pilotage numérique du premier étage d'ARIANE IV. Synthèse des lois de commande robustes, Contrat CNES, Rapport CERT/DERA No. 1/7340, February 1984

N. Imbert and A. Piquereau, Etude du pilotage numérique du premier étage d'ARIANE IV. Synthèse des lois de commande robustes, Contrat CNES, Rapport CERT/DERA No. 2/7340, June 1984

J.C. Doyle and G. Stein, Robustness with observers, IEEE Automatic Control, v. 24, No. 4, August 1979

J.C. Doyle and G. Stein, Multivariable feedback design: concept for a classical modern synthesis, IEEE Automatic Control, v. 24, No. 1, February 1981

R.L. Stapleford, L.G. Hofmann, et al., Transfer function approximations for large highly coupled elastic boosters with fuel slash, NASA, CR 464, April 1966

## *CHAPTER 10*

## *REDUCTION OF DISTURBANCES: APPLICATION TO PASSENGER COMFORT ON AIRCRAFT*

### *10.1 – INTRODUCTION*

Wind gust alleviation, or turbulence absorption in general, forms one of the main applications of active control in aeronautics. Aircraft are naturally subject to atmospheric disturbances which have undesirable effects not only on manoeuvring but also on the aircraft structures; turbulence absorption is the reduction of these effects by the co-ordinated operation of the control surfaces which produce aerodynamic forces or moments, in other words by an active control system.

The systems may be classified according to their objectives, which are as follows:

– Gust alleviation to improve the flight qualities and the maintenance of the aircraft trajectory, this being particularly important for safety reasons at the time of take-off and landing.

– Turbulence absorption, to reduce the acceleration at different positions of the aircraft and thus to improve the comfort of the pilot and passengers.

– The control of loads induced by turbulence, making it possible to reduce the loads and the structural fatigue, resulting in either a lightening of the structure or a lengthening of the wings, and in any case producing a cost saving.

These systems also differ in the ranges of frequencies within which they operate, and, consequently, in the control models used. The model considered may be based on:
– the conventional rigid body dynamics,
– or on a separation of the effects of turbulence on the fuselage, the wings and the tail,
– or on aeroelastic dynamics with structural and non-stationary aspects.

There are many similarities in the principles and the definitions of these anti-turbulence systems; this chapter, however, will be limited to the study of turbulence absorption for the improvement of passenger comfort in the case of a rigid aircraft.

This application is particularly relevant to aircraft of the "commuter" type such as the ATR 42, which have a low wing loading, are typically used for short journeys, and fly at lower altitudes where the turbulence is greater.

## *10. 2 – PRINCIPLE OF TURBULENCE ABSORPTION FOR THE IMPROVEMENT OF PASSENGER COMFORT*

Many studies have been carried out, particularly in the United States, on the comfort and satisfaction of airline passengers, and these have shown that passengers are primarily sensitive to the vertical and lateral acceleration to which they are subjected in an aircraft. Fig. 1 shows the criterion of comfort as a function of the acceleration, and demonstrates that passengers are twice as sensitive to lateral acceleration as to vertical acceleration.

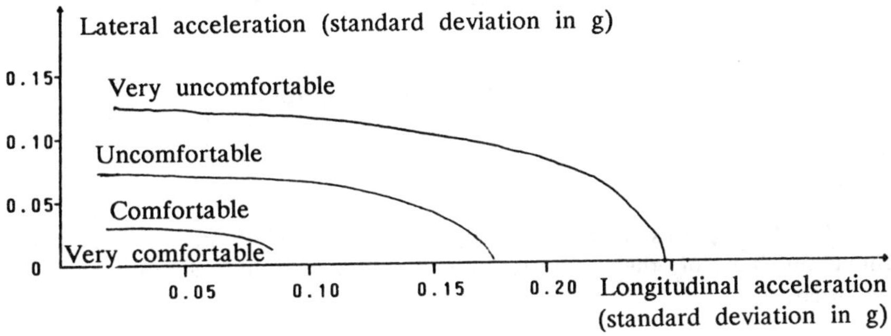


Passenger comfort as a function of accelerations

Fig. 1

When an aircraft is manoeuvred in a turbulent atmosphere, the aerodynamic forces and moments fluctuate, resulting in accelerations which are unpleasant for the pilot and passengers. To improve comfort and reduce these accelerations, the forces and moments created by the turbulence must be compensated by forces and moments created by the control surfaces. This must be achieved without modifying the flight performance in relation to the pilot inputs. The control surfaces generally used are, in the longitudinal direction, the ailerons or spoilers operating symmetrically and the elevators, and, in the lateral direction, the ailerons operating normally and the rudder. Depending on their inclination, the canards can generate lateral or vertical forces or a combination of the two. In order to be effective, these control surfaces must be actuated by sufficiently rapid servo control systems.

In the case of longitudinal motion, which will be studied in more detail, the direct lift force is created by the symmetrically operated ailerons and the moment about the centre of gravity is created by the conventional elevator. Since the aircraft is assumed to be a rigid structure, the degree of comfort is characterised by two accelerations along the fuselage, or by the equivalent values of the acceleration at the centre of gravity and the pitch acceleration.

There are two main techniques for the control of the surfaces, as follows:

- The open loop technique which requires a turbulence measuring device from which the control law for the control surfaces is generated (Fig. 2):

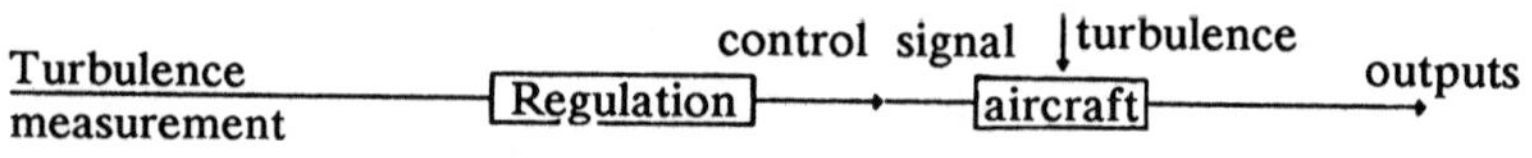


Open loop control

Fig. 2

In theory, this technique has the advantage of not changing the flight qualities of the aircraft; only its response to the turbulence is modified.

- The closed loop technique, in which the control is based directly on measurements made on board the aircraft (Fig. 3).

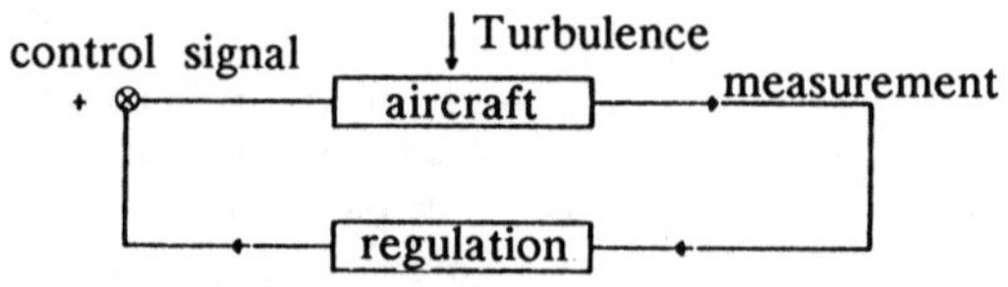


Closed loop control

Fig. 3

A hybrid structure may be obtained by combining the two above systems.

The determination of the corresponding control laws requires the modelling of the behaviour of the aircraft in a turbulent atmosphere and a representation of atmospheric disturbances.

## 10.3 – MODEL OF A RIGID AIRCRAFT IN A TURBULENT ATMOSPHERE

For the study of a system which absorbs vertical gusts, it is sufficient to examine the small longitudinal movements of the aircraft over a short period, in other words the equations of the oscillation of the angle of attack obtained when the longitudinal velocity is assumed to be constant. In fact, the variations of velocity and the phugoid mode are slow enough to be controlled either by the pilot or by the automatic pilot.

The equations for the aircraft in a calm atmosphere are obtained by writing the lift and moment equations, in the conventional form:

$$\dot{\gamma} = \frac{1}{2}\,\rho\,\frac{SV_0}{m}\left[\,C_{z\alpha}\,\alpha + C_{z\dot{\alpha}}\,\frac{1}{V_0}\dot{\alpha} + C_{zq}\,\frac{lq}{V_0} + C_{z\delta q}\delta q + C_{z\delta A}\,\delta_A\,\right]$$

$$\dot{q} = \frac{1}{2}\,\rho\,\frac{SV_0 l}{B}\left[\,C_{m\alpha}\,\alpha + C_{m\dot{\alpha}}\,\frac{1}{V_0}\dot{\alpha} + C_{mq}\,\frac{lq}{V_0} + C_{m\delta q}\delta q + C_{m\delta A}\,\delta_A\,\right]$$

$$\dot{\theta} = q$$

$$\gamma = \theta - \alpha$$

where $V_0$ represents the velocity of the aircraft, m is the mass, S is the surface, l is the standard mean chord, B is the moment of inertia, $\rho$ is the density of the air, $\gamma$ is the inclination, $\theta$ is the attitude, q is the pitch rate, $\alpha$ is the angle of attack, $\delta q$ is the inclination of the elevator, and $\delta_A$ is the inclination of the symmetrical ailerons. $C_{zx}$ and $C_{mx}$ are the aerodynamic derivatives which express the variations of the lift and moment coefficients for variations of the parameter x.

The control surfaces are actuated by servo control signals modelled by a first order transform:

$$\frac{\delta}{\delta_c} = \frac{1}{1 + T_S s}$$

In general, the atmospheric disturbances are introduced as variations of the angle of attack, of its derivative, and of the pitch rate in the expression of the aerodynamic forces:

$$\alpha_a = \alpha + \alpha_{vc}$$

$$\dot{\alpha}_a = \dot{\alpha} + \dot{\alpha}_{vcf}$$

$$q_a = q - \dot{\alpha}_{vcf}$$

with $\alpha_{vc} = W_c/V_0$, where $W_c$ is the vertical velocity of the wind at the centre of gravity of the aircraft, and $\alpha_{vcf}$ is the angle of attack due to the wind, output of the following filter:

$$\alpha_{vcf} = \frac{\alpha_{vc}}{1+T_\alpha s}$$

The variation of the pitch rate and the time constant $T_\alpha$ show the effect of turbulence penetration.

This classical modelling of the effect of turbulence, which is used for the study of automatic pilots, is found to be inadequate at high frequencies. It is preferable for the study of turbulence absorption systems to consider the variations of the velocity of the disturbances along the aircraft and to model separately their effects on the wings and on the tail unit, or, better still, to examine separately the nose of the aircraft, the central part with the wings, and the tail with the horizontal plane.

In this case, the respective local angles of attack are:

$$\alpha_1(t) = \alpha - \frac{ql_1}{V} + \frac{W_1}{V}$$

$$\alpha_2(t) = \alpha - \frac{ql_2}{V} + \frac{W_2}{V}$$

$$\alpha_3(t) = \alpha - \frac{ql_3}{V} + \frac{W_3}{V} - \epsilon\,\alpha_2(t-\tau)$$

where $w_i$ and $l_i$ are the vertical velocity of the wind to which the element i is subjected, $l_i$ is the distance of the element from the centre of gravity, and $\epsilon$ is the wing deflection. The corresponding aerodynamic coefficients are $C_{z\alpha i}$ and $C_{m\alpha i}$.

The turbulence itself is assumed to be homogeneous, isotropic, stationary, and characterised by its spectrum. Two forms of spectrum are used, namely the Karman and the Dryden spectra; the latter has the advantage of being in a rational analytic form.

For the vertical turbulence, the Dryden spectrum of the vertical velocity is given by

$$\Phi_W(\omega) = \sigma_W^2 \frac{L_W}{V_0} \frac{1 + 3\left(\frac{L_W}{V_0}\omega\right)^2}{\left[1 + \left(\frac{L_W}{V_0}\omega\right)^2\right]^2}$$

where

$$\sigma^2 = \frac{1}{2\pi} \int_{-\infty}^{+\infty} \Phi_W(\omega)\, d\omega$$

and where $L_W$, the length, and $\sigma_W$, the standard deviation of the turbulence, are functions of the altitude.

Anti-turbulence systems are also tested on gusts, particularly the cosine type expressed by

$$W(t) = \frac{W_M}{2}\left(1 - \cos \frac{2\pi V t}{d_M}\right)$$

where $W_M$ and $d_M$ represent the amplitude and length of the gust.

To these models must be added the measurement equations, consisting of the local angle of attack, $\alpha_S(t)$, the pitch rate, q, and the normal accelerations, $N_{ZP}$.

$$N_{ZP} = \frac{V_0}{g} \dot{\gamma} + \frac{X - X_{CG}}{g} \dot{q}$$

where X and $X_{CG}$ are the positions of the accelerometer and of the centre of gravity respectively.

The complete model integrating the aircraft, the disturbances, and the measurements has an analytic form including the effects of delays which are a function of the aircraft velocity. For the frequency range in question, these time delays can be approximated with a good degree of precision by the first order Padé approximation:

$$e^{-\tau s} \# \frac{1 - \tau s/2}{1 + \tau s/2}$$

Note: A discrete model would enable these delays to be taken into account, but would require a higher-order representation which would also be variable, since the delays depend on the velocity of the aircraft.

Finally, with the approximation of the delays, the model can be written in the standard form:

$$\dot{X} = A X + B u + v$$

$$S = C X + D u$$

where

X represents the overall state of the system, consisting of the state of the aircraft ($\alpha$, q), the state of the servo controllers ($\delta q$, $\delta A$), and the state of the wind, the dimension of which depends on the number of elements chosen for the representation;

u denotes the control signals to the ailerons, $\delta_{AC}$, and to the elevator, $\delta_{Pc}$;

v is a white noise process;

S is the vector of the outputs, comprising the measurements $\alpha_{aS}$, q, $N_{ZCG}$, and the accelerations at different points of the aircraft, for the evaluation of the degree of comfort.

## *10.4 – DETERMINATION OF THE CONTROL LAWS*

Among the different control techniques, that of stochastic optimal control with a quadratic criterion (linear quadratic Gaussian: LQG) appears to be particularly well suited to this problem of regulation of an aircraft subjected to random atmospheric disturbances.

The objective of reducing the accelerations along the fuselage, whilst complying with the constraints on the inclination and on the speed of inclination of the control surfaces, may be formulated by a criterion of deviation in the following form:

$$C = \lim_{T\to\infty} \frac{1}{2T} \int_{-T}^{+T} (\lambda_1 \dot{q}^2 + \lambda_2 N_{ZCG}^2 + \lambda_3 \delta q_c^2 + \lambda_4 \delta \dot{q}_c^2 + \lambda_5 \delta A_c^2$$

$$+ \lambda_6 \delta \dot{A}_c^2) dt$$

The choice of the parameters $\lambda_1 \ldots \lambda_6$ enables a compromise to be made between the degree of comfort (characterised by the acceleration at the centre of gravity, $N_{ZCG}$, and the angular acceleration, $\dot{q}$) and the amount of control surface movement (characterised by the controlled inclinations and speeds of inclination, $\delta q_c$, $\delta \dot{q}_c$, $\delta A_c$, $\delta \dot{A}_c$.

Although gust alleviation systems are implemented in digital form, as is usual with aeronautical control systems, they are designed in the continuous domain and subsequently digitised for implementation. This solution is justified by the fact that the sampling rates are relatively high.

Conventionally, the control laws are determined either by the Wiener method or by the linear quadratic Gaussian method, depending on whether the open loop or closed loop approach is used. In fact, as section 10.5 shows, the LQG method may be used for either of these two systems.

In the absence of constraints on the derivatives of the control signals, the minimisation of C can be expressed as a standard LQG problem as follows:

$$C = \lim_{T\to\infty} \frac{1}{2T} \int_{-T}^{+T} (X^T Q_1 X + u^T Q_2 u)dt$$

In steady state conditions, the control signal is a constant gain state feedback:

$$u = -L \hat{X}$$

whose matrix L is obtained by the solution of a Riccati equation, and where $\hat{X}$ is the estimate of X provided by a Kalman filter.

If it is found to be necessary to introduce constraints on the derivative of the control signals, the standard LQG method may be used, assuming a velocity control of the system:

$$\begin{bmatrix} \dot{X} \\ \dot{u} \end{bmatrix} = \begin{bmatrix} A & B \\ 0 & 0 \end{bmatrix} \begin{bmatrix} X \\ u \end{bmatrix} + \begin{bmatrix} 0 \\ 1 \end{bmatrix} \dot{u} + \begin{bmatrix} 1 \\ 0 \end{bmatrix} v$$

The optimal control signal is then

$$\dot{u} = -(L_1 \; L_2) \begin{bmatrix} \hat{X} \\ u \end{bmatrix}$$

in other words a filtered control signal:

$$\frac{u}{\hat{X}} = \frac{-L_1}{p+L_2}$$

The use of a Kalman filter to estimate the state in LQG control requires the use of regulators, of an order at least equal to that of the aircraft, which appear to be unnecessarily complex. These regulators can be considerably simplified without any degradation in performance by means of a technique of identification or model reduction. Since the criterion used for the simplification is generally different from the quality criterion of the gust alleviation system, the reduced order regulator must be a good approximation to the optimal regulator.

The ideal solution is to find directly the reduced fixed order optimal regulator or the dynamic output feedback which minimises the quality criterion. This technique, which ultimately forms a problem of optimal non-linear optimisation, has the disadvantage of being more difficult to implement than LQG.

## *10.5 – CONTROL STRUCTURES*

With the control law and performance of the system determined from theoretical considerations, various control structures may be envisaged.

LQG is generally used in a closed loop control system consisting of a Kalman filter together with a state feedback gain matrix (Fig. 4):

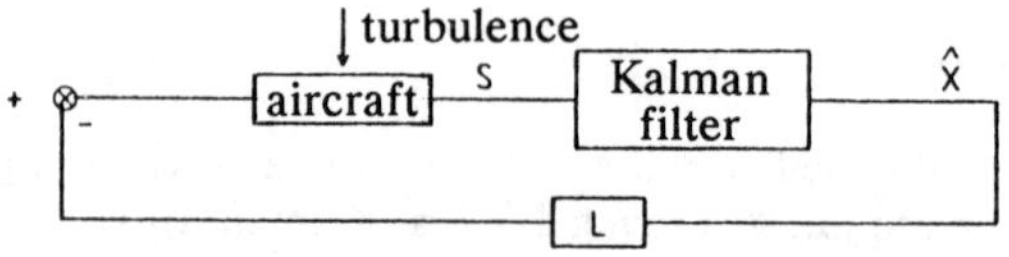


Linear quadratic Gaussian closed loop control

Fig. 4

This structure is modified or simplified according to the measurements available and the values of the various disturbances. If all or part of the state can be measured with a negligible measurement noise, the Kalman filter is of a reduced order. If the measurement of turbulence is available and the aircraft is stable, the solution is an open loop control structure. In this case, the Kalman filter is reduced to a simulation of the aircraft subjected to the turbulence (Fig. 5).

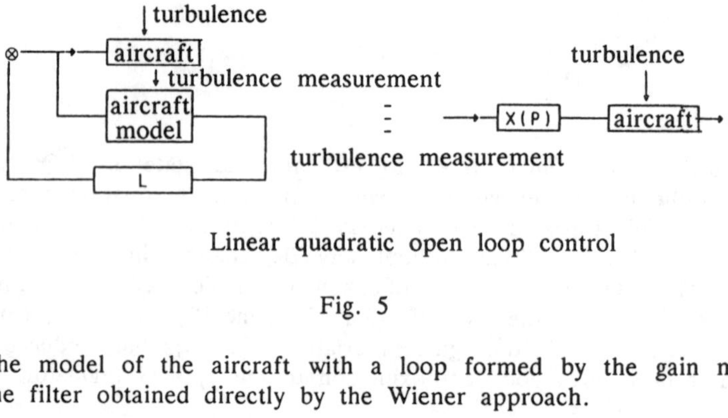


Linear quadratic open loop control

Fig. 5

The model of the aircraft with a loop formed by the gain matrix L is the filter obtained directly by the Wiener approach.

In theory, this open loop controller does not modify the modes of the aircraft and therefore does not alter its behaviour in relation to other inputs, in particular the pilot input. This interesting property enables the problem of turbulence absorption to be decoupled from the pilot control of the aircraft.

In practice, the turbulence estimation is based on measurements made on the aircraft, and the system is not really open-loop. To retain the open-loop properties, the disturbance measurement must be independent of the dynamics of the aircraft. For this purpose, the measurement of the longitudinal turbulence is obtained from a dynamic combination of measurements of the normal acceleration $N_Z$, the pitch rate q, and the angle of attack $\alpha_a$, at one point of the aircraft. The equations for measuring these parameters at the point S situated at a distance $X_S$ in front of the centre of gravity are thus written as:

$$N_{ZS} = \frac{V_0}{g} (q - \dot{\alpha}) + \frac{X_S}{g} \dot{q}$$

$$\alpha_a = \alpha - \frac{X_S}{V_0} q + \alpha_v$$

The estimate of the vertical velocity of the turbulence $W_V$, or the equivalent variation of the angle of attack $\alpha_V = W_V/V_0$ due to the wind, may be obtained with a Kalman filter which allows for the measurement noises of the different sensors and for the spectrum of disturbances.

The gust alleviation system and the pilot control can also be decoupled by a total closed loop control system as shown in Fig. 6.

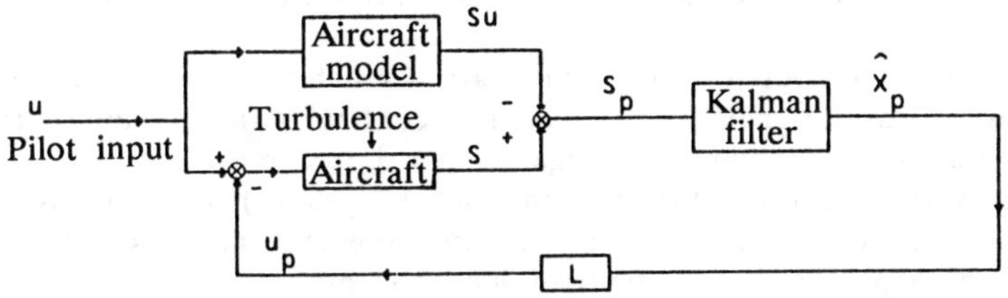


Closed loop turbulence control, decoupled from pilot input

Fig. 6

In this structure, where $S_u$ and $S_p$ represent the outputs of the aircraft due respectively to the input u and the disturbances p, in the absence of atmospheric disturbances the closed loop controller of the disturbances has no effect; moreover, it has no effect on an input, such as that from the pilot, which is independent of the disturbances.

Even if a measurement of turbulence is available, it is useful to add a closed loop controller, using, for example, the structure of a model reference system (Fig. 7), to the open loop controller to compensate for imperfections in modelling.

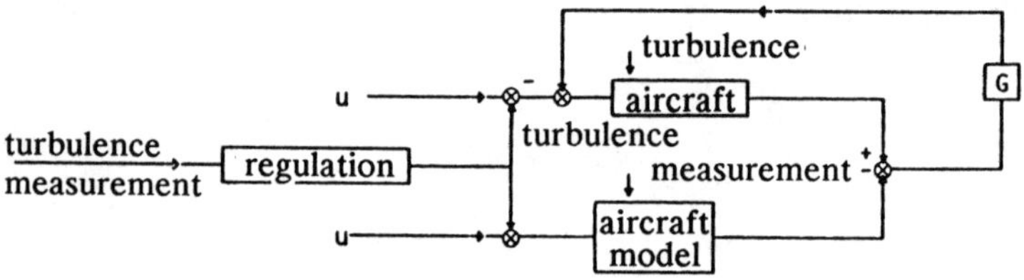


Controller with reference model

Fig. 7

If, from a theoretical viewpoint, all these controllers are equivalent and provide equal performance in response to turbulence, the structure is chosen in accordance with considerations of complexity, sensitivity to uncertainties of the model, and measurement noises.

## *10.6 – DIGITAL IMPLEMENTATION OF THE CONTROLLER*

Numerous methods of digitisation are available for transforming a continuous control signal into a digital control signal. The bilinear algebraic transform has the advantage of not introducing any time delay (which is very important for anti-turbulence systems) and of enabling the different systems (such as the gust measurement and the controller in the open loop case) to be digitised independently.

The bilinear algebraic method comprises the replacement of the continuous controller p in the expression by (2/Dt)(Z–1/Z+1), or the integration of the state of the controller by the trapezoidal rule, which is equivalent.

Thus, starting with the following open loop controller:

$$\left| \begin{array}{l} \dot{X}_R = A_R X_R + B_R \alpha_{vm} \\ u = C_R X_R + D_R \alpha_{vm} \end{array} \right.$$

where $\alpha_{vm}$ is the measurement of the gusts, the trapezoidal rule integration leads to:

$$X_R(n+1) = X_R(n) + A_R \frac{\Delta t}{2} [X_R(n+1)+X_R(n)] + B_R \frac{\Delta t}{2} [\alpha_{vm}(n+1)+\alpha_{vm}(n)]$$

$$u(n) = C_R X_R(n) + D_R \alpha_{vm}(n)$$

$$X_R(n+1) = A_1 X_R(n) + B_1[\alpha_{vm}(n+1) + \alpha_{vm}(n)]$$

$$u(n) = C_R X_R(n) + D_R \alpha_{vm}(n)$$

where

$$A_1 = (I - A_R \frac{\Delta t}{2})^{-1} (I + A_R \frac{\Delta t}{2})$$

and

$$B_1 = (I - A_R \frac{\Delta t}{2})^{-1} B_R \frac{\Delta t}{2})$$

It is preferable to write this equation in the state form:

$$Y(n+1) = A_{RD} Y(n) + B_{RD} \alpha_{vm}(n)$$

$$u(n) = C_{RD} Y(n) + D_{RD} \alpha_{vm}(n)$$

specifying that

$$Y(n) = X_R(n) - B_1 \alpha_{vm}(n)$$

## *10.7 - EXAMPLE OF APPLICATION - CONCLUSION*

The proposed method has been tested in a simulation for an aircraft of the NORD 262 type whose ailerons, when symmetrically inclined, produced a direct lift force. The ailerons and elevator are actuated by servocontrol systems with a time constant of 0.125 seconds, their inclination is limited to $10^o$, and their rate is limited to $50^o$/s. This aircraft has an $\alpha$ vane at the front of the fuselage, a pitch rate sensor, and an accelerometer, enabling the vertical velocity of the wind to be measured at the nose of the aircraft and allowing an open loop controller to be used.

Figs. 8 and 9 show the time response behaviour of the aircraft subjected to vertical turbulence with a standard deviation of 1.8 m/s with and without absorption. ALFV represents the variation of the angle of attack due to the wind, in degrees, NZCG represents the vertical acceleration at the centre of gravity in g, QP represents the angular pitch acceleration in degrees per second$^2$, and DAC and DQC are the controlled aileron and elevator angles.

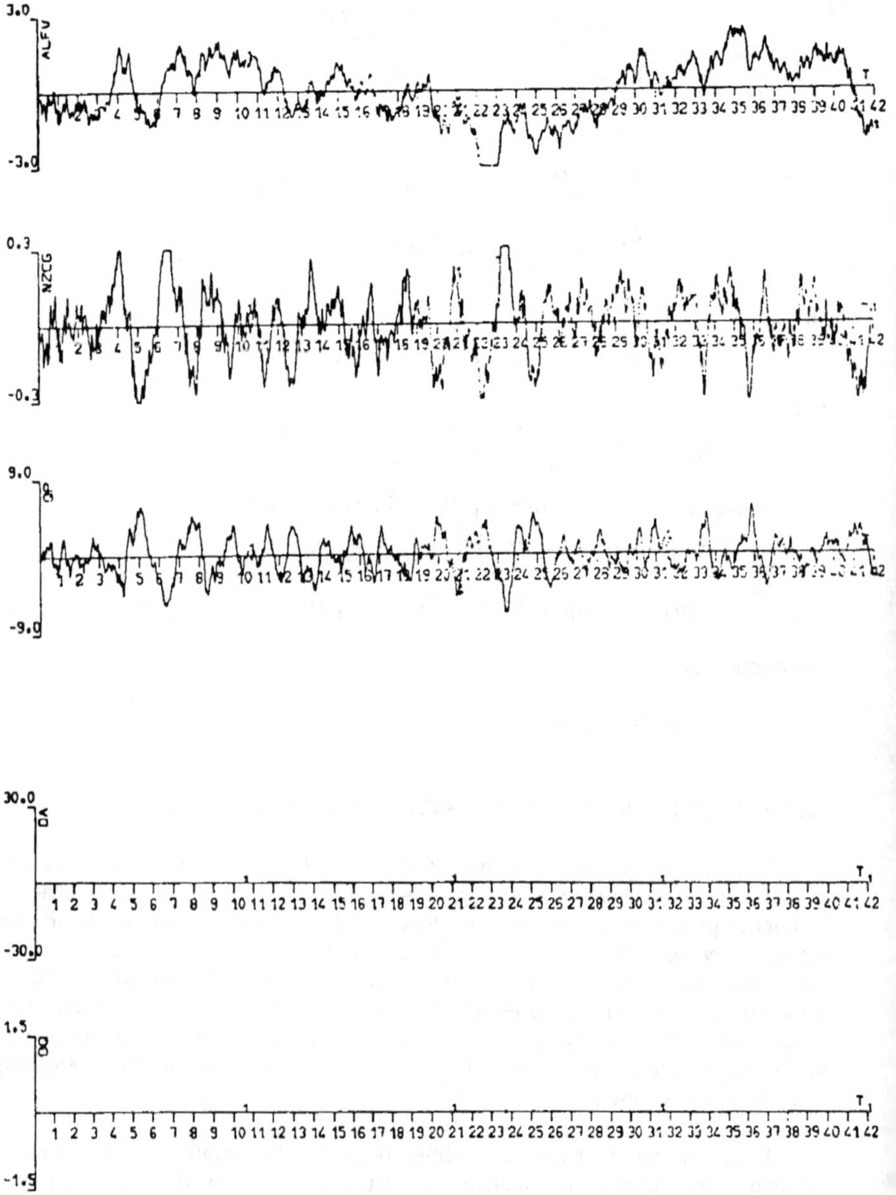


Behaviour of the aircraft without anti-turbulence system

Fig. 8

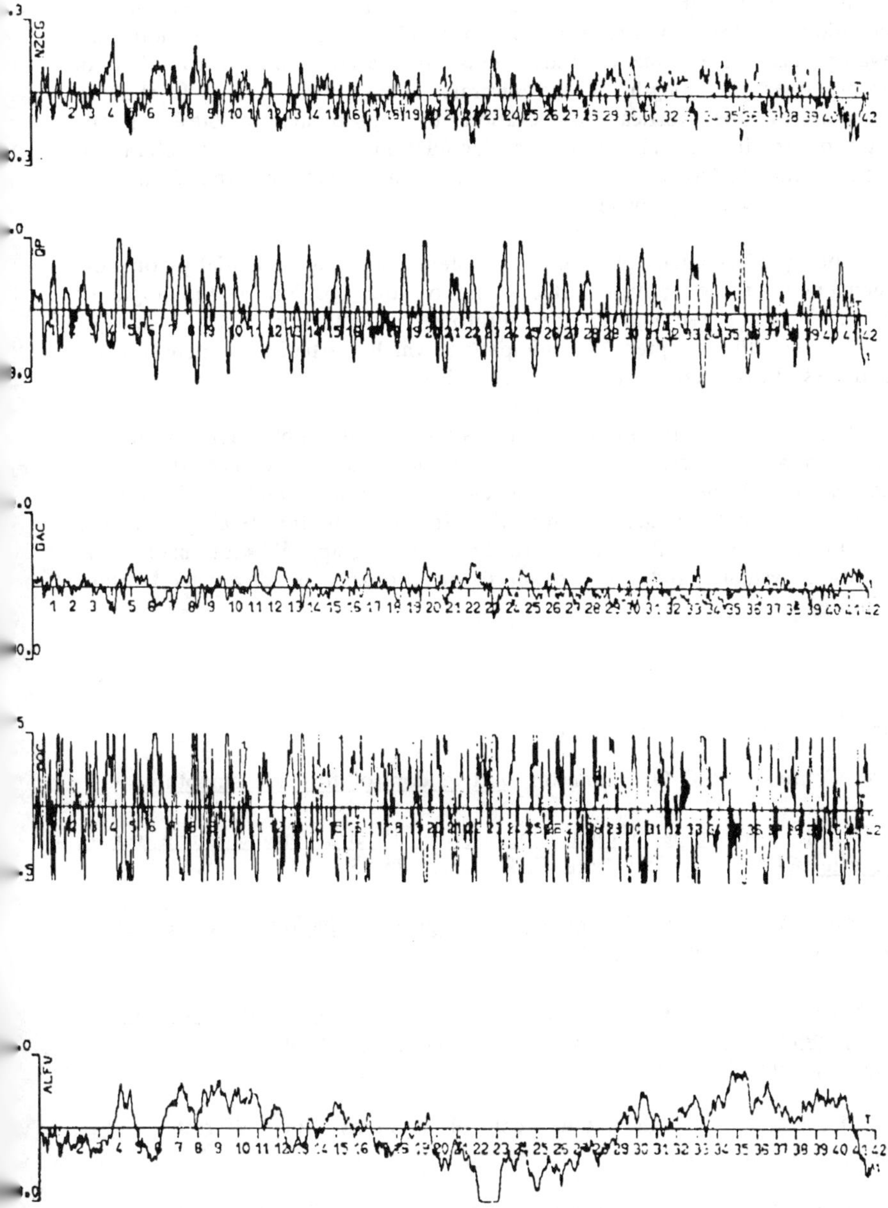


Behaviour of the aircraft with an anti-turbulence system

Fig. 9

The measurement of gusts and the controller are both digital, with a sampling frequency of 30 Hz. In the development of the controller, the most constraining factors were found to be the limitations on the inclination of the ailerons and, in particular, the rate limitation; the elevator has much smaller movements and affects the distribution of comfort over the length of the aircraft. Thus Figs. 8 and 9 show a reduction of the vertical acceleration by approximately 50% at the centre of gravity, associated with a degradation in the pitch acceleration, which is manifested in a difference in comfort between the front and the rear, but which provides a better overall performance.

The performance of such a system depends essentially on the effectiveness of the control surfaces and on the position of the angle of attack sensor in relation to the aerofoils, providing a measurement of disturbance with a degree of anticipation which partially compensates for the delays introduced by the control devices.

Thus the absorption of gusts is a useful application of active control in aeronautics, to the extent that it does not require any major modification of the aircraft and enables the comfort and satisfaction of passengers to be increased. Some aircraft, such as the Boeing 747 and the Dornier 28 TNT, have already been equipped with such gust alleviation systems, and these systems will probably be developed further in future.

## *REFERENCES*

D.W. Conner and G.O. Thomson, Active control for ride smoothing

O. Herail, Commande optimal stochastique d'ordre réduit. Etude sur la maquette ATA, Technical Note CERT, No. 85/83

B. Krag, Active control technology for gust alleviation, Lecture series, Von Karman Institute, Dec. 1978

G. Krag and H. Wunneberg, Olga, a gust alleviation system for improvement of passenger comfort of general aviation aircraft, ICAS Congress, Munich, 1980

D. McLean, Gust alleviation control systems for aircraft, IEEE Proceedings, July 1978

J.L. Pac, J.L. Boiffier, and M. Labarrère, Amélioration du confort passager. Etude sur maquette ATA, Technical Note CERT DERA No. 228/82 and report STPA/DERA No. 67/83

Quinton P. Poisson and J.C. Wanner, Evolution de la conception des avions grâce aux commandes automatiques généralisées, Congrès SEE, Grenoble, 1977

E.G. Rynaski, Gust alleviation - criteria and control laws, AIAA Conference - Boulder, Colorado, August 1979

E.G. Rynaski, D. Andrisani, and B.J. Eulrich, Gust alleviation using direct turbulence measurements, AIAA Conference - Boulder, Colorado, August 1979

# CHAPTER 11

# APPLICATION OF INTERNAL MODEL CONTROL TO THE AUTOMATIC STEERING OF A SHIP

## 11.1 - INTRODUCTION

This chapter is concerned with the application of methods of process control by internal model, and more particularly with an application to the automatic steering of a ship.

This application (*) begins with a feasibility study of the control law used in a realistic simulation of the ship, and this is followed by the design and construction of the automatic pilot and finally the sea trials.

## 11.2 - GENERAL DESCRIPTION

The purpose of this study was to define the characteristics of an automatic pilot suitable for the special case of ships required to carry out special missions in a highly constrained environment. An example of this is the case of surface ships which must be manoeuvred very precisely for surface or submarine research, exploration, and operations.

Since the performance of currently available automatic pilots was inadequate, the decision was made to study a new form of automatic pilot, whose main claim to originality was that it made maximum use of an a priori knowledge contained in the mathematical model representing the system to be steered.

---

* The development and scope of this application provide an example of fruitful co-operation among the Technical Departments concerned (DCAN, STCAN), a designer and manufacturer of automatic systems (GCA/ALCATEL), a research laboratory (ADERSA/GERBIOS), and the Direction des Recherches et Etudes Techniques (DRET).

This a priori knowledge, when carefully used, makes it possible to achieve optimal operation of the ship. The principal feature of the model is that it permits the prediction at any moment of the behaviour of the ship at any point in the future, and the calculation of the permissible instantaneous control signals enabling this development to be controlled.

Since the modelling of the ship will probably be imperfect, it is advisable to guard against differences in behaviour between the ship and its model. A control algorithm having this property will be described as "robust".

The two principal characteristics of the automatic pilot will therefore be that it makes use of prediction and provides robustness.

The first application was to the tripartite (*) programme of mine hunter construction. The mine hunter developed by co-operation between France, Belgium and the Netherlands is a ship of 550 tonnes, with a length of 50 metres, and provided, for reasons of noise reduction and manoeuvrability, with two independent propulsion systems as follows:

- a "main" propulsion system with a diesel engine for movement up to a maximum speed of 15 knots, and for mine sweeping at 8 knots;

- an "auxiliary" propulsion system for mine hunting, usable from 0 to 7 knots, comprising two active rudders.

A purely transverse-acting bow propeller complements the action of the active rudders in mine hunting.

The objective of the automatic pilot of the tripartite mine hunter (PACT) is to control the movement of the ship during the hunting, identification, and destruction of mines, as well as in free travel and in various manoeuvres.

To perform the different functions required of it, PACT must provide the following modes of operation:

---

* A programme for the construction of forty units, distributed among the three navies (French, Belgian, and Dutch) participating in the project.

1 – Aided manual operation: the operator inputs the force and torque references to the ship from a portable control panel. In this mode, PACT formulates and transmits to the actuators the commands which will enable the ship to follow the references given.

2 – Course keeping: in this mode, PACT must enable the ship to maintain its course in accordance with the reference given, using only the rudder. The ship is powered by the main propulsion system whose control is provided by conventional means external to PACT.

3 – Track keeping: in this mode, PACT must control the active rudders to enable the ship to follow an imaginary rail in relation to the sea bed. This mode is used in the mine detection phase.

4 – Fixed point: in this mode, PACT must control the active rudders and the bow propellers in such a way as to keep the ship in a fixed position at a particular point defined in relation to the sea bed.

PACT is designed as a multiple microprocessor system, assisted by rapid calculation units. It is modular and functionally independent of the other navigation aids. The manufacture, installation, and maintenance of the system have been undertaken by CGA-ALCATEL.

## *11.3 – DESCRIPTION OF THE PACT SYSTEM*

### 11.3.1 – FUNCTIONAL CHARACTERISTICS OF PACT

The mathematical model of the ship, described in the form of multivariable non-linear state equations, is developed in accordance with the characteristics of the ship and of the propulsion systems. The structure of the model is also valid for any operating mode with the auxiliary propulsion system.

The formulation of the constraints which affect the actuators, together with the operational constraints which determine the objectives, will enable us to completely define the problem which the control law is required to solve.

#### 11.3.1.1 – **Characteristics of the ship and its propulsion system**

Detailed calculations (see Reference No. 6) show that the ship can be represented by the following system of state equations:

$$X_{n+1} = F(X_n, PERT_n) + G \cdot u_n \quad (1)$$

$$Y_{n+1} = H(X_{n+1}) \quad (2)$$

$$U_n = K(X_n, ACT_n) \quad (3)$$

a) - **The state vector**

The choice of the state vector X makes it possible to follow the motion of the ship in longitudinal and lateral translation and in rotation. It should be noted that the study is carried out for the horizontal plane, the compensation for the effects of rolling and pitching being provided at the level of measurement at this stage of the study. The modelling of the pitching and rolling forms part of the planned future developments (see section 5).

The components of the state vector are:

X(1) = x, abscissa of the centre of gravity in an earth reference system in metres

X(2) = y, ordinate of the centre of gravity in an earth reference system, in metres

X(3) = K, course of the ship in radians

X(4) = u, longitudinal component of the speed in relation to the water, in m/sec

X(5) = v, lateral component of the speed in relation to the water, in m/sec

X(6) = $K^{\circ}$, angular velocity in rad/sec.

Fig. 1 shows the location of the state variables and of the induced variables which can be calculated from the instantaneous values of the state vector.

b) - **The control variables**

The control variables $ACT_n$ are:

ACT(1) = DELTA, the angle of the active rudders in relation to the longitudinal axis of the ship, in radians (see Fig. 1)

ACT(2) = VGA, the speed of the active rudders in r.p.s.

ACT(3) = VPE, the speed of the bow propellors in r.p.s.

The positions of the control variables develop lateral and longitudinal forces composing the vector U(n), the input term of the inertial model (equation (1)):

U(1) = FX, the longitudinal force in kilonewtons (kN)

U(2) = FYAV, the lateral force developed at the position of the bow propellors, in kN

U(3) = FYAR, the lateral force developed at the position of the active rudders, in kN.

It should be noted that $U_n$ acts in a linear way on the state, since the matrix G has constant coefficients. This is important for the choice of the general structure of PACT.

c) - **The disturbances**

The components of the disturbance vector $PERT_n$ are the terms characterising the current, the wind, and the swell.

The current, characterised by its components in an absolute frame of reference, intervenes in an additive way in the first two state equations.

The wind intervenes through the forces created by it and acting on the ship. These are in the form $\alpha V^2$, where V represents the relative velocity of the wind with respect to the ship. The term $\alpha$ depends on the superstructures and is specific to each ship; it may be obtained from tables.

The swell has an effect comprising low-frequency forces and high-frequency movements (pitching, yawing). The high-frequency movements act as noises and are not taken into account in the model, unlike the low-frequency forces. The swell is characterised by its relative mean bearing, its amplitude, and its period.

d) - **The output variables**

The output variables must completely define the position of the ship.

Given the navigation system, the position of the ship is determined by the distance and the azimuth of the objective as well as by the course of the ship. In these conditions, the observation equation is non-linear.

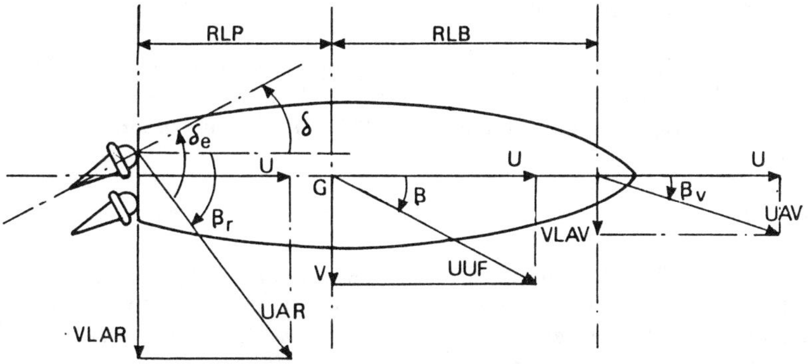


Figure 1- a

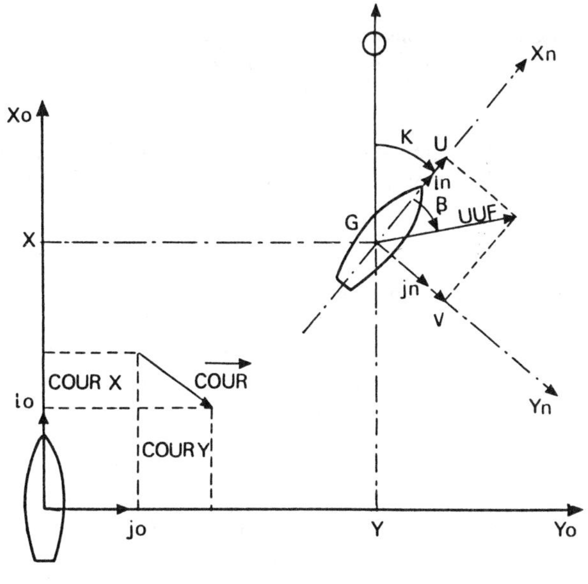


Fig. 1-b: Positioning

Fig. 1: Description of the state variables of the ship

### 11.3.1.2 - Constraints on the actuators

**a) - Inherent constraints**

Technical constraints associated with the nature of the propulsion units. Rotation speed limited (upper limit and lower limit) and limited gradient of change of regime.

Since the screws of the bow propellors and active rudders are driven by variable-frequency motors, they have an uncontrollable operating range corresponding to ±20% of the maximum speed about zero.

**b) - Noise constraints**

The noise limitation affects the nominal thrust of the active rudders, which is a function of the state of the ship (drift, velocity in relation to the water, helm angle). This limitation may form an additional constraint on the actuators.

**c) - Power constraints**

The power limitations affect each propulsion unit, as well as the total power demand; the total installed power is not sufficient to supply the propulsion units at their nominal power.

Table 1 shows the various constraints, emphasising the deliberate under-powering of the ship, which makes the task of PACT even more difficult.

### 11.3.1.3 - Operational constraints

Allowing for the performance of the instrumentation, some examples of which are given in Table 2, the performance of PACT must be as follows:

**a) - For course keeping:**

For a maximum sea state of 3, a wind of 7, and a current of 3 knots, the course of the ship must be maintained for more than 60% of the time within 1.5 degrees of the reference course and for more than 90% of the time within 4.5 degrees of the reference course.

In this mode, the ship is driven by the main propulsion unit. Its speed is between 7 and 15 knots. The rudder is the only available actuator (see Fig. 2.a).

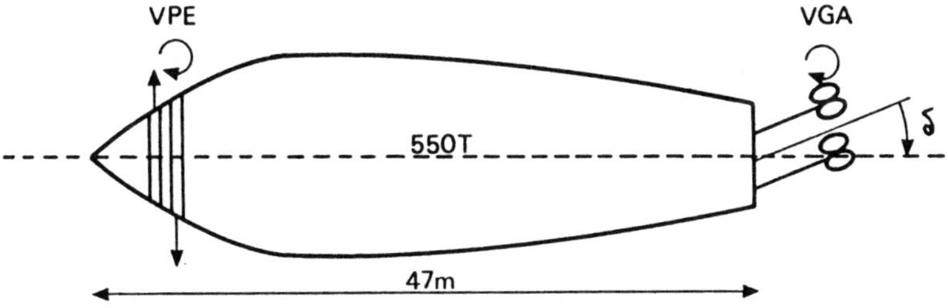


PROPULSION: (1) main propulsion unit 1400 Kw
(2) active rudders 2 x 90 Kw
(3) bow propellor 35 Kw

CONSTRAINTS: (1) acoustic 7 limit on power and acceleration

(2) limitation of installed power

| Maximum attainable forces: | | capable of being produced by a Force 7 wind | capable of being produced by a Force 5 swell |
|---|---|---|---|
| longitudinal | $F_X$ = 35 KN | 80 KN | 20 KN |
| lateral | $F_Y$ = 28 KN | 80 KN | 20 KN |
| torque | $F_N$ = 800 KN | 1000 KN | 100 KN |

(3) actuator constraints

| constraint | min | max | dead zone | velocity constraint |
|---|---|---|---|---|
| screw velocity, active rudders | - 3.75 | + 3.75 | -.833 à +.833 | .583 t/s$^2$ |
| screw velocity, bow propellor | - 7.083 | + 7.083 | -1.283 à + 1.283 | 1.133 t/s$^2$ |
| helm angle | - 1.18 | 1.18 | - | .136 rad/sec |

Table 1:

b) – **For track keeping:**

In the same sea conditions, the variation of position with respect to the imaginary rail and speed with which the rail is followed must be as specified in Fig. 2.b.

c) – **For the fixed point mode:**

The reference point of the ship must not move outside a circle with a radius of 35 m, centred at 145 m from the contact to be investigated. The course of the ship must never be outside the range defined by $+15^{o}$ and $-15^{o}$ from the reference course orientated in the direction defined by the centre of the circle to be maintained and the contact position (see Fig. 2.c).

### 11.3.1.4 – **Characteristics of the control law**

At this stage, the ship has been modelled, and the operating domain has been specified with respect to the constraints and the objectives.

The essential characteristics of the control law are as follows:

– Prediction: the control law uses the model to predict the behaviour of the ship in the near future.

– Resetting: the prediction is renewed at each sampling instant n from the measured state of the ship.

The model has already been described. In this form, equation (1) represents the inertial model, and equation (3) represents the model of the actuators. This separation between the system of force creation and the system showing the effects of these forces means that:

– it is only necessary to make partial updates when the propulsion system is changed;

– use can be made of the linear dependence between the forces and the state: X is a linear function of U.

However, this has the disadvantage of requiring the transformation of the well-known physical constraints affecting the actuators into force constraints. This transformation requires the use of approximations to avoid exceeding the actuator constraints, and therefore the constraints on the forces are more restrictive than necessary.

1) location systems

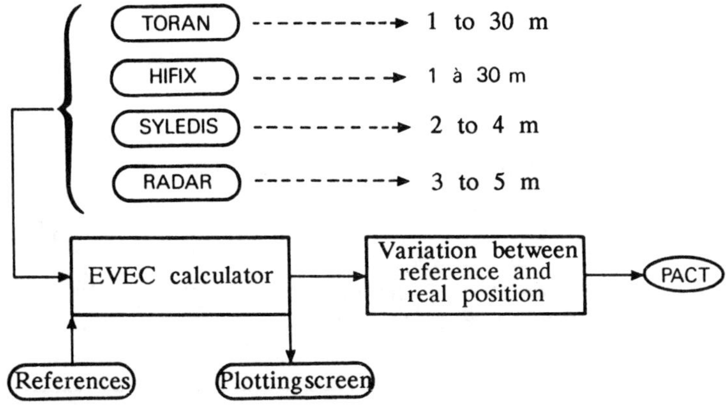


2) Measurement systems

| Value | Instrument | Specified accuracy |
|---|---|---|
| surface speed, speed relative to sea bed | LOCH DOPPLER | ± 0.05 à ± 0.22 nd |
| course | gyrocompass | ± 75′ à ± 101′ |
| relative roll, absolute pitch | SONAR | 1,5° |
| wind direction | wind vane | ± . 30° |
| relative wind velocity | anemometer | ± 2 m/s |
| actuator positions | submerged detectors | $\delta$ = ± 1°<br>$\omega, \Omega$ = ± 2% – |

Table 2: Performance of the instruments

OPERATIONAL CONSTRAINTS

CURRENT ⩽ 3 knots
WIND ⩽ Force 7 (28 to 33 knots)
SEA ⩽ Force 3 (0.5 to 1.5 m)
sea depth varying between 12.5 m and 80 m

**Course maintenance (a)**

Main propulsion ile
Surface speed: $7 \leqslant V \leqslant 15$ knots

Maintenance of reference course $\hat{K}$
à $\pm 1.5°$ for 60% of the time
à $\pm 4.5°$ for 90% of the time

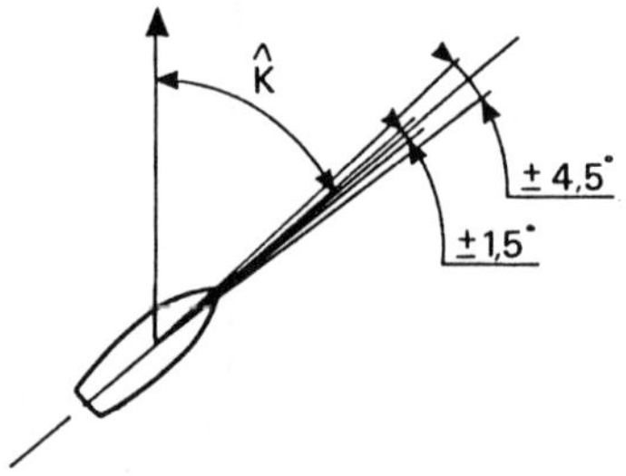


**Rail following (b)**

Propulsion: active rudders

Rail following: in position
à $\pm 15$ m for 60% of the time
à $\pm 45$ m for 90% of the time

Rail following: when moving
reference speed from 1 to 6 knots
maintenance at less than 0.2 knots (60%)
0.6 knots (90%)

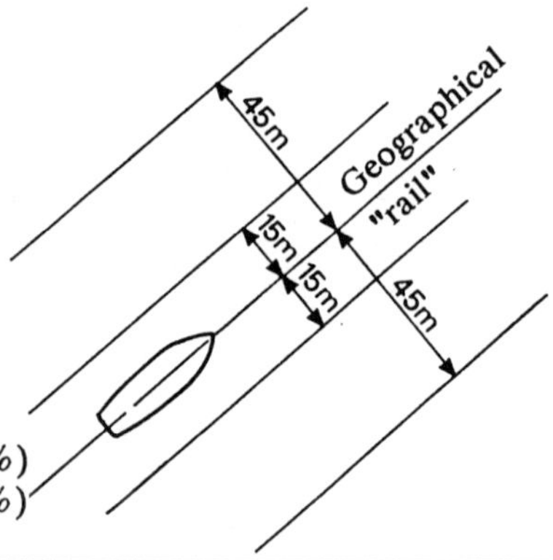


**Fixed point (c)**

Auxiliary propulsion
- active rudders
- bow propellors

Position maintained within the 35 m circle
Course maintained ,15°

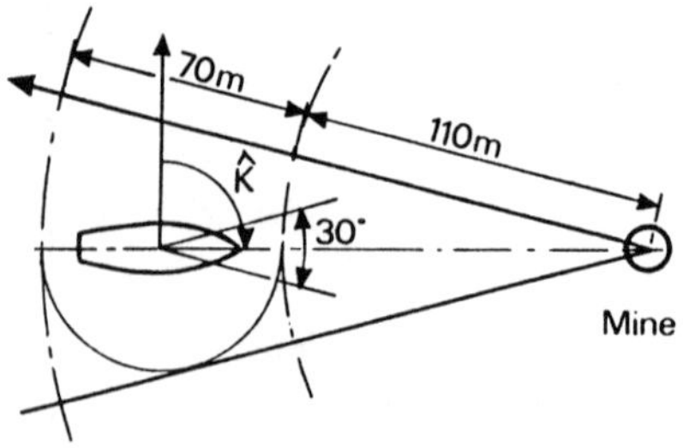


Fig. 2: Operational constraints in the different modes

When the control problem is formulated in this way, its solution requires the performance of four separate tasks, namely simulation, calculation of trajectories, calculation of the commands, and distribution of the commands.

a) - **Simulation**

Given the current measured state (at instant n) of the ship, the disturbances, the actuators, and the commands sent to the actuators, this procedure determines the probable state of the ship and of the actuators at the instants n+1 and n+2.

$U_n$ is determined by the measurements of $ACT_n$ and $X_n$, and equation (3). $X_{n+1}$ is determined by the measurements of $PERT_n$ and $X_n$, the calculation of $U_n$, and equation (1).

Since it is known that the local feedback loops of the actuators have a response time which is shorter than the sampling period T, it may be assumed that $ACT_{n+1}$ corresponds to the reference sent to the actuators at the instant n. According to equation (3), $U_{n+1}$ corresponds to $ACT_{n+1}$ and $X_{n+1}$, which were determined previously. The estimation of the development of the disturbances at the instant n+1 according to the measurement of $PERT_n$, combined with $X_{n+1}$ and $U_{n+1}$ and equation (1), makes it possible to calculate $X_{n+2}$.

As seen from the instant n, the first controllable state is $X_{n+2}$, the state vector determined by the reference commands sent to the actuators at the instant n. This is equivalent to a time delay of two sampling periods for the control law.

b) - **Calculation of the trajectory**

The characteristics required of the state trajectories are:

- They must converge on the objectives, regardless of the starting point; in other words, regardless of $X_{n+2}$ and therefore $X_n$, $ACT_n$, and $PERT_n$;

- They must be optimal in the sense of the constraints affecting the state. The constraints on the state are deduced from the constraints on the forces which themselves result from the physical constraints on the actuators. This two-stage relationship, and the approximations that it necessitates, explain the sub-optimality and also the robustness of a controller of this type.

- They must take into account the hierarchical structure of the objectives; they must observe the priorities when the constraints on the state variables are reached.

The method of calculating the trajectories is as follows: after initialisation on the calculated state of the ship for the instant n+2, and in accordance with a predetermined prediction horizon, the sequence of longitudinal, lateral and angular velocities which the ship must follow to comply with the reference is calculated.

The velocities and corresponding accelerations must be in accordance with the capacities of the ship in the following modes:

*** In fixed point mode**

After the determination of a favourable reference course, in the sense that it permits good control of the distance to the fixed point, the first objective is to maintain this course.

The second objective is to maintain the distance. It may be momentarily sacrificed for the sake of the first objective, while complying with the maximum operational limits (mine safety circle). The converse is not possible.

The change in the velocities is entirely dependent on the change in the course and the distance. For the sake of simplicity, the dynamics of the reference models for the output variables (distance and course) are of the first-order lag type. Since the two objectives may be attained separately, the time constants of the first-order lag type may be different and may depend on the initial state or on the capacities of the ship.

After the modification of the velocities in accordance with the constraints, the values used to calculate them are restored, up to and including the distance and the course found in the next sampling period.

The longitudinal, lateral and angular velocities at the centre of gravity are memorised so that they can be processed by the control module. This calculation is performed over the whole length of the prediction horizon.

*** In track keeping mode**

The first objective is to maintain the course. The second objective is to maintain the track keeping velocity. The absence of the bow propellor subordinates the maintenance of the distance directly to the maintenance of the course. The second objective may be momentarily sacrificed for the sake of the first. The converse is not possible. The velocity vector must satisfy the reference command for the rail following velocity.

The first order reference model for the course can be used to calculate the requisite angular velocity. The lateral velocity is directly related to the angular velocity, since the bow propellors are inactive.

The first order reference model for the velocity can be used to calculate the longitudinal velocity when the lateral velocity is known.

The constraints and the procedures required by them are identical to those for the fixed point mode.

*** In course keeping mode**

The only objective is to maintain the course. The course reference model is a first order system whose time constant depends on the velocity of the ship. In this mode, the ship is driven by the main propulsion system; consequently, the lateral and longitudinal velocities are measured and are assumed to be constant over the whole prediction horizon.

The rear lateral velocity is deduced from the variation described by the reference model for the angular velocity. The constraints and procedures for this velocity are identical to those for the fixed point mode.

Regardless of the mode, the trajectory calculation module transforms the operational commands from the bridge or from the operation control point into the desired components of the state vector, designated SMR(I) for future instants, for I varying from 1 to NHOR (number of sampling points on the prediction horizon).

Why is it necessary to choose a prediction horizon with a range greater than the sampling period? To put it another way, why not use a simple following point control method to calculate $U_n$ such that $X_{n+2}$ is attained? There are a number of reasons for this, as follows:

- Because of the inertia of the system, it is advisable to provide a suitable anticipation time to guard against violations of the operational constraints due to unforeseen variations in the environment. In other words, there must be a sufficiently anticipatory reaction to predicted overshoots. If compliance with the operational constraints cannot be ensured, because of the constraints on the actuators, the operation control point will be warned of this in sufficient time to enable it to take the necessary measures (changing the reference course, returning to the main propulsion system, etc.).

- The existence of non-minimum phase differences in the model makes it necessary to control the predicted states beyond these phase differences.

- A beneficial effect of control signal filtering can be obtained from calculation following a horizon. The control signals are moderated as the horizon lengthens.

c) - **Calculation of the control signals**

For all modes of operation, and regardless of the initial state and force conditions, the trajectory calculation module can determine a sequence of future states capable of being followed by the ship while complying with the set of constraints as closely as possible.

The problem to be solved by the control signal calculation module is that of determining the sequence of forces which the actuators must create so that the desired state (SMR of the reference model) is followed as closely as possible.

Under the effect of previously applied forces, the ship will move in a certain way along the prediction horizon (this movement will be called the natural change of the state vector during the prediction horizon). The control signal calculation module determines the increments of force that must be applied at future instants so that the natural change is as close as possible to the desired change.

Let the state model be as described previously:

$$X_{n+1} = F(X_n, PERT_n) + G \cdot u_n \tag{1}$$

$$Y_{n+1} = H(X_{n+1}) \tag{2}$$

$$U_n = K(X_n, ACT_n) \tag{3}$$

where:

- all the past, up to and including the instant n, is known, including, in particular, $X_n$, $ACT_n$, and $PERT_n$;

- the required behaviour is also known, and is denoted $SMR_{n+i}$ for i=2, NHOR; SMR(n+1) is entirely predetermined at the instant n and has the value of X(n+1);

- the change in the disturbances is estimated:

  - flat prediction: $PERT_{n+i} = PERT_n \ \forall i = 1, NHOR-1$

  - use of a predictor: $PERT_{n+i} = f(PERT_{n+i-1}, \ldots, PERT_n, \ldots)$

In these conditions, and if the desired behaviour is acceptable for the ship, there is a vector of force increments $dU_{n+i}$, i = 0, NHOR−2, such that the natural change is in accordance with the desired change.

In the case of the tripartite mine hunter, owing to the linear relationship between the state and the forces $U_n$ (the matrix G having constant coefficients), the problem is reduced to the solution of a constrained linear system which is written thus:

$$X_{n+2} - F(X_{n+1}, PERT_{n+1}) - G \cdot U_n = G \cdot dU_N$$

$$X_{n+3} - F(X_{n+2}, PERT_{n+2}) - G \cdot U_n = G \cdot (dU_n + dU_{n+1})$$

$$X_{n+NHOR} - F(X_{n+NHOR-1}, PERT_{n+NHOR-1}) - G.U_n$$

$$= G.(dU_n + dU_{n+NHOR-1})$$

When the state X is replaced by the desired state SMR, the system is written thus:

$$SMR_2 - F(X_{n+1}, PERT_{n+1}) - G \cdot U_n = G \cdot dU_N$$

$$SMR_3 - F(SMR_{n+2}, PERT_{n+2}) - GU_n$$

$$= G \cdot dU_n + G \cdot dU_{n+1}$$

$$SMR_{NHOR} - F(SMR_{NHOR-1}, PERT_{n+NHOR-1}) - G.U_n$$

$$= G.dU_n + \ldots + G.dU_{n+NHOR-1}$$

The whole of the left-hand term is determined at the instant n, it is then possible to calculate the vector DU with the component $dU_{n+i}$ for i=0, NHOR-1.

A number of solutions to this problem can be envisaged. The one adopted here is an iterative procedure of solution by minimisation of the structure distance which does not require a matrix inversion. With this approach, it is possible to modify, at the end of each iteration, the solution vector DU in accordance with the constraints on the forces.

If the exact solution does not lie within the domain formed by the constraints, the procedure finds the solution which complies with the constraints and is closest to the exact solution.

Note:

This incremental approach is equivalent to determining the operations required from the actuators to follow a given state trajectory by acting in addition to the forces naturally induced by the disturbances.

d) – **Distribution of the commands**

This is a matter of calculating the commands to be sent to the actuators (helm angle, DELC; bow propellor speed, VPEC; stern propellor speed, VGAC) from the forces to be developed at the instant n+1 ($U_n$ + $DU_n$) which correspond to the reference forces at the instant (n) issued by the controller in automatic mode or requested by the pilot in assisted manual mode.

In this solution, which is equivalent to the inversion of equation (3) (finding $ACT_n$ when $U_n$ and $X_n$ are known, under the constraints stated in 11.3.1.2), a hierarchic procedure, of the same type as that used previously, is implemented. Therefore, if a constraint on one of the actuators is exceeded, the corresponding actuator command is clipped and this limitation is applied to the other commands in order to maintain, as far as possible, the required torque.

The components of $U_n$ (front lateral force FYAV and rear lateral force FYAR) can easily be transformed into the lateral forces and torques applied to the centre of gravity when the corresponding lever arms are known.

11.3.1.5 – **Implementation of PACT in the three modes**

Fig. 3 shows the general structure of PACT. Its three main functions can be described as follows:

– **guidance:** defining the state reference commands (u, v, K) as a function of the objectives set either by the operation control unit or from the bridge or the subsidiary panel (a smaller portable control unit);

– **piloting:** calculating the forces enabling the state references to be followed;

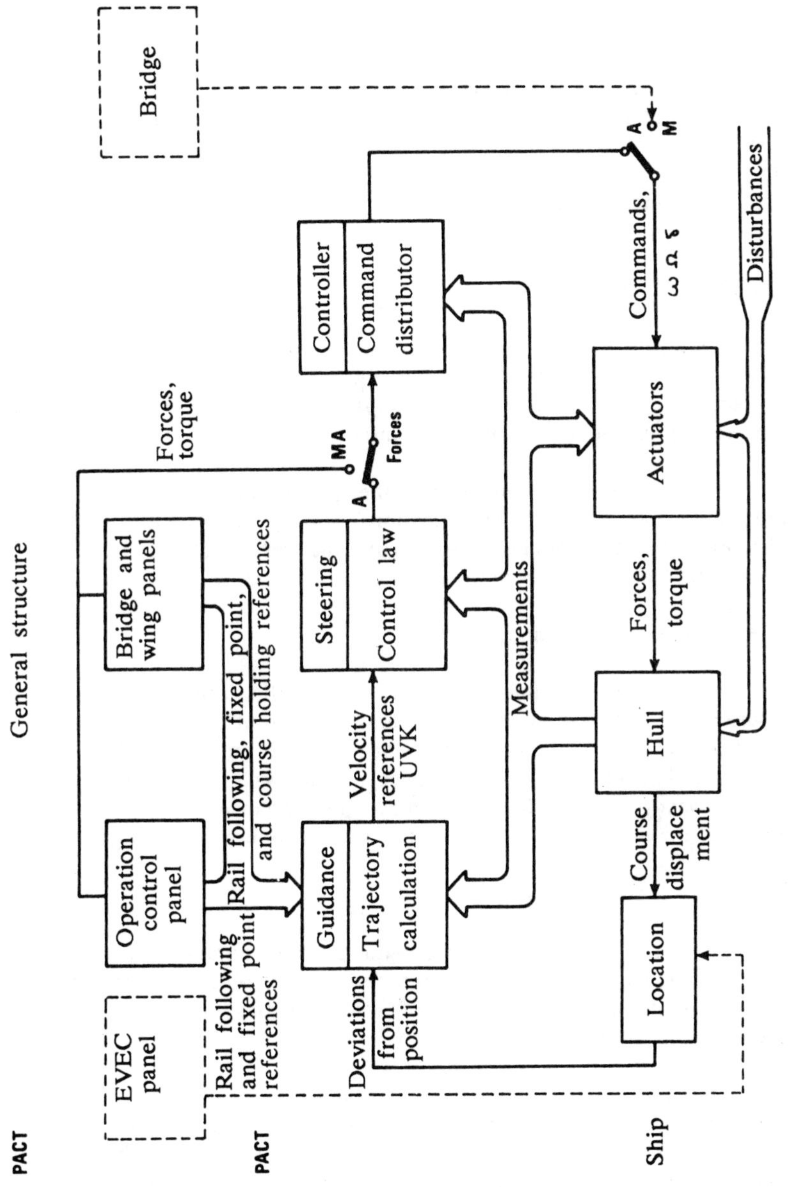


Fig. 3: Structure of PACT system

– **control**: calculating the commands to be sent to the actuators either from the piloting module in automatic mode (A) or from the assisted manual control panels (MA).

The commands are issued to the actuators either directly from the bridge in the manual operation mode, or from the command distribution unit. These commands produce forces which result in displacements of the ship. These displacements will be measured by the set of location systems which will send the information to PACT. The disturbances act on the ship, and PACT receives measurements associated with these at each sampling instant.

## 11.3.2 – PERFORMANCE

Numerous tests of the automatic pilot for the tripartite mine hunters have been carried out in the first mine hunter, "Eridan", and subsequently in four other ships (Alkmaar, Delfzuil, Cassiopée, Dordrecht) in the series of forty units.

These tests have shown that the performance of PACT falls within the limits set by the specifications. Some individual results for the different operating modes, taken from the results of official trials of the "Alkmaar", are given below.

1 – **Course keeping mode**:

This mode was used frequently in specific tests and in outward and return travel phases, with a number of different reference courses and a number of modifications made while in this mode, and with variable speeds and orientations with respect to the waves. The behaviour of the system was judged to be satisfactory, the variance of deviations being $1^{o}$ and the maximum deviation $3^{o}$, for several hours of testing.

2 – **Track keeping mode**:

Ten tests were performed, with a total duration of several hours. These following tests were preceded by manoeuvres in the assisted manual mode and by a phase of rejoining the rail from a distance of several tens of metres.

The following performance was judged to be good: distance read from the cabinet: variance 2 metres, maximum error ±7 m. The distance observed at the EVEC screen was within the thickness of the luminous trace. The velocity of following, 3 and 6 knots, was maintained except in the case of VRAIL = 6 knots with noise limitation (acoustic alarm, maximum possible speed $< 5$ knots).

**3 - Fixed point mode**

Ten tests were performed, with the reference point located in the safety circle around an imaginary mine. The conditions were:

sea 2 to 3, wind force 4, current 1 knot

The performance was as follows: mean distance <5 m, distance variation 1.5 m, max. ±5m, corresponding to maximum distances of 0 to 10 metres from the fixed point. Course variation <10°.

All the tests were considered to be satisfactory.

**4 - Assisted manual mode:**

This mode was judged to be very satisfactory and was frequently used, both for specific tests (entering and leaving harbour, various manoeuvres) and in its function as a transitional mode between one automatic mode and another (a contractual obligation).

## 11.3.3 - DEVELOPMENTS AND PROSPECTS

PACT, in the form in which it has been developed, provides the expected services for ships and missions corresponding to the operation of tripartite mine hunters. However, a certain number of improvements may reasonably be proposed:

a) With respect to the control law, a more global approach, without decoupling between the propulsion model and the inertial model, appears promising as regards the performance.

With respect to the problems posed by the inversion of a non-linear system and the allowance to be made for non-linear constraints affecting the intermediate variables (such as the propulsion forces in the case of the mine hunter), the control problem can be solved as a simple problem of parameter optimisation.

At each sampling instant, a hypothesis on the control sequence ($\delta$, $\omega$, $\Omega$ on a horizon of length L) to be applied in accordance with the constraints on these control signals enables the behaviour of the model to be simulated whether or not it is linear. The comparison between the predicted

portable wing steering panel

detail of the PACT cabinet

PACT cabinet

central operation control panel

Components of the PACT system

Fig. 4

and desired behaviour (reference model or command) makes it possible to construct a scalar criterion which can be minimised by an optimisation procedure of the non-linear programming type, by modification of the control scenarios.

At the sampling instant n, only the control signals determined by optimisation of the criterion and corresponding to this instant will be applied; the other control signals will act as initial values for the calculation to be performed at the instant n+1.

*** Disadvantage of this type of approach**

* There are numerous possible combinations of control signals in the multivariable case if there is a large prediction horizon.

*** Remedy:**

Parametrisation of the control signals to make the length of the prediction horizon independent of the number of control signals.

Use of the model in open loop form which assumes certain precautions (introduction of loops) in order to ensure that the output variables are adjusted towards the references even if the model is imperfect.

*** Advantages of this type of approach**

- Direct verification of the constraints on the control variables.
- The method does not require any special model representation mode and in particular does not require any hypothesis of linearity.
- There are numerous methods of optimisation available.

b) With respect to the determination of the model of the ship and its updating, various methods may be proposed. These methods will be useful for the development of automatic pilots for ships other than the tripartite mine hunter.

The robustness of the control laws make it possible to propose the use of an automatic pilot which only has access to rudimentary information concerning the ship (tonnage, dimensions, power, etc.). A supervision system can be used to supplement and refine this information by making judicious use of the data gathered during the operation of the automatic pilot.

***REFERENCES***

J. Boudon, G. Bouthinon, et al., Automatic pilot for tripartite mine hunter: design studies and hardware implementation - Symposium on Ship Steering Automatic Control, Genoa, 1980

G. Bouthinon and G. Charbonnier, Pilotage automatique des chasseurs des mines type Eridan, Navigation, No. 107

J. Richalet, A. Rault, and R. Pouliquen, Identification des processus par la méthode du modèle, Gordon and Breach, London, 1971

J. Richalet, A. Rault, J.L. Testud, and J. Papon, Model predictive heuristic control: applications to industrial processes, Automatica, v. 14, 1978

Notices Techniques (ADERSA GERBIOS) – Modèle dynamique du bâtiment, 1978

Chasseur de mines tripartites – Pilote Automatique – Spécifications techniques STCAN 1979

## *CHAPTER 12*

# *CONTROL BY REFERENCE MODEL: APPLICATION TO DECOUPLING AND MANOEUVRABILITY IN A HELICOPTER*

## *12.1 - INTRODUCTION*

An examination of control theory makes it possible to imagine the effects of the introduction of computers and electronic flight controllers in helicopters.

The problem facing the designer of the control law is not a simple problem of stability. It is also necessary to ensure that the combination of the helicopter and the control law has good manoeuvrability and "pure" behaviour, so that, for example, there is no lateral movement in response to a pilot command for longitudinal movement.

The totality of these objectives may be represented by a reference model describing an ideal stable, decoupled, and manoeuvrable form of behaviour.

It is, of course, also desirable for the resulting system to be operational in the widest possible flight domain. The control law must be robust.

## *12.2 - GENERAL DESCRIPTION OF THE HELICOPTER*

The helicopter is kept airborne by a rotating aerofoil. This characteristic introduces a very high degree of complexity into the aerodynamic behaviour.

The mean angle of incidence of the blades governs the modulus of the lift force. The direction of this force is governed by a sinusoidal modulation whose two components in quadrature (longitudinal and lateral pitch) are determined by the pilot by means of a mechanical chain of command (joystick - swash plate - blades); a supplementary collective pitch control can be used to vary the position of the swash plate and consequently the lift force. The reaction torque of the main rotor is counteracted, in the case of a single-rotor helicopter, by a tail rotor to control the yaw. The pitch of the tail rotor is controlled by the pilot, using the rudder bar.

The helicopter has four controls and a large number of degrees of freedom. However, if the rotor dynamic is disregarded, the standard six degrees of freedom are present. The difficulties concerning the flying of the helicopter are connected with the non-decoupling of the longitudinal and lateral movements, the instability of certain flight configurations, and the extreme variation of the aerodynamic coefficients within the range of flight velocities (0 – 300 km/hr).

The design of the control law was supported by a series of linear models with eight states for different values of the longitudinal velocity.

For each flight case, we may write:

$$\dot{X} = AX + BU; \qquad Y = EX \tag{1}$$

The components of the state vector are:

| | |
|---|---|
| u | longitudinal aerodynamic velocity along the helicopter axis |
| w | normal velocity |
| q | pitch rate |
| $\theta$ | pitch attitude |
| v | lateral velocity |
| p | roll rate |
| $\varphi$ | roll attitude |
| r | yaw rate |

The components of U are:

| | |
|---|---|
| $\theta_2$ | longitudinal cyclic pitch |
| $\theta_o$ | collective pitch |
| $\theta_1$ | lateral cyclic pitch |
| $\theta_{ar}$ | pitch of the tail rotor |

The vector Y denotes the vector of measurements.

## *12.3 – PRINCIPLE OF DETERMINATION OF THE CONTROL LAW*

### 12.3.1 – GENERAL THEORY

The ideal behaviour of the helicopter is described by the equations of a reference model with the subscript m:

$$\begin{aligned} \dot{X}_m &= A_m X_m + B_m U_m \\ Y_m &= E_m X_m \end{aligned} \tag{2}$$

The minimisation of a criterion of the type

$$J = \int_0^\infty [(X-X_m)^T Q(X-X_m) + (U-U_m)^T R(U-U_m)]dt$$

results in a law of the form

$$U = K_u U_m + K_1 X + K_2 X_m$$

which may be written in the following forms:

$$U = K_u U_m + K X_m + (K_m + K)(X_m - X) \quad (3)$$

or

$$U = K_u U_m + K X_m + K_m(X_m - X) \quad (4)$$

The law will be implemented in form (4) but analysed and synthesised in form (3). The reference model and the gain K are calculated in such a way that

$$\begin{aligned} A_m &\simeq A + BK \\ B_m &\simeq B K_u \end{aligned} \quad (5)$$

for particular flight conditions (stationary flight, for example).

The gain $K_m$ is calculated in such a way that the eigenvalues of $(A-BK_m)$ (closed loop control of helicopter) have a higher modulus than those of $A_m$ for different flight conditions, for example 150 km/hr.

In these conditions, at 0 km/hr, the open loop controller (K) is sufficient to provide satisfactory behaviour of the helicopter. At 150 km/hr, this controller is poor, but the feedback controller with the reference model is then very effective and makes the behaviour of the whole system very satisfactory.

However, the supporting feedback controller must be stable for all the intermediate flight conditions, and this constraint makes it necessary to introduce parameter optimisation.

### 12.3.2 - PARAMETER OPTIMISATION

The coefficients of $K_u$, K and $K_m$ are optimised by a algorithm for finding the function minimum with constraints.

The advantage of this formulation is that it makes it possible to take into account explicitly the fact that the feedback controller must be effective for a large number of operating points.

### 12.3.2.1 – The performance indices

It is difficult to introduce a single criterion describing the objective of the control law. However, it is possible to define this objective with the aid of a number of performance indices, which may be related. The first indices are standard quadratic criteria calculated by a discrete method (k indices) for a number of different sets of flight conditions (i indices).

$$J_i = \sum_{k=0}^{N-1} [X_m(k\Delta t) - X_i(k\Delta t)]^T Q[X_m(k\Delta t) - X_i(k\Delta t)]$$

$$+ \sum_{k=o}^{N-1} [U_m(k\Delta t) - U(k\Delta t)]^T R[U_m(k\Delta t) - U(k\Delta t)]$$

Other criteria may be introduced:

$$J' = \lambda \, ||K_m|| + \mu ||K||$$

where

$$||K|| = \sup \, |K_{ij}|$$

Frequency indices for the placement of the poles of the closed loop system may be envisaged:

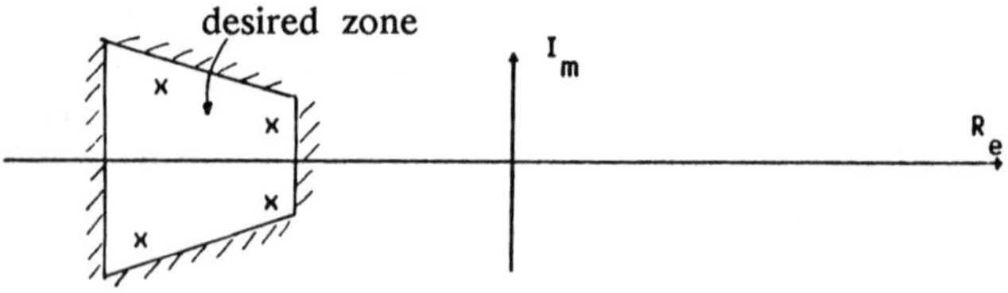


Approximate pole placement

Fig. 1

These indices may be:

$$J_i'' = \sup |R_e(s)| \qquad J_i''' = \sup |-R_e(s)| \qquad J_i'''' = \sup \left| \frac{I_m(s)}{R_e(s)} \right|$$

and must be calculated for all the poles in the closed loop system and for all the operating points taken into account in the design of the control law.

### 12.3.2.2 – Management of the indices

A non-linear programming algorithm can only be used with a scalar criterion. A simple linear combination of the indices $J_i$, $J'$, $J_i''$,

$J_i'''$, $J_i''''$ cannot be used because of the difficulty inherent in the interactive choice of the weighting functions.

However, it is very easy to choose one criterion (such as a combination of the $J_i$) and to use the other indices as constraints. The maximum thresholds are established by an interactive method.

## *12.4 - CHOICE OF THE REFERENCE MODEL*

The behaviour of the helicopter–control law system will be more satisfactory if the equations of the reference model can be obtained from a simple state feedback:

$$A_m \simeq A + BK$$

However, the reference model can be constructed on the basis of an approximation of a helicopter model controlled by a state feedback system. It is sufficient for the model to be decoupled and manoeuvrable.

### 12.4.1 - MANOEUVRABILITY STANDARD

The NASA standards take the form of tables specifying the following:

$$q_{max},\ \dot{q}_{max},\ \frac{1}{\tau_o} = \frac{\dot{q}_{max}}{q_{max}} \quad \text{for } \Delta\theta_2$$

$$p_{max},\ \dot{p}_{max},\ \frac{1}{\tau_1} = \frac{\dot{p}_{max}}{q_{max}} \quad \text{for } \Delta\theta_1$$

$$r_{max},\ \dot{r}_{max},\ \frac{1}{\tau_r} = \frac{\dot{r}_{max}}{r_{max}} \quad \text{for } \Delta\theta_r$$

$\Delta\theta_2$, $\Delta\theta_1$, $\Delta\theta_r$ are fixed at a value of 10% of the full activity of the controllers. The relation $\tau = \dot{q}_{max}/q_{max}$ shows that the response of the helicopter in terms of angular velocity is modelled by a first order tangential to the high frequencies.

For the present purposes, only a point in space $\dot{q}$, q(or $\dot{p}$,p or $\dot{r}$,r) was chosen, for example the point corresponding to flight without visibility.

| | | |
|---|---|---|
| $\Delta\theta_2 = 2.6^o$ | $\dot{q}_{max} = 15.3^o/s^2$ | $q_{max} = 7^o/s$ |
| $\Delta\theta_1 = 1.3^o$ | $\dot{p}_{max} = 30.5^o/s^2$ | $p_{max} = 7.6^o/s$ |
| $\Delta\theta_{ar} = 6.75^o$ | $\dot{r}_{max} = 152.6^o/s^2$ | $r_{max} = 19.1^o/s$ |

The following objective of vertical manoeuvrability was arbitrarily added to these data:

$$\Delta\theta_o = 1.4^o \qquad \dot{W}_{max} = 8.4 \text{ m/s}^2 \qquad W_m = 4.2 \text{ m/s}$$

### 12.4.2 - DESIGN OF THE REFERENCE MODEL

The definition of a reference model is a problem which is not often dealt with explicitly, but which will be formalised here. The reference model must be decoupled to meet the objectives of manoeuvrability defined in 12.4.1. But it is also desirable for the model response not to be too different from that of a helicopter, especially as regards the static gains in attitude and lateral and longitudinal velocities.

The state equations of the reference model are of the type:

$$\dot{X}_m = \underbrace{\begin{vmatrix} x & 0 & x & 0 & 0 & 0 & 0 & 0 \\ 0 & x & 0 & 0 & 0 & 0 & 0 & 0 \\ x & 0 & x & x & 0 & 0 & 0 & 0 \\ x & 0 & x & x & 0 & 0 & 0 & 0 \\ 0 & 0 & 0 & 0 & x & x & x & 0 \\ 0 & 0 & 0 & 0 & x & x & x & 0 \\ 0 & 0 & 0 & 0 & 0 & 0 & 0 & x \end{vmatrix}}_{A_m} X_m + \underbrace{\begin{vmatrix} x & 0 & 0 & 0 \\ 0 & x & 0 & 0 \\ x & 0 & 0 & 0 \\ x & 0 & 0 & 0 \\ 0 & 0 & x & 0 \\ 0 & 0 & x & 0 \\ 0 & 0 & 0 & x \end{vmatrix}}_{B_m} U_m$$

where the x's show the positions of the non-zero terms.

The values of the terms $A_m(2,2)$, $A_m(\theta,\theta)$, $B_m(2,2)$, $B_m(8,4)$ is entirely defined by the manoeuvrability coefficients defined in 12.4.1, since the transfer functions

$$\frac{W_m}{\theta_{o_m}}, \quad \frac{r_m}{\theta_{arm}}$$

are of the first order. The subsystems $\{u_m \; q_m \; \theta_m \; \theta_{2m}\}$ and $\{v_m \; p_m \; \varphi_m \; \theta_{1m}\}$ are of order 3.

The calculation technique used in the example of the longitudinal subsystem will now be explained. The starting point is the equations of the helicopter in stationary mode, in which the lateral and normal terms are eliminated:

$$\begin{aligned} \dot{u} &= -0.044\, u + 0.0066\, q + 0.171\, \theta - 0.192\, \theta_2 \\ \dot{q} &= -1.89\, u - 1.22\, q \qquad\qquad\qquad + 15.4\, \theta_2 \\ \dot{\theta} &= q \end{aligned} \tag{6}$$

This imaginary system is governed by the control law

$$\theta_2 = \mu\theta_{2m} + \lambda_\theta\ \theta + \lambda_q\ q + \lambda_u\ u \tag{7}$$

The problem can be solved by a number of simulations for different values of $\mu$, $\lambda_\theta$, $\lambda_q$, $\lambda_u$. The value $\mu$ is chosen in such a way as to obtain the desired $q_{max}$ for a $\theta_{2m}$ of 2.6°. $\lambda_\theta$ and $\lambda_q$ govern the value of $q_{max}$ and determine the modes of the response, but have little effect on the static gains:

$$G = \inf\left(1, \frac{12}{u}\right) \tag{8}$$

$\lambda_u$ acts independently on the static gains ($\theta/\theta_{2m}$ and $u/\theta_{2m}$ in the standard mode).

The interactive determination of $\lambda_u$, $\lambda_\theta$, $\lambda_q$ is therefore very simple. Finally the system (6), governed by the established control law (7), provides a suitable longitudinal reference model. The same type of procedure is followed for the lateral subsystem.

But the perfect decoupling defined by this model has no meaning except in stationary flight. In directional flight, the natural relationship between $r_m$ and $v_m$ must be taken into account (sideslip).

If T is the lift force of the rotor, m is the mass of the machine, and $\Psi$ is the heading, the following equations can be written (approximate or in standard mode):

$$T\cos\varphi = mg \qquad \dot{\beta} = -\dot{\Psi} = -r = \frac{v}{u}\,\frac{180}{\pi}$$

$$T\sin\varphi = m\dot{v}$$

After all the calculations have been performed, the reference model equations are written thus:

$$\begin{aligned}
\dot{u}_m &= -0.0267\ u_m + 0.0185\ q_m - 0.159\ \theta_m - 0.0734\ \theta_{2m}\\
\dot{q}_m &= 0.504\ u_m - 1.37\ q_m - 0.924\ \theta_m + 0.88\ \theta_{2m}\\
\dot{\theta}_m &= q_m\\
\dot{w}_m &= -2\ w_m - 6\ \theta_m\\
\dot{v}_m &= -0.091\ v_m - 0.0785\ p_m + 0.162\ \varphi_m + 0.0942\ \theta_{1m} - u\,\frac{\pi}{180}\,r_m\\
\dot{p}_m &= -0.916\ v_m - 3.05\ p_m + 1.90\ \varphi_m + 23.5\ \theta_{1m}\\
\dot{\varphi}_m &= p_m\\
\dot{r}_m &= -8\ r_m - 22.6\ G_r\ \theta_{arm}
\end{aligned} \tag{9}$$

The term u in the equations denotes the true velocity of the helicopter (and not the deviation with respect to the equilibrium point). The requirement for the yaw velocity is too high for directional flight, and has been attenuated by introducing a multiplicative gain $G_r$ into the yaw reference model.

## *12.5 – ADJUSTMENT OF THE CONTROL LAW*

The control law is implemented in the form of equations (4), to which are added integral terms to allow for the fact that the reference model, unlike the real helicopter, always operates around the equilibrium point (0,0). To obtain the complete structure of the control loop shown in Fig. 2, it is necessary to add a filter F, designed to filter the noise returned to the control surfaces, the servo control block S, a decoupling matrix K'u, and a matrix C which is used to calculate the model inputs from the commands of the human pilot.

The different blocks can be adjusted or defined separately.

### 12.5.1 – BLOCKS $K_U$, T, F, S

The gain $K'_U$ is used to directly compensate the yawing movements arising from the control of the main rotor. The reading of the Dauphin control derivatives shows that it is permissible to assume that

$$K'_u = \begin{vmatrix} 1. & 0. & 0. & 0. \\ 0. & 1 & 0. & 0. \\ 0. & 0. & 1 & 0. \\ 0. & 4. & 0. & 1. \end{vmatrix} \qquad (10)$$

T is a diagonal transfer function matrix:

$$T = (1 + \frac{0.1}{s}) \; I \qquad (11)$$

F and S are a transfer functions matrix:

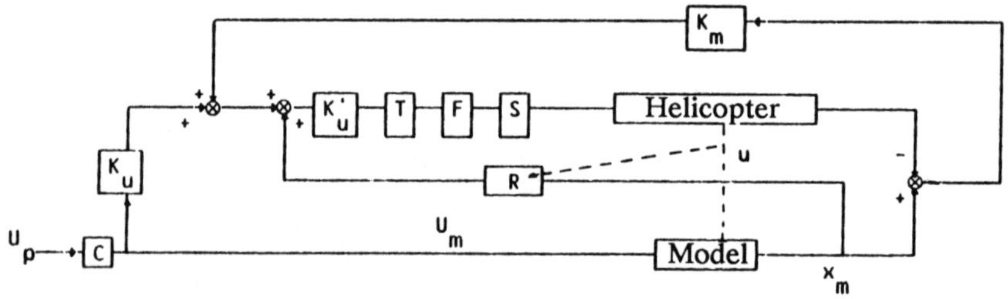


Structure of the control law – the control surface velocities are limited to 50 °/s

Fig. 2

$$F = \frac{\omega_F^2}{s^2+1.6\omega_F s+\omega_F^2} I$$

$$\omega_F = \omega_s = 20 \text{ rad/s}$$

$$S = \frac{\omega_s^2}{s^2+1.2\omega_s \Delta+\omega_s^2} I$$

## 12.5.2 - BLOCK $K_m$

This is obtained by assuming a zero reference ($U_m$ = 0). The chosen gain $K_m$ has a diagonal tendency so that it can be applied to the most unvarying subsystems of input variables and state variables. It is not necessary to refer to the complete state nor to provide decoupling of the resetting modes, but it is necessary to provide the following:

- rapid resetting modes;
- stability over a wide range of operating points;
- a sufficient phase margin to enable the filters of the control surfaces, the servo-operated control surfaces, and any parasitic delays to be tolerated. The whole set of these constraints is manifested in optimisation by a delay of 0.15 s in the loop.

The control law chosen is of the type ($X_m$=0):

$$\theta_2(n) = k_2\ q(n) + k_2'\ \theta(n)$$

$$\theta_o(n) = k_o\ w(n)$$

$$\theta_1(n) = k_1\ p(n) + k_1'\ \varphi(n) + k_1''\ v(n)$$

$$\theta_{ar}(n) = k_r\ r(n) + k_r'\ v(n) + k_r''\ v(n)$$

The terms k are chosen and their values are obtained by minimising a time criterion based on resetting to 0, with a correct dynamic of the state of the helicopter, on the basis of initial non-zero conditions on $\theta$, w, $\varphi$, r.

The frequency constraints used are:

$$\sup_i \frac{I_m(s_i)}{R_e(s_i)} \leqslant 2$$

$$\sup_i R_e(s_i) \leqslant 0.2$$

The following resetting law is obtained:

$$\theta_2 = 0.3\,(q_m - q) + 0.4\,(\theta_m - \theta)$$

$$\theta_o = -(w_m - w)$$

$$\theta_1 = \boxed{1.5}\,(v_m - v) + 0.1\,(p_m - p) + 0.3\,(\varphi_m - \varphi)$$

$$\theta_{ar} = \boxed{1.}\,(v_m - v) - 0.7\,(r_m - r) + \boxed{0.5}\,(\dot{v}_m - \dot{v}) \tag{13}$$

The terms in the boxes can be reset to zero if necessary for certain proposed versions of the closed loop controller. These results were obtained by an interactive linear method, using the non-linear programming algorithm and the management of the criteria and constraints.

The following procedure is recommended:

- the resetting in $\theta$ of the helicopter alone is optimised with the aid of two feedback coefficients in $\theta$ and q and a given resetting mode;
- the resetting in w, in $\varphi$, and in r are optimised in turn;
- the resettings in u and v are obtained by assuming slower resetting modes;
- the delay of 0.15 s is then introduced and the damping is optimised;
- the maximum value of the poles in the closed loop configuration and the damping are improved in an interactive way by varying the nature of the constraints and their values.

## 12.5.3 - BLOCK K

The block K was simplified to the maximum extent. The following law was adopted:

$$\theta_o = \frac{0.4\,u}{0.275\,u + 252}\,q_m$$

This provides normal longitudinal coupling and may be suppressed if required. It was obtained by reducing the equation in $\dot{w}$ to:

$$\dot{w} = Z_q\,q + Z_{\theta_o}\,\theta_o$$

The values of $Z_q$ and $Z_{\theta o}$ are available in the form of curves which are functions of u, the longitudinal velocity of the helicopter. The term in $\dot{w}$ is therefore cancelled directly in the open loop configuration. The maximum values of the parameters were monitored and introduced, if necessary, into the constraints.

## 12.5.4 - BLOCK C

This is an equation co-ordination block:

$$\theta_{2m} = \theta_{2p} \quad \theta_{om} = \theta_{op} \quad \theta_{1m} = \theta_{1p} \quad \theta_{arm} = G_r\ \theta_{arp} + \lambda\theta_{1p}$$

A weakening of the gain in yaw velocity for high velocities is provided, and $\lambda$ can be used to co-ordinate the sideslip. Calculations of the maximum lateral acceleration and the equilibrium in sideslip show that

$$G_r = \inf\ |1.,\ \frac{12}{u}| \qquad \lambda = \inf\ |49,\ 5u|$$

The unco-ordinated side-slip option corresponds to $\lambda = 0$.

### 12.5.5 - BLOCK $K_u$

An adjustment in the form $K_u \neq 0$ would not make any apparent difference, because of the very high speed of action of the resetting law, and also because of the limitation of the velocities of the commands entering the chain to 50°/s. However, the conditions of manoeuvrability in $\dot{q}_{max}$, $\dot{p}_{max}$ and $\dot{r}_{max}$ are poorly complied with in the absence of a direct transmission of $\theta_{2m}$, $\theta_{1m}$ and $\theta_{arm}$ to the input of the helicopter.

The following will be used:

$$K_u = \begin{vmatrix} 0.4 & & & \\ & 0. & & \\ & & 0.5 & \\ & & & 4. \end{vmatrix}$$

## *12.6 - RESULTS*

### 12.6.1 - FREQUENCY RESULTS

The resetting law was tested as regards frequency by replacing the blocks S, F and the time delay for the application of the law by a rational fraction,

$$\frac{1 - 0.07\ s}{1 + 0.07\ s}$$

The poles found in the closed loop configuration are shown in the following table for two operating points and reference model 1.

| 0 km/hr | 150 km/hr |
|---|---|
| -11.33 | -11.52 |
| - 5.43 ± j 4.54 | - 7.77 |
| - 5.08 ± j 7.58 | - 5.87 ± j 7.6 |
| - 4.61 ± j 5.89 | - 5.33 ± j 9.17 |
| - 1.80 | - 1.98 ± j 4.32 |
| - 1.26 | - 0.97 |
| - 1.14 ± j 1.12 | - 0.69 ± j 1.55 |
| - 0.18 | - 0.17 |
| - 0.10 | - 0.10 |
| - 0.09 | - 0.07 |
| - 0.08 | - 0.06 |
| - 0.01 | 0. |

These poles only concern the resetting feedback control. The poles with very low values correspond to the trim modes. The poles with high values are introduced in an imaginary way by the phase shifter responsible for providing the phase margin. The pole virtually equal to zero is due to the absence of longitudinal velocity feedback in the control law.

### 12.6.2 – TIME RESULTS

Figs. 3 to 6 show the results obtained for operating points 0 km/hr (curve 1) and 150 km/hr (curve 2) with the following pilot inputs:

Fig. 3: $\theta_{2p} = 2.6^{o}$

Fig. 4: $\theta_{op} = 1.4^{o}$

Fig. 5: $\theta_{1p} = 1.3^{o}$

Fig. 6: $\theta_{arp} = 6.75^{o}$

Fig. 7 shows the result obtained with $\theta_{1p} = 1.3^{o}$ in the co-ordinated sideslip option for the points 50 km/hr (1), 100 km/hr (2), and 150 km/hr (3).

$\theta_2$, $\theta_o$, $\theta_1$, and $\theta_{ar}$ in these figures concern the commands actually sent to the rotor of the helicopter.

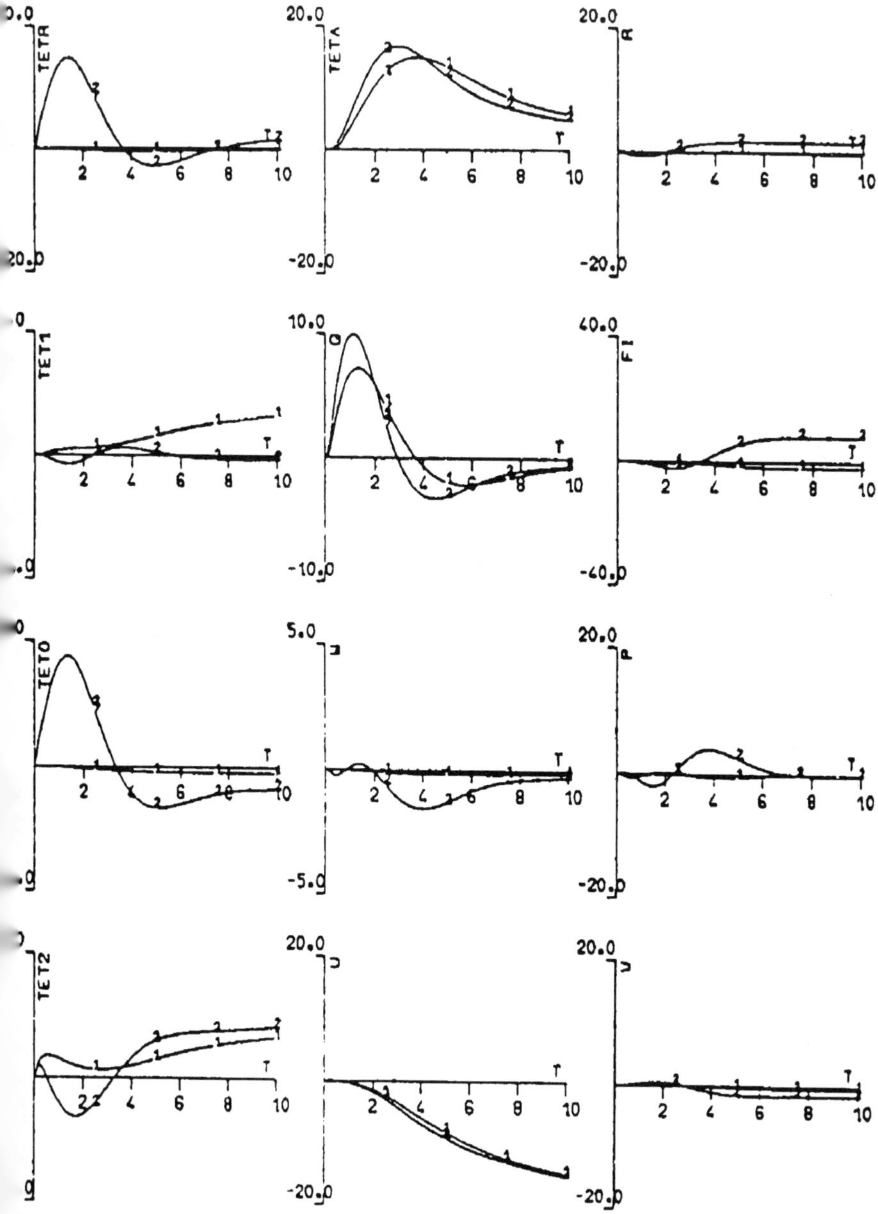


Longitudinal pitch ramp, 0–150 km/hr

Fig. 3

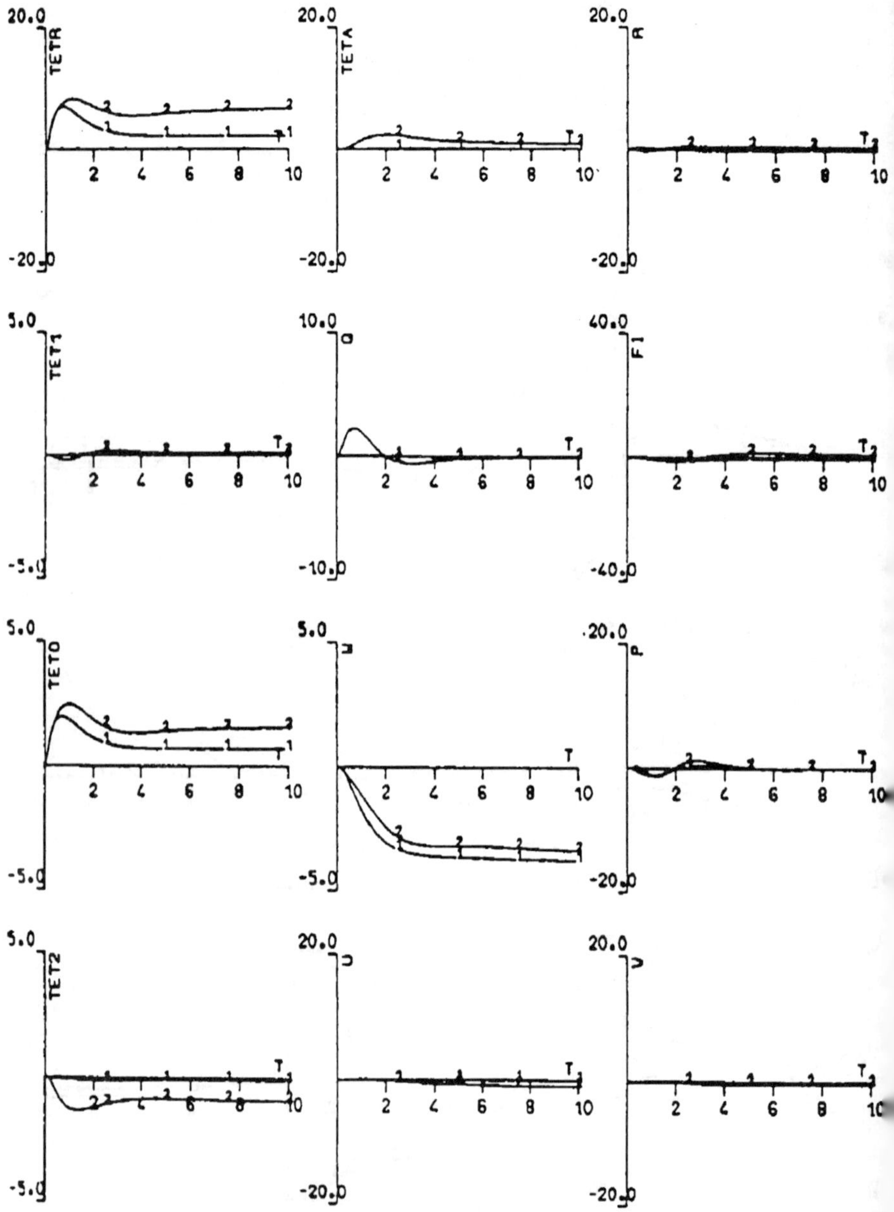


Collective pitch ramp, 0–150 km/hr

Fig. 4

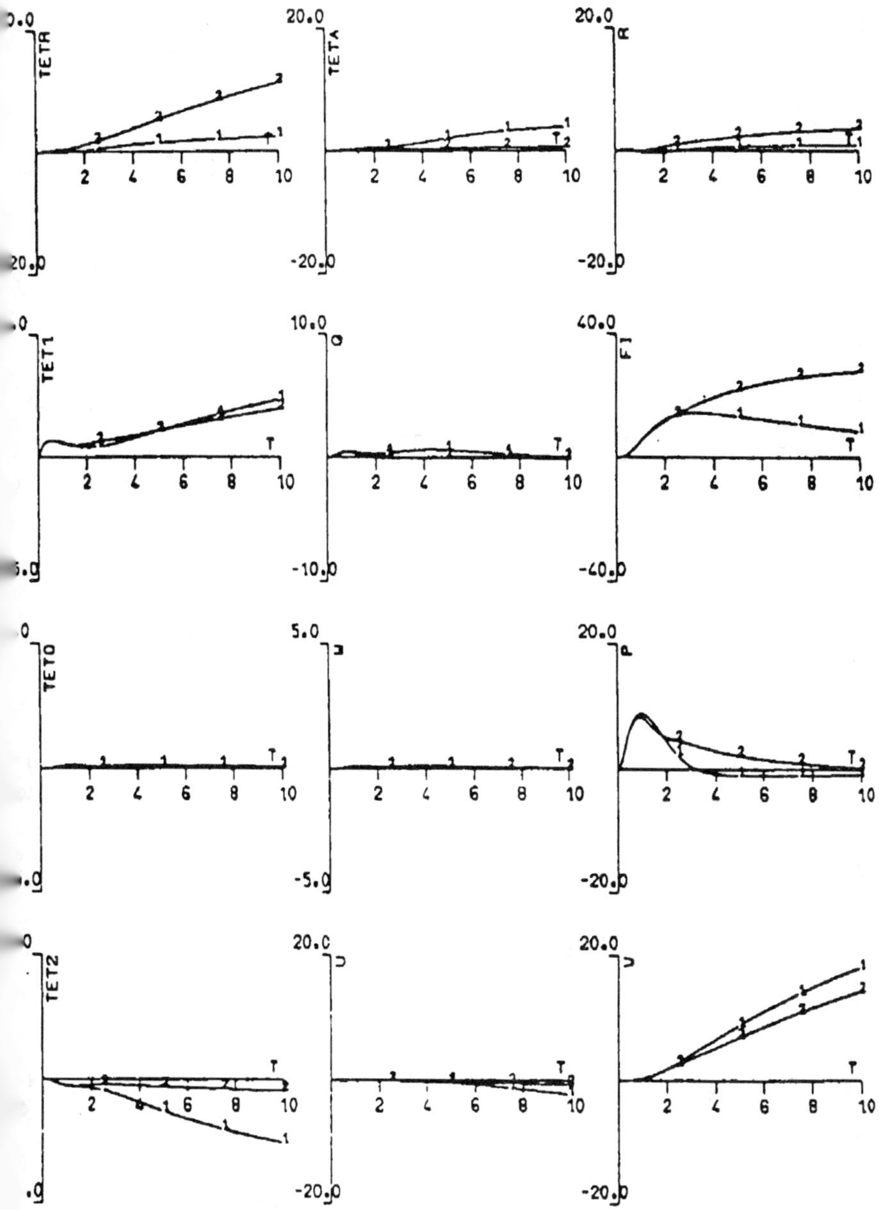


Lateral pitch ramp, 0–150 km/hr

Fig. 5

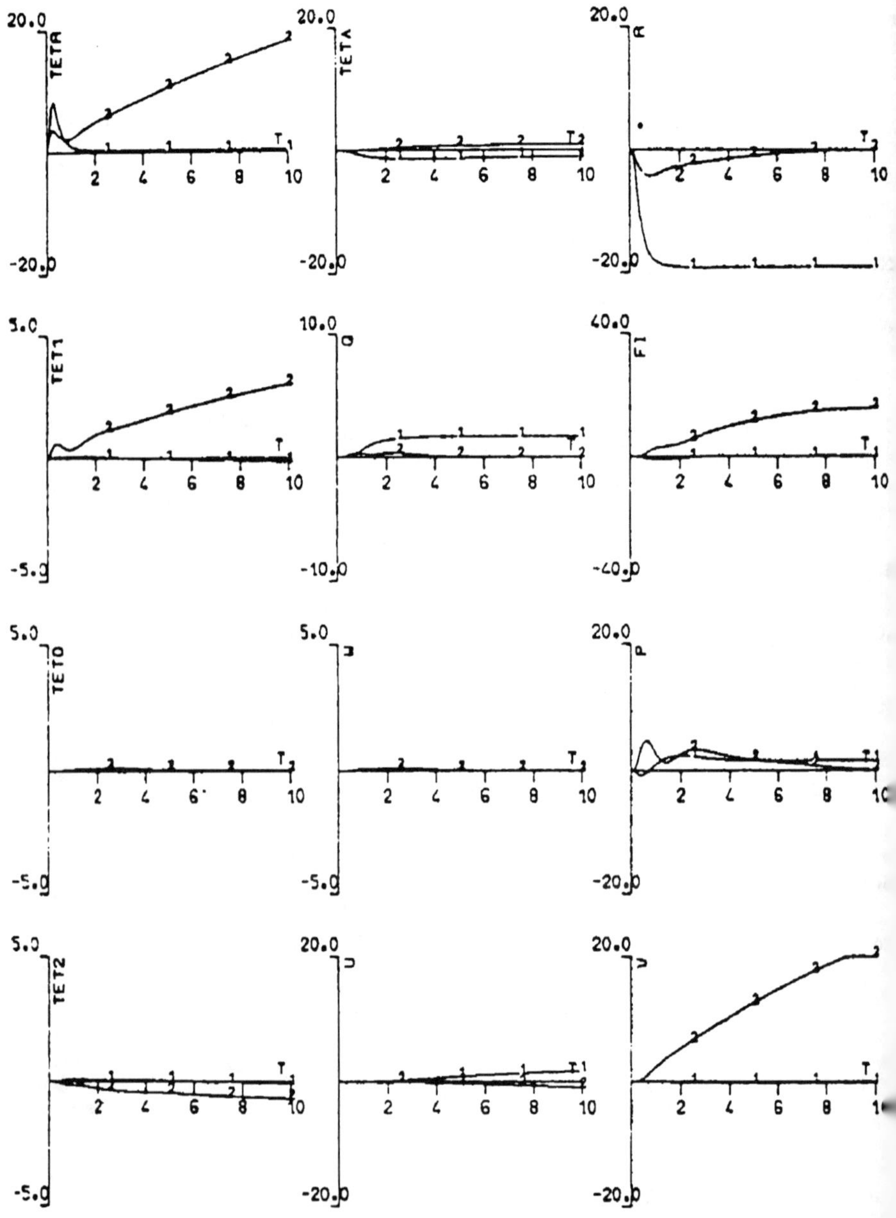


Tail rotor pitch ramp, 0–150 km/hr

Fig. 6

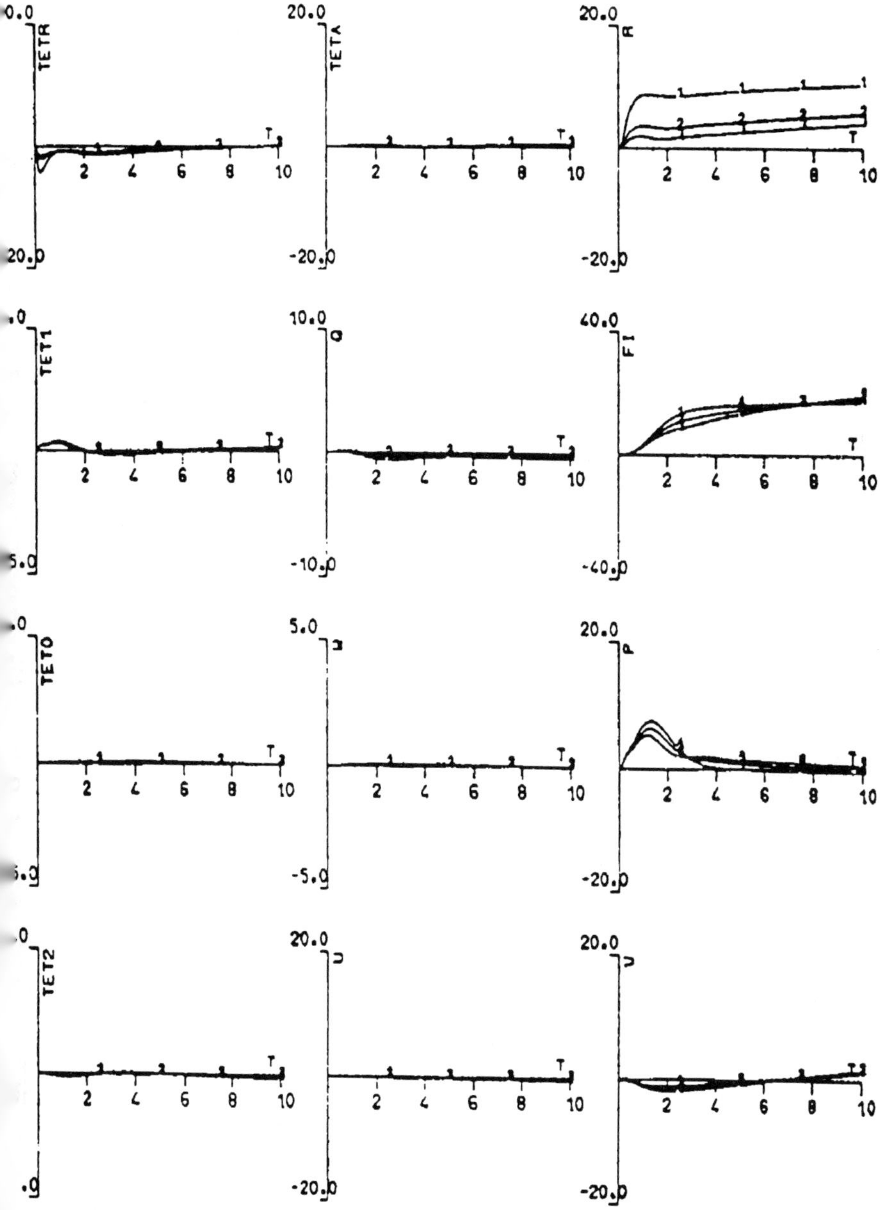


Lateral pitch ramp, co-ordinated sideslip, 50, 100, 150 km/hr

Fig. 7

## *12.7 – CONCLUSION*

In the adjustment of this law, allowance was made for the implementation constraints by introducing a small number of coefficients, by reducing to a minimum the values of the feedback control gains, and by introducing control filters to minimise the effect of the noise returned to the control surfaces. The resulting manoeuvrability is satisfactory except as regards the yawing movement, for which the manoeuvrability constraint is very severe.

The procedure for finding control laws described above demands considerable experience from the user. However, the fully debugged program now stored in the Aerospatiale computer at Marignane is a useful tool for computer–assisted control law design. It can be used to establish a dialogue between the researchers and those responsible for the automatic pilot and the electronic flight control systems.

Admittedly, the analytical aspect has to some extent been replaced by the algorithmic, interactive procedure, but in compensation for this the resulting control explicitly allows for a number of sets of flight conditions and parasitic phase shifts. Consequently the robustness with respect to stability and performance is particularly well developed.

## *REFERENCES*

L.R. Anderson, Preliminary demonstration of a robust controller design method, N 81 198 22, November 1980

H. Bourles, Qualification et amélioration de la robustesse des régulateurs multivariables avec applications au pilotage d'un avion, Doctoral Thesis

C.B. Chato, Design of optimally robust control system, AD A 085 762, January 1980

B. Gimonet, Etude des lois de pilotage d'un système stabilisé trois axes, Rapport DERA AS No. 7242, August 1980

B. Gimonet, Définition de nouvelles lois de pilotage, Rapport DERA AS No. 7282, December 1981

B. Gimonet, Définition d'une loi de pilotage robuste pour hélicoptère assurant le découplage, Rapport DERA AS No. 7382, August 1982

G. Kreisselmeier et al., In flight test of a parameter insensitive controller, N 80 21358, November 1979

D.H. Owens et al., Simple models for robust control of unknown or badly defined discrete multivariable systems, N 81 30860, December 1980

D.H. Owens et al., Robust control of unknown discrete multivariable systems, N 81 30853, January 1981

Rontani, De Lestrade, Dérivées aérodynamiques du modèle S 80, Notes de travail Aérospatiale, 1980

G.N. Wanderplaats, A fortran program for constrained function minimization, NASA Ames Research Center, August 1979

*CHAPTER 13*

# *ADAPTIVE CONTROL APPLIED TO THE REDUCTION OF VIBRATIONS IN HELICOPTERS*

## *13.1 - INTRODUCTION*

Problems of vibration have always been a source of concern in helicopters, where the rotor is the basic cause of vibrations for the machine as a whole. In level flight, the blades are subject to asymmetric aerodynamic loads, as a result of which the rotor is submitted to alternating forces which are multiples of the fundamental frequency.

Althrough vibrations were considerably reduced in the earliest helicopters by rotor balancing and use of passive devices, such as hanging dampers, most present-day helicopters have a level of vibration in excess of the specifications of international organisations. The civil and military organisations have set a level of $\pm$ 0.02 g which can only be attained by active self-adaptive methods which allow for variations in the flight conditions of the helicopter.

The work presented here concerns the development of a multivariable self-adaptive controller which generates suitable control signals to reduce the vibrations directly at their basic source, in other words at the rotor. After establishing the procedure for modelling the system, the adaptive technique itself, which implies the principle of separation, will be described, and the experimental performance of the controller for significant flight configurations will be analysed. This work is an application of the theoretical results of Chapter 6, and is based on the work of Taylor, Farrar, and Miao (1980).

## *13.2 - HELICOPTER MODEL*

The purpose of multicyclic control is to minimise the vibrations at pulsation $b\Omega$ where b and $\Omega$ are respectively the number of blades and the rotation pulsation of the rotor, by means of control signals sent to the fixed plate.

In practice, it is also possible to minimise the vibration level by means of a control signal at or close to $b\Omega$ pulsation.

The effect of the multicyclic components $\theta$ on the vibratory components, Z, of the helicopter is represented in steady state, assuming that the system is linear, by the deterministic relationship

$$Z = Z_o + S\ \theta \tag{1}$$

where

$$\theta^T = [\theta_{b-1,c},\ \theta_{b-1,s},\ \theta_{bc},\ \theta_{bs},\ \theta_{b+1,c},\ \theta_{b+1,s}]$$

The vector $\theta$ of dimension 6 consists of cosine and sine components of the modes $(b-1)\Omega$, $b\Omega$ and $(b+1)\Omega$.

$Z_o$ is the vector of dimension 6 formed by the vibratory components without multicyclic control, and S is a matrix representing the sensitivity of the vibratory components to the multicyclic control signals.

It is very simple to deduce from relationship (1) the recursive relationship for small movements which forms the basis of this modelling:

$$Z_{k+1} = Z_k + S\ \Delta\theta_k + \eta_k \tag{2}$$

In this vector expression, $\Delta\theta k$ is the variation of the control vector and $\eta_k$ is the measurement noise which is assumed to be white and to have a zero mean. The variance, $R_k$, of this noise forms one of the parameters in the experimental analysis.

Notes

With regard to equation (2), some preliminary comments are necessary.

1. The dynamic matrix is a unit matrix whose eigenvectors are all equal to unity. Consequently, the controllability of the system is related to the rank of the matrix $[S, SI, \ldots, SI^{2n-1}]$, which must be equal to or less than 2n (2n = 6, corresponding to the control vector dimension). It is therefore important that the matrix S should have a rank equal to 2n. Controllability is generally achieved by making the number of measurements equal to 3, thus providing a vector Z of dimension 6.

2. Equation (2) shows that the identification of the parameters of the matrix S, which act as control parameters, will be necessary in the adaptive phase. These parameters are time-variable.

3. System (2) always has a minimum phase.
   The stochastic criterion to be minimised in relation to the general equation (2) has the following form:

$$J = E(Z_{k+1}^T\ W_Z\ Z_{k+1} + \Delta\theta_k^T\ W_\theta\ \Delta\theta_k) \tag{3}$$

where the expected value is calculated for all the random variables affecting the criterion. This one step criterion permits an immediate rigorous analytical solution of the control problem. The criterion could be enriched, but this would result in the much more sophisticated controllers known as dual controllers.

## *13.3 – OPTIMAL CONTROLLER SYNTHESIS*

Taking (2) and (3) into account, we find that

$$J = E[(Z_k^T + \Delta\theta_k^T S^T + \eta_k^T) W_Z (Z_k + S\Delta\theta_k + \eta_k) + \Delta\theta_k^T W_\theta \Delta\theta_k]$$

At the instant k, the measurements $Z_k$ are known, and the control signals $\Delta\theta_k$ to minimise the above criterion are to be found. The above equation may be written thus:

$$J = Z_k^T W_Z Z_k + \Delta\theta_k^T E(S^T) W_Z Z_k + Z_k^T W_Z E(S)\Delta\theta_k + \Delta\theta_k^T E(S^T W_Z S)\Delta\theta_k + E(\eta_k^T W_Z \eta_k) + \Delta\theta_k^T W_\theta \Delta\theta_k \qquad (4)$$

At this level of calculation, it is possible, without loss of generality, to assume that $W_Z = I$, the matrix $W_\theta$ being used to weight the effect of the control signals in the criterion. It will also be assumed that the estimates of the parameters of the 2n lines of the matrix S are provided by 2n Kalman filters which provide the estimates and the associated covariance matrices.

To anticipate a later section of this chapter, $\hat{S}$ denotes the estimated matrix of S and $P^j_k$ denotes the error covariance matrix of the parameters of the j-th line of S (denoted $S_j$), i.e.

$$P^j_k = E(S_j^T - \hat{S}_j^T)(S_j - \hat{S}_j)$$

Observing that $P^j_k = P_k$ for any j, we may make the following deductions from the notation above:

$$E(S^T) = \hat{S}_k^T \qquad E(S) = \hat{S}_k$$

$$E(S^T W_Z S) = E(S^T S) = \hat{S}_k^T \hat{S}_k + 2n P_k$$

Taking these expressions into account, and minimising relationship (4) with respect to $\Delta\theta_k$, we obtain the optimal controller:

$$\Delta\theta_k^* = -(\hat{S}_k^T \hat{S}_k + 2nP_k + W_\theta)^{-1} \hat{S}_k^T Z_k \qquad (5)$$

In the above development, the parameters of matrix S are identified by a Kalman filter method which will be described in the next section. Relationship (5) expresses the optimal controller when the parameters of matrix S are unknown but constant. Additionally, if the principle of separation is applied, equation (5) takes the following form:

$$\Delta\theta^*_k = -(\hat{S}^T_k \hat{S}_k + W_\theta)^{-1} \hat{S}^T_k Z_k \tag{6}$$

The performances of these two types of control were tested experimentally as described in a subsequent section.

## *13.4 – ESTIMATION OF THE PARAMETERS OF THE MATRIX S*

### 13.4.1 – FIRST CASE: WITH CONSTANT PARAMETERS

Let $S_j$ denote the j–th line of the matrix A. At any instant k, the following recursive equations are true:

$$S^T_{j,k+1} = S^T_{j,k} \tag{7}$$

$$\Delta Z^T_{j,k+1} = \Delta Z_{j,k+1} = \Delta\theta^T_k S_{j,k} + \eta_{j,k} \tag{8}$$

for j = 1,..., 2n
k = 0, 1, 2

with

$$\Delta Z_{j,k+1} = Z_{j,k+1} - Z_{j,k}$$

Equation (7) expresses that the parameters are constant; this relationship can also be modified by adding a white noise of variance $Q_k$ to allow for parameter variations (9).

$$S^T_{j,k+1} = S^T_{j,k} + v^T_{j,k} \tag{9}$$

The system of equations (8) and (9) is presented in the conventional form of state estimation by the Kalman filter method (the state vector here represents the parameter vector). The identified parameters of the matrix S are therefore found by means of the following equations:

$$H_k = \Delta\theta^T_k \tag{10}$$

$$P_k = P_{k-1} + Q_k \tag{11}$$

$$K_k = P_k H^T_k (H_k P_k H^T_k + R_k)^{-1}$$

$$P_{k+1} = P_k - P_k H^T_k (H_k P_k H^T_k + P_k)^{-1} H_k P_k \tag{12}$$

$$\hat{S}^T_{j,k+1} = \hat{S}^T_{j,k} + K_k(\Delta Z_{j,k+1} - \Delta\theta^T_k \hat{S}^T_{j,k}) \quad j=1,\ldots,2n \tag{13}$$

Equations (10), (11) and (12) are used to calculate the corrected covariance matrix $P_k$ of the parameter estimates as well as that of the Kalman gain. These values are independent of the line of matrix S being estimated.

Equation (13) provides the parameter estimate for the vector $S^T_{j,k}$ which is required for the synthesis of the optimal controller. The fact that the matrix $\Delta\theta_k$ is variant prevents off-line calculation of the Kalman gain of the matrix $P_k$. The volume of calculation is limited, however, since only equation (13) is required for each value of j.

### 13.4.2 – SECOND CASE: WITH VARIABLE PARAMETERS

When there is a change of configuration, the parameters of the matrix S vary more or less rapidly. To adapt the identification method to this new constraint, it is possible to give exponential weights to the information received, in other words to filter it by a first order network.

This gives a new recursive formulation:

$$\hat{S}^T_{j,k+1} = \hat{S}^T_{j,k} + K_k(\Delta Z_{j,k+1} - \Delta\theta^T_j \hat{S}^T_{j,k}), \quad j = 1,\ldots,2n$$

$$P_k = \frac{1}{\gamma}(P_{k-1} - K_k H_k P_{k-1}) \qquad (14)$$

$$K_k = P_{k-1} H^T_k (H_k P_k H^T_k + \gamma)^{-1} \qquad (15)$$

where $\gamma$ is the weighting coefficient.

It is now possible to describe the self-adaptive multicyclic control system, consisting of a minimal variance control estimator, in the form of a block diagram. The accelerometers installed on the fuselage detect the vibrations, and a harmonic analyser restores the components Z of the modes $(b-1)\Omega$, $b\Omega$, and $(b+1)\Omega$. On the basis of these data, the self-adaptive controller generates the control signals $\Delta\theta^*$ for the actuators, these control signals being limited to $2^o$ in amplitude. This constraint determines the choice of the parameters of the weighting matrix $W_\theta$.

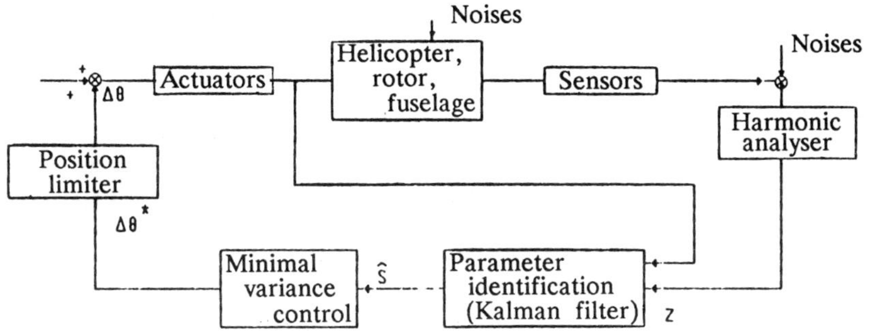


Multicyclic control system.

### *13.5 - EXPERIMENTAL RESULTS*

To demonstrate the performance of multicyclic control systems, a preliminary input–output simulation was made of the vibratory behaviour of the helicopter. The parameters of the different matrices S provided by Aérospatiale for the set of configurations (V = 200, 230, 250, and 280 km/hr; load factor 1.25, 1.5, and 1.7) formed the basis for this digital simulation initialised by the given vibration vector $Z_0$.

The identification of the parameters of S in real time was applied to the vibration measurements, and the optimal control signal was then applied, thus enforcing the principle of separation. In the case of speed variation, a linear or polynomial extrapolation was made from the above nominal configurations. Some of the more significant results will now be described [Gauvrit and Chaves (1982)].

#### 13.5.1 - CONSTANT SPEED CONFIGURATION (V = 280 km/hr and V = 200 km/hr)

Fig. 1 shows the development of the vibration level with a self-adaptive multicyclic controller and the optimal control signals to the three actuators, which are each limited to $2^{o}$. Fig. 2 shows the development of three identified parameters of the matrix S. Three conclusions immediately follow:

1 - The identification is rapid and very good. The convergence of the estimated parameters to the nominal parameters confirmed in Fig. 2 ensures that the performance is very good.

2 – The three control signals for an optimal choice of the matrix $W_\theta$ rarely reach saturation.

3 – The total vibration level is reduced very rapidly (in 12 revolutions of the rotor, i.e. 2 seconds) from its initial level (2.9 $m/s^2$) to a level of the order of 0.20 $m/s^2$, which is compatible with the requirements of specifications.

Figs. 3 and 4 show the variations with time of the vibration level, the control signals, and the parameter estimates in the case of the configuration for 200 km/hr. The reduction of the vibration level and the convergence of the estimates confirm the quality of the performance of the self–adaptive controller.

### 13.5.2 – VARIABLE SPEED CONFIGURATION

VARIATION FROM V = 200 km/hr to 280 km/hr WITH AN ACCELERATION OF $\gamma = 4$ $m/s^2$

We shall now consider a speed variation of the helicopter from 200 km/hr to 280 km/hr with a constant acceleration of $\gamma = 4$ $m/s^2$. During this operation, the parameters of the matrix S vary in a large range; the estimated parameters must therefore follow these variations to avoid any degradation of the control system performance. Fig. 5 shows the variation of the total vibration level in this configuration. This is kept below 0.4 $m/s^2$; without multicyclic control, it would have reached 3 $m/s^2$. Fig. 6, which shows the real variations of the parameters, demonstrates that three parameters are correctly followed by the Kalman filters. The results are identical for the whole set of 36 parameters. The performance of this real–time parameter identification confirms the importance of the reduction of the vibration level.

The performance is excellent for the whole set of flight configurations investigated.

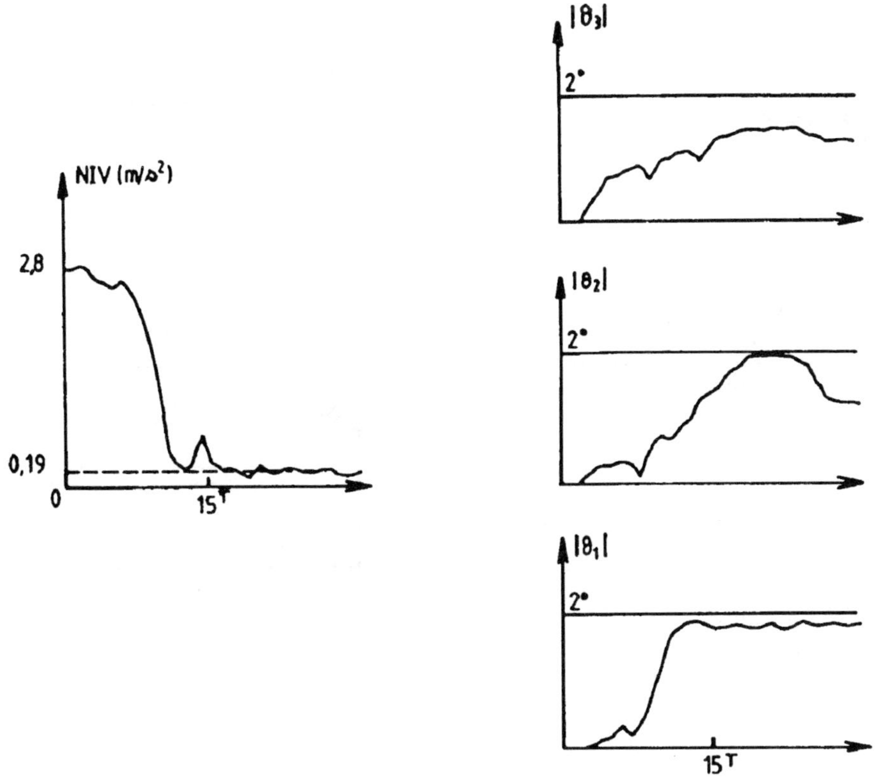


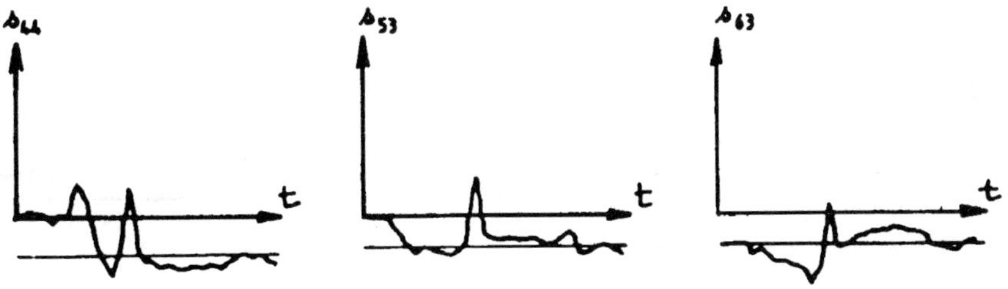


Figs. 1 and 2

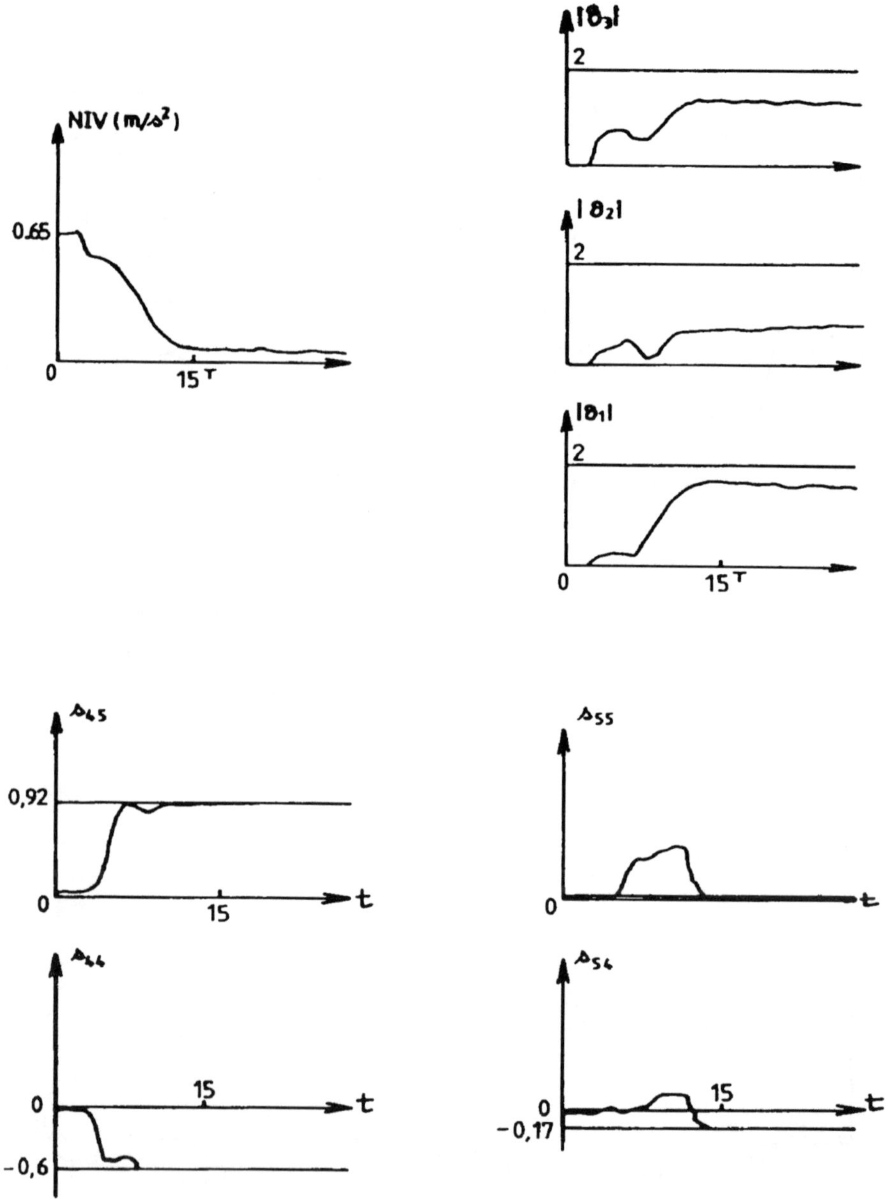
NIV (m/s²)
0.65
0
15 T
|θ3|
2
|θ2|
2
|θ1|
2
0
15 T
s45
0,92
0
15
t
s55
0
t
s44
0
15
t
-0,6
s54
0
-0,17
15
t

Figs. 3 and 4

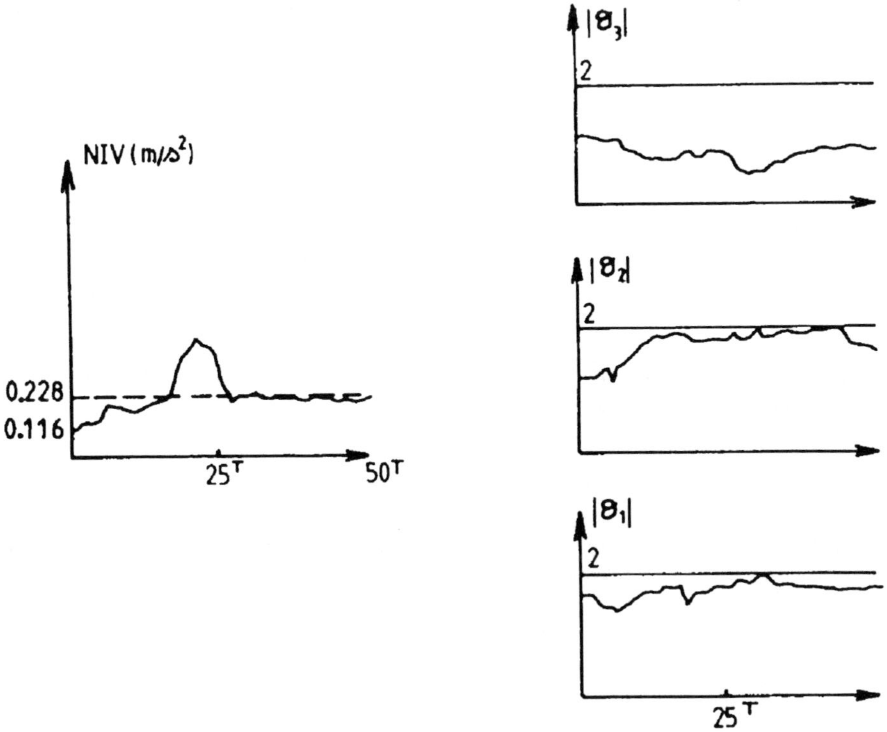


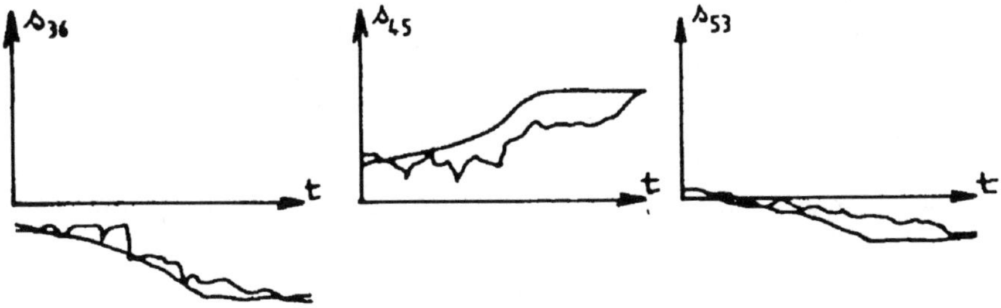


Figs. 5 and 6

## *13.6 – CONCLUSION*

This study, carried out in close co–operation with Aérospatiale [Achache (1982)], led to the design of a self–adaptive multivariable controller for the reduction of vibrations in a helicopter.

A number of important comments must be made:

1. The behaviour of the controller produces performances which on the whole are good, as confirmed by the results of digital simulation. The parallel connection of the Kalman filters enables the mathematical formulation to be radically simplified, thus reducing the storage required in the on–board computer. The controller, based on constant parameters, was enabled to follow the variation of the parameters by introducing a sliding memory into the Kalman filter.

2. The simulation of the helicopter was based on experimental results provided by Aérospatiale, thus guaranteeing the quality of the chosen modelling procedure.

3. If parameter variations do exceed the capacity of the controller (as in the case of an increased load factor), supplementary measurements (of speed, for example) may be used to enrich and complement the structure of the self–adaptive controller.

We consider that the future stages of this implementation, namely the bench tests and flight tests, will provide valuable information for refining the adjustment of the controller parameters.

## *REFERENCES*

Achache, Conduite multicyclique – algorithme adaptatif, Document Aérospatiale, 1982

A. Chaves, Réduction des vibrations sur un hélicoptère par une commande adaptative multicyclique, ENSAE Thesis, Toulouse, 1983M. Gauvrit and A. Chaves, Commande multicyclique des pales d'hélicoptères – algorithme adaptatif stochastique, Rapport DERA, No. 1/7296, 1982

Taylor, Farrar, Miao, An active control system for helicopter vibrations, 36th Annual Forum of the American Helicopter Society, Washington, Dec. 1980

## *CHAPTER 14*

## *MINIMISATION OF THE OPERATING COST OF AN AIRCRAFT FLIGHT BY OPTIMISATION OF THE TRAJECTORY*

### *14.1 - INTRODUCTION*

The problem of the optimisation of aircraft trajectories is not new. Interest was shown in this problem thirty years ago [Rutowski (1954)], and hundreds of studies have been carried out since then, mostly in contexts which are restricted in various ways, as follows:

i) the optimisation is "off line" and therefore intended more for the definition of nominal procedures than for effective implementation;

ii) the criteria are related to time (minimum), or to consumption, but rarely to both at once;

iii) the optimisation only affects one flight phase, such as climbing, acceleration, or turning;

(iv) the modelling is very simplified, especially as regards the aerodynamics (polar parabolic models) and consumption (taken to be proportional to the thrust).

Moreover, these studies were generally much more concerned with military aviation (and therefore with highly manoeuvrable aircraft for which the use of aggregated variables such as total height was natural and for which procedural constraints could be ignored) than with civil aviation, for which the need for better management of flight profiles only became evident after various "oil crises".

In the civil context, where Performance Management Systems (PMS) are used, the following requirements are encountered:

- the trajectory as a whole must be optimised from the beginning of the climb to the end of the descent at the time of the capture of the ILS;
- the criterion must include both time and consumption, which are both essential components of the direct operating cost;
- highly developed models must be used, since the percentage gains are small;
- the optimisation conditions must be applied in real time, with

allowance made for uncertainties in the system and the environment;
- allowance must be made for constraints affecting not only the aircraft but also the ground control procedures;
- and the whole system must be developed with limited computing facilities.

Consequently, even if the problem is easily formulated as "minimising a criterion weighting the flight time and consumption between a given initial state (position, altitude, velocity, inclination) and a given final state, while satisfying the constraints on the aircraft and on the procedures", the solution of this problem by the standard optimisation techniques is hardly practicable. The difficulty is particularly due to the fact that the system is characterised by a non-linear dynamic model which is highly complex, not so much at the level of the formal writing of the flight mechanics equations and the number of states as at the level of the modelling of the aerodynamics and the operation of the propulsion systems. Difficulties also arise from the atmospheric environment and the disturbances encountered, which means that what is required is not an "optimal control trajectory" but an "optimal control law". In the first case, the control signal is specified as a function of time, whereas in the second case it is a function of the state, and it can be applied in closed loop form with the advantages of self-correction.

After the description of the problem, and in particular the criterion and mathematical model of the system (14.2), the reasons for the choice of the method of solution are given (14.3). The general mathematical formulation is described in (14.4), and section (14.5) is concerned with the presentation of the application itself and the results.

## *14.2 - CRITERION AND MATHEMATICAL MODEL*

### 14.2.1 - CRITERION

As stated in the introduction, most of the studies of trajectory optimisation only deal with one flight phase and use a time criterion. The objective in these is to minimise the time for performing a turn [Hedrick (1972), Peterson (1981), Lin (1982)] or for reaching a given height [Ardema (1976)].

In the case of commercial aircraft and PMS, the objective is to optimise a direct operating cost over the whole trajectory, with a weighting for the price per minute of flight and the consumption, in the form:

$$Q = \int_{0}^{t_f} (C_t + C_f\ f')\ dt \qquad (1)$$

where $C_t$ and $C_f$ are the costs for each minute of flight and for each kilo of fuel, f' is the instantaneous consumption, and $t_f$ is the time taken for the stage concerned.

## 14.2.2 - DYNAMIC MODEL

The dynamic model must meet three partially conflicting requirements: it must be simple, since its complexity affects that of the algorithms, and real-time implementation is required; it must be suitable for the optimisation method adopted; and naturally it must be sufficiently precise to achieve the required objective. Obviously, apart from any explicit optimisation, the air transport companies wish to fly their aircraft in the correct conditions, and the percentage gains which may be expected from an optimisation are low. A range of 1 to 3% is generally accepted (a considerable amount in terms of volume: the amount of fuel consumed in one year by an Airbus is estimated to be 3.7 million gallons, corresponding to a cost of 38 million francs). A saving of 1 to 3% in fuel consumption will repay the cost of an optimisation computer and its interfaces in two to three years.

### 14.2.2.1 - **Flight mechanics equations**

In view of the context of the problem (civil aircraft which are less manoeuvrable), it is possible to deal only with movements in the vertical plane and the movement of the centre of gravity.

In these conditions, the flight mechanics equations are written thus:

$$\dot{X} = V \cos \gamma \qquad (a)$$

$$\dot{m} = -f' \qquad (b)$$

$$\dot{h} = V \sin \gamma \qquad (c) \qquad (2)$$

$$\dot{V} = (F-X)/m - g \sin \gamma \qquad (d)$$

$$\dot{\gamma} = (Z - mg \cos \gamma)/mV \qquad (e)$$

where V is the velocity in the trajectory, $\gamma$ is the slope, m is the mass, F is the thrust, and X and Z are the drag and the lift respectively. The control signals are F and Z.

This apparently simple system must include:

- The aerodynamic model required for the calculation of the drag and lift coefficients $C_X$ and $C_Z$ which, together with the atmosphere, intervene in the calculation of X and Z:

$$X = \frac{1}{2} \rho S V^2 C_X, \qquad Z = \frac{1}{2} \rho S V^2 C_Z$$

where $\rho$ is the air density, and S is a reference aerofoil surface.

- The propulsion model for the calculation of f' and F.

### 14.2.2.2 – **Aerodynamic model**

The relationship between $C_x$ and $C_z$ is complex, being partly non-linear and partly a function of the Mach number. The phenomena of compressibility must also be taken into account. These polars $C_z = F(C_x)$ are represented in Fig. 1.

a) – Modelling of the parabolic type:

$$C_x = C_{x_o} + k_1 C_z + k_2 C_z^2 \tag{3}$$

where $k_1$ and $k_2$ are simple functions of the Mach number (M), is very commonly used in theoretical studies [Mehra (1979), Calise (1976)]. In the present case, in view of the "break" appearing in the polars, especially for $M \geqslant 0.8$, such modelling was found to be inadequate. For $k_1$ and $k_2$, for example, which are polynomials of the 3rd order in M, the error is 3.5%.

b) – It is therefore necessary to reconstruct the polars directly from the precise data which are available in the form

$$C_x = C_{x_o} + C_{x_i} + C_{x_c} \tag{4}$$

where $C_{xo}$, the drag coefficient at zero lift, is dependent on $\eta = V/\nu$. The coefficient $\nu$ denotes the kinematic viscosity of the air, which is a function of temperature and pressure, and $C_{xi}$ is the induced drag coefficient, which is a function of $C_z$ and M:

$$C_{x_i} = f_1 (C_z, M) C_z^2 \tag{5}$$

and $C_{x_c}$ is the drag coefficient due to compressibility:

$$C_{x_c} = f_2 (M - f_3 (C_z)) \tag{6}$$

All these data are provided by the aircraft manufacturer, point by point, in a form which is difficult to use directly. To obtain sufficient precision, $C_{xo} = f(\eta)$, for example, must be approximated by a polynomial of a minimum order of 4, and $f_2$ must be approximated by a polynomial of order 5.

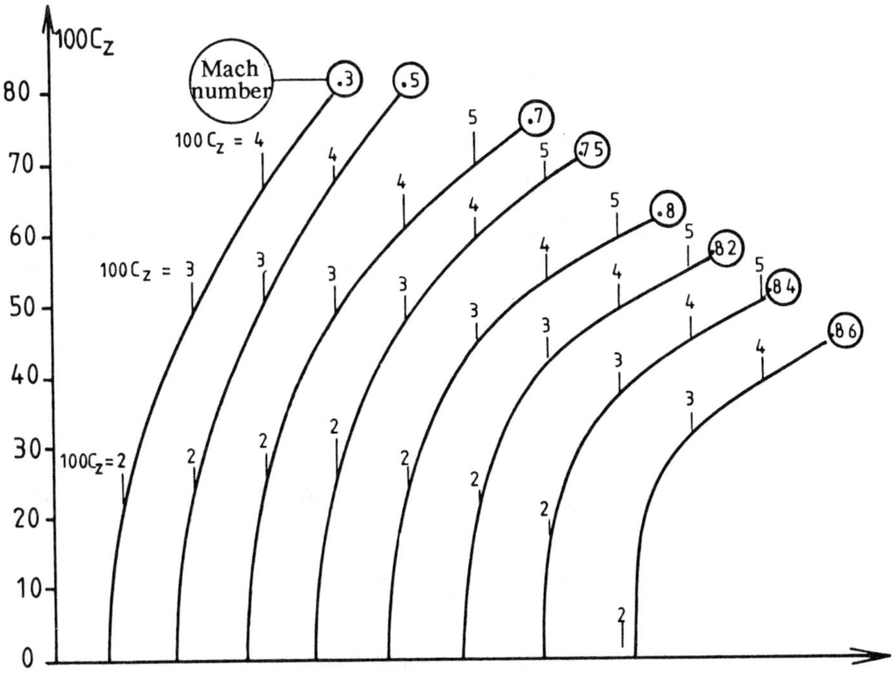


Polars for $C_Z = F(C_X)$

Fig. 1

c) – In order to comply with the imperatives related to the precision and analytical nature of the model, it was therefore necessary to use spline functions with two dimensions in the following form:

$$C_x(M,C_z) = t_1 M + t_2 C_z + t_3 + \sum_{i=1}^{N} S_i \omega_i \log \omega_i \qquad (7)$$

$$\omega_i = (M - M_i)^2 + (C_z - C_{z_i})^2$$

on the basis of a set of triplets, M, $C_Z$, $C_X$. The error was thus reduced to 0.15%.

## 14.2.2.3 – Modelling of propulsion

f' and F are also complex functions, depending not only on the current engine regime but also on the temperature, the density, the Mach number, etc.

The normalised output, $f'_n = f'/\delta_2 \theta_2^{0.61}$, and the reduced thrust, $F/\delta$, are given by the engine designer in the form of numerous tables of values, as a function of M and $F/\delta$ for $f'_n$, and as a function of the reduced engine regime $N_1/\sqrt{\theta_2}$ for $F/\delta$.

Polynomial expressions of the type

$$f'_n = \sum_{i=0}^{2} \sum_{j=0}^{3} |C_{ij} M^i| \left(\frac{F}{\delta}\right)^i \tag{8}$$

have been found to be inadequate, having a precision of only 1.26%, and therefore spline modelling (with a precision of 0.38%) had to be used in this case also.

### 14.2.2.4 – Modelling of the atmosphere

This is mentioned simply as a matter of interest. It corresponds to the modelling of the standard atmosphere, with the possibility of displaced initial conditions at sea level and non-standard temperature gradients modelled by altitude layers.

Table I summarises the various points made above.

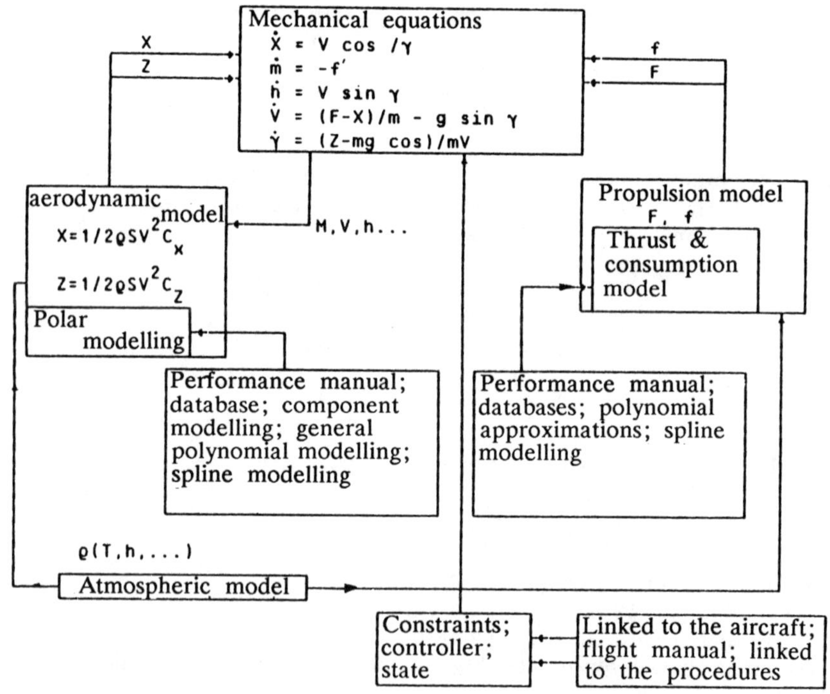


Table 1

### 14.2.3 - CONSTRAINTS

Finally, to complete the definition of the problem, there are certain constraints, some connected with the aircraft (and not affecting the present problem except for those concerning the maximum thrust, but essential in the case of fighter aircraft with a minimum time criterion), and some connected with the procedures (constant-level cruising, etc.).

These will be considered in Section 14.4.

## *14.3 - CHOICE OF THE OPTIMISATION METHOD*

As indicated in the introduction, the term "trajectory optimisation" actually implies two problems, namely the determination of the optimal trajectories and the associated controls, and the application of these, in other words their implementation in real time. Owing to the complexity of the current applications which was demonstrated in 14.2, these two problems cannot be considered independently, even though they can be treated sequentially or in parallel.

### 14.3.1 - SEQUENTIAL PROCEDURE

The procedure is sequential if the determination of the trajectories and optimal controllers is performed initially, independently of their implementation and therefore off-line, followed by application by means of a simplification and/or parametrisation of the resulting laws. In the first stage, the trajectory and controls are defined as a function of time, and the methods used are generally derived from the calculus of variations.

i) It is possible to solve the problem directly by using the maximum principle and the Hamilton formula; this makes it necessary to integrate not only the direct system, but also the adjoint system, with the added difficulty of having to correctly initialise the adjoint variables.

ii) It is also possible to use projected gradient techniques which consist in starting from a "nominal control law" (and a nominal trajectory) and modifying this in such a way as to reduce the criterion and to satisfy the constraints. This method is frequently used in aeronautics for the determination of trajectories in the medium term, particularly for turning [Blank (1982), Well (1982), Huynh (1983)].

In all these cases, the control law is defined as a function of time. Therefore, independently of the calculation difficulties associated with finding the law, the problem of its application still remains, and its solution requires the analysis and characterisation of the laws which have been found.

### 14.3.2 – PARALLEL PROCEDURE

As stated above, it is also possible to proceed in parallel. For this it is necessary either to find the control law, as a function not of time but of the state of the system, or to find a sub-optimal solution which is sufficiently simple to be calculated in real time at any instant.

Three modes of action are possible, depending on whether the designer is dealing with:

- the dynamic model;
- the structuring of the controllers; or
- the method of optimisation

i) The simplification of the dynamic model is a very widely used method. In addition to what has already been stated in 14.2 (disregarding the motion about the centre of gravity, decoupling the longitudinal and transverse movements), this method can make use of aggregated variables such as the total altitude $E = h + V^2/2g$ [Ardema (1976), Calise (1981), Mehra (1979)]. An extreme case of the use of this "aggregation", where the system dynamic is reduced to the single variable E, is described by Erzberger (1975). Assuming that for a passenger aircraft the trajectory consists of three segments (climb, cruising, descent), in the course of which E is respectively monotonically increasing, constant, and decreasing, it is possible to use this variable in place of time as the independent variable [Erzberger (1975)].

In the problem considered here, the simplification of the dynamic model is not a realistic proposition. It has been shown that the complexity of the model is due less to the number of states than to the complexity of the functions relating their variations and that, taking into account the criterion and the relative smallness of the expected gains, it is necessary to use a precise model of the system.

ii) Another possibility is to parametrise the required controls as a function of a certain number of parameters whose optimal values are to be found. In this way a dynamic optimisation problem is reduced to a static optimisation problem, but evidently the optimal trajectories found by this method are only optimal in the class of solutions corresponding to the chosen parametrisation. For example, if it is specified, according to current commercial procedure, that the climb is made with constant $V_i$ and Mach number, the problem of optimisation is reduced to the search for only two parameters. A climb profile $\hat{V}_i$, $\hat{M}$ is determined, corresponding to an improvement of the criterion with respect to the present situation (where $V_i$, M are fixed once for all, regardless of the weight of the aircraft, state of the atmosphere, etc.), but this may be far from the theoretical optimum. In fact, there is a possible compromise between an excessive restriction of the "acceptable" controls and complete freedom, consisting in the characterisation of the control by the N values $u_k$ (considered as parameters) taken by u for k values

$x_k$ of a variable x, which may be time or a system variable (altitude above ground level, distance to the runway, etc.). This provides a trajectory control in the sense stated above, the controller u(t) being calculated as a function passing through these N points in the sense of a spline or a Chebyshev polynomial. This procedure was used, for example, to define the optimal flight procedures for an approaching aircraft encountering large wind shears [Jacob (1982)]. It is currently used for the determination of optimal helicopter trajectories in the case of failure of an engine, using other parametric optimisation algorithms.

It will be noted that, although this last method can be used to calculate the control signal as a function of the state, and therefore in a directly applicable form, the calculations required by the parametric optimisation are such that the calculation cannot be performed on board the vehicle.

## *14.4 - METHOD OF FORCED SINGULAR PERTURBATIONS*

The third method, which will be adopted here, is to find a sub-optimal solution giving a good compromise between the quality of the criterion and the difficulty of solution, by using the concept of the separation of the set of variables into distinct dynamic classes. It is known that, in this case, the overall control problem can be divided into sub-problems each associated with a different dynamic [Fossard and Magni (1982), Kokotovic (1983)]. The method, known as the forced singular perturbations method, will be described first of all, before its practical application is examined in the next section.

Let the system have two time scales, defined by the equations

$$\begin{aligned} \dot{x}_1 &= f_1\,(x_1\ x_2\ u) \qquad x_1(0) = x_{10} \quad x_1(t_f) = x_{1f} \\ \epsilon\dot{x}_2 &= f_2(x_1\ x_2\ u) \qquad x_2(0) = x_{20} \quad x_2(t_f) = x_{2f} \end{aligned} \tag{9}$$

The problem is to find the control signal u which makes the following criterion extreme:

$$\int_{t_o}^{t_f} L(X\ u)\ dt \qquad \text{with } X = \begin{vmatrix} x_1 \\ x_2 \end{vmatrix} \tag{10}$$

In equations (9) the parameter $\epsilon$ recalls the conventional terminology of systems with singular perturbations and simply indicates that the variables $x_2$ change in a time scale which is rapid by comparison with that associated with $x_1$.

It is known that a uniformly valid approximation in the interval ($t_o$ $t_f$) (which in fact is only the first term of an asymptotic approximation of the optimal trajectory in a series of $\epsilon$) may be written as follows:

$$u^*(t, t_o, x_o) = [\hat{u}(t)-\bar{u}(t_o)] + \bar{u}(t) + [\tilde{u}(\sigma)-\tilde{u}(t_f)] \qquad (11)$$

where $\bar{u}$ is the reduced solution and $\hat{u}$, $\tilde{u}$ are the solutions in the initial and final layers, calculated in the following time scales:

$$\tau = \frac{t - t_o}{\epsilon} \quad \text{and} \quad \sigma = \frac{t - t_f}{\epsilon}$$

If it is required to calculate the "optimal" control signal (11) at the initial instant, it is only necessary to calculate:

$$u^* (t_o, t_o\ x_o) = \bar{u}(t_o) + |\tilde{u}(\sigma) - \bar{u}(t_f)|$$

where

$$\lim_{\sigma \to -\infty} \tilde{u}(\sigma) = \bar{u}(t_f)$$

In fact, it can be shown that, in general, the convergence in the preceding equation is exponential:

$$|\tilde{u}(\sigma) - \bar{u}(t_f)| < e^{-b|\sigma|}$$

and consequently [Mehra (1979)] it is found that

$$u^*(t_o, t_o\ x_o) = \hat{u}(t_o) + 0\ (e^{-b\frac{(t_f-t_o)}{\epsilon}})$$

$0(f/\epsilon)$ is a function which decreases as $f(\epsilon)$.

The determination of the "optimal" control signal apparently only requires the calculation of $\hat{u}(0)$ which, as will be seen, requires the calculation of the reduced solution $\bar{u}$.

a) The reduced solution is calculated by making $\epsilon=0$ in equations (9). This gives

$$\begin{aligned} \dot{x}_1 &= f_1\ (\bar{x}_1\ \bar{x}_2\ \bar{u}) \\ 0 &= f_2\ (\bar{x}_1\ \bar{x}_2\ \bar{u}) \end{aligned} \qquad (12)$$

and the Hamilton formula is written

$$\bar{H} = L\ (\bar{x}_1, \bar{x}_2\ \bar{u}) + \bar{\lambda}_{x_1}\ f_1(\bar{x}_1, \bar{x}_2\ \bar{u}) + \bar{\lambda}_{x_2}\bar{f}_2$$

with

$$\dot{\bar{\lambda}}_{x_1} = - \frac{\delta\bar{H}}{\delta\bar{x}_1} \qquad \epsilon\dot{\bar{\lambda}}_{x_2} = 0 = - \frac{\delta\bar{H}}{\delta\bar{x}_2} \qquad (13)$$

It will be noted in particular that, as a result of equations (12) and (13), $\bar{x}_2$ appears here as a pseudo-control, and the optimal values $\bar{x}_2^*$ $u^*$ are defined thus:

$$\bar{x}_2^*, \; \bar{u}^* = \min_{\bar{x}_2 \bar{u}} \; \bar{H}\,(\bar{x}_1, \bar{x}_2, \bar{u}, \bar{\lambda}_{x_1}) \tag{14}$$

$\bar{x}_2^*$ and $\bar{u}^*$ are therefore functions of $x_1$ $\lambda_{x_1}$

$$\bar{x}_2^* = \bar{x}_2^*\,(\bar{x}_1, \bar{\lambda}_{x_1}), \quad \bar{u}^* = u^*\,(\bar{x}_1, \bar{\lambda}_{x_1})$$

The system of equations (12) and (13) is then written thus:

$$\begin{aligned} \dot{\bar{x}}_1 &= f_1\,(\bar{x}_1, \bar{x}_2^*\,(\bar{x}_1, \bar{\lambda}_{x_1}), \; \bar{u}^*\,(\bar{x}_1, \bar{\lambda}_{x_1})) \\ \dot{\bar{\lambda}}_{x_1}^* &= -\frac{\delta \bar{H}}{\delta \bar{x}_1}\,(\bar{x}_1, \bar{x}_2^*\,(\bar{x}_1, \bar{\lambda}_{x_1}), \bar{u}^*\,(\bar{x}_1, \bar{\lambda}_{x_1})) \end{aligned} \tag{15}$$

with $x_1(o)$ and $x_1(t_f)$ as terminal conditions.

Although $\bar{x}_1$ satisfies the given terminal conditions, the same is not true of $\bar{x}_2$, since $\bar{x}_2$ is calculated from $\bar{x}_1$ and $\bar{u}^*$ by the second equation (12) (see Fig. 2).

It will also be noted that, if the final state $x_{1f}$ is fixed but the (current) initial state is assumed to be free,
$\lambda_{x_1}^*$ and $x_2^*$ are functions of $x_{10}$ and $t_f$:

$$\bar{\lambda}_{x_1}^*\,(x_{10}, t_f); \qquad \bar{x}_2^*\,(x_{10}, t_f) \tag{15b}$$

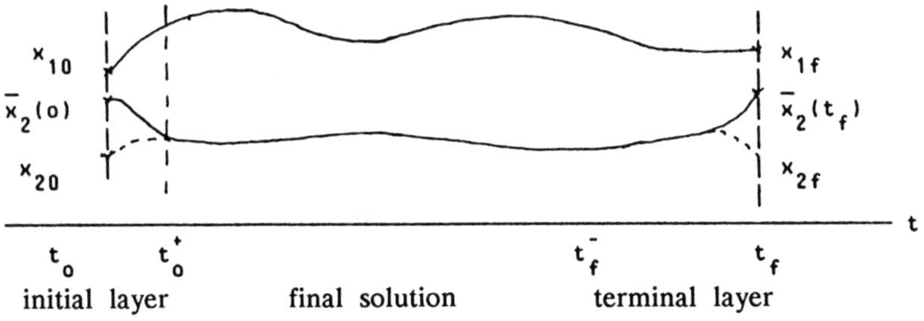


Fig. 2

b) The solution $\hat{u}$ in the initial layer is calculated in the time scale $\tau = t/\epsilon$. With respect to $\tau$, the equations are written as follows, if $\epsilon$ is again made equal to 0:

$$\dot{\hat{x}}_1 = 0 \qquad \dot{\hat{x}}_2 = f\ (\hat{x}_1, \hat{x}_2, \hat{u}) \tag{16}$$

$\hat{x}_1$ is therefore constant (equal to $x_{10}$) as is the adjoint variable:

$$\hat{\lambda}_{x_1}(\tau) = \lambda_{x_1}(o) = \overline{\lambda}^*_{x_1}\ (x_{10}, t_f)$$

The problem to be solved is therefore defined by

$$\min\ [\hat{L}\ (\hat{x}_2, \hat{u}, \hat{\lambda}^*_{x_1}, x_{10})] \tag{17}$$

where

$$\hat{L}^* = L\ (x_{10}\ \hat{x}_2\ \hat{u}) + \overline{\lambda}_{x_1}^*\ (0)\ f_1\ (x_{10}\ \hat{x}_2\ \hat{u})$$

with the second equation (16) and the terminal conditions satisfied.

$$\hat{x}_2(o) = x_{20}, \quad \lim_{t\to\infty} \hat{x}_2(\tau) = \overline{x}_2\ (x_{10}, t_f) \tag{18}$$

The optimal solution $\hat{u}$ therefore depends explicitly on $x_{20}$ and indirectly on $x_{10}$ and $t_f$, since $\overline{\lambda}^*_{x1}$ depends on these.

At first sight, this solution does not appear to provide any major advantage. In fact, if it is necessary to solve two problems with smaller dimensions, the second may be very complicated, since on the one hand the associated criterion (17) becomes complicated and on the other hand the terminal condition (18) is variable ($x_{10}$ being considered to be a current initial condition).

However, without discussing the theory any further, since this chapter is essentially concerned with a practical application, it will be seen in the next section that the problem is actually considerably simplified if each successive variable $x_i$ can be considered as belonging to a separate, more rapid dynamic class. The problem in each layer is thus reduced to a problem of static optimisation.

## *14.5 - APPLICATION TO THE PRESENT PROBLEM*

Consider the system of equations (2) presented in section 14.2 (assuming that the weight is constant for the instant) and rewritten in the form:

$$
\begin{aligned}
&\dot{x} = V \cos \gamma && x(o) = x_o && x(t_f) = x_f \\
&\epsilon \dot{h} = V \sin \gamma && h(o) = h_o && h(t_f) = h_f \\
&\epsilon^2 \dot{V} = (F-X)/m - g \sin \gamma && \text{"} && \text{"} \\
&\epsilon^3 \dot{\gamma} = (Z - m\ g \cos \gamma)/m\ V && \text{"} && \text{"}
\end{aligned}
\qquad (19)
$$

$$Q = \int_{t_o}^{t_f} (C_f\ f' + C_t)\ dt$$

where the variables $\epsilon$ indicate that, in the order used, the variables X, h, V, $\gamma$ belong to classes of increasing rapidity. Although this is based on a physical reality, this complete decoupling is somewhat artificial at first sight, as regards h and V, but will be justified subsequently by the results obtained.

The total problem would consist in writing the Hamilton formula:

$$H = C_f\ f' + C_t + \lambda_x V \cos \gamma + \lambda_h V \sin \gamma + \lambda_V \left[\frac{F-X}{m} - g \sin \gamma\right] + \lambda_\gamma (Z - mg \cos \gamma)/mV \qquad (20)$$

and finding the "control signals" F and Z which minimise H under the "constraints" specified by equations (19) and the adjoint equations:

$$\dot{\lambda}_x = -\frac{\delta H}{\delta x}, \quad \epsilon \dot{\lambda}_h = -\frac{\delta H}{\delta h}, \quad \epsilon^2 \dot{\lambda}_V = -\frac{\delta H}{\delta V}, \quad \epsilon^3 \dot{\lambda}_\gamma = -\frac{\delta H}{\delta \gamma} \qquad (21)$$

where $H = 0$ if $t_f$ is free.

### 14.5.1 - REDUCED SOLUTION (CRUISING PHASE)

If it is specified that $\epsilon = 0$ in the above equations, to obtain the reduced solution, the only remaining dynamic equation is the first one. The Hamilton formula is reduced to:

$$\bar{H} = \mathring{H} = C_f\ f' + C_t + \lambda_x\ V \cos \gamma \qquad (22)$$

and the new constraints are:

$$\gamma = 0 \qquad F = X \qquad Z = mg \qquad (23)$$

In physical terms, it may be seen that the reduced solution corresponds to the phase of stable cruising (constant altitude, drag balanced by thrust). It will also be seen that, according to equation (20), the initial control signals F and Z are replaced here by the pseudo-control signals h and V. Finally, since H does not depend on x, $\lambda_x$ is constant.

If $\mathring{\eta}$ denotes the variable $\eta$, to indicate that the layer concerned is reduced (layer 0), the problem becomes:

$$\mathring{h}^*,\ \mathring{V}^* = \underset{h.V}{\arg\min}\ |C_f\ f' + \mathring{\lambda}_x\ \mathring{V}\ |$$

which, since the Hamilton formula only contains a single adjoint variable, may be reduced to:

$$\mathring{h}^*,\ \mathring{V}^* = \underset{h.V}{\arg\min}\ \frac{C_f\ f' + C_t}{V}\ \left| \begin{array}{l} F = X \\ Z = mg \\ \gamma = 0 \\ m \ \text{given} \end{array} \right. \qquad (24)$$

Also, since the final time is free, $\dot{H} = 0$, and therefore

$$\mathring{\lambda}_x = -\frac{C_t + C_f\ f'}{\mathring{V}} \qquad (25)$$

The solution obtained from equation (24) obviously depends on:

- the costs $C_f$ and $C_t$ (often expressed in relation to $q = C_t/(C_f+C_t)$;
- the mass of the aircraft (by the constraint Z = mg);
- the atmosphere (standard or otherwise).

Figs. 3 and 4 show the variations of $\mathring{h}^*$ and $\mathring{V}^*(M)$ as a function of the mass, for q = 1 (minimum consumption) and q = 0.66 (direct operating cost).

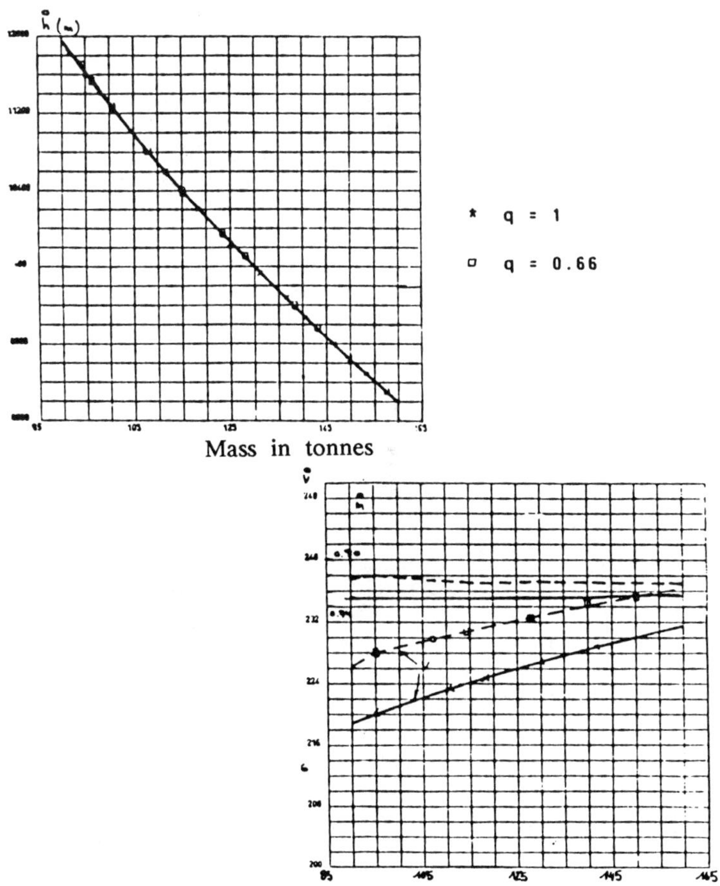


Figs. 3 and 4

In particular, since the optimal altitude increases as the mass decreases, then in theory the altitude increases as the aircraft becomes lighter. Since this is incompatible with the procedural constraints which oblige the aircraft to cruise at a constant flight level, equation (24) is replaced by

$$\dot{V}^* \ (\text{or } \dot{M}^*) = \underset{V(\text{or } m)}{\arg\min} \frac{C_f \, f' - C_t}{V} \left| \begin{array}{l} F = X \\ Z = mg \\ \gamma = 0 \\ m \ \text{given} \\ h = \text{const.} \end{array} \right.$$

The optimal Mach number is shown in Fig. 5 as a function of the mass and flight level chosen. At any given flight level, therefore, it decreases slightly in the course of the flight. The fuel economy obtained by adopting a Mach number which varies in the course of cruising, rather than a constant M, is small if the constant Mach number is carefully chosen (for an initial mass of 130 t at a flight level of 300, the optimal Mach number varies from 0.748 to 0.716. If a constant Mach number of 0.732 is adopted, the gain is only 0.94%. However, for a higher Mach number of 0.8, the difference becomes considerable and may be as much as 4%.

## 14.5.2 - SOLUTION FOR THE FIRST INITIAL LAYER

### a) - Climb

In the first initial layer, as has been demonstrated (14.5), there is a change in the variable $\tau = (t-t_0)/\epsilon$. Equations (19) are then written:

$$\dot{X}=\epsilon V \cos\gamma,\ \dot{h}=V\sin\gamma,\ \epsilon\dot{V}=(f-x)/m-g\sin\gamma,$$
$$\epsilon^2\dot{\gamma}=(Z-mg\cos\gamma)/mV \quad (26)$$

and for $\epsilon=0$ the constraints are:

$$F = X + mg\sin\gamma, \quad Z = mg\cos\gamma$$

The Hamilton formula (20) is written:

$$H^1 = (C_f\, f' + C_t) + \lambda_x^1 V^1 \cos\gamma^1 + \lambda_h^1 V^1 \sin\gamma^1 \quad (27)$$

(the indices 1 show that the solution is evaluated in the first layer), where

$$\dot{\lambda}_x^1 = 0, \quad \frac{\delta H^1}{\delta V^1} = 0, \quad \frac{\delta H^1}{\delta\gamma^1} = 0 \quad (28)$$

This first initial layer corresponds to the climbing phase, from the initial altitude h(o) to the altitude at the beginning of the cruising phase $\dot{h}(o)$, and the "control signals", in accordance with (28), are velocity and slope.

Also in accordance with (28), $\lambda_x^1$ is constant in this layer, being equal to $\dot{\lambda}_x(o)$.

Thus the optimal solution is obtained at each altitude by the equation

$$V^{1*},\ \gamma^{1*} = \underset{V,\ \gamma}{\arg\min}\ \frac{C_f\ f' + C_t + \dot{\lambda}_x(o)\ V\cos\gamma}{V\sin\gamma}\ \Bigg|\begin{array}{l} F = X + mg\sin\gamma \\ Z = mg\cos\gamma \\ h = \text{const.}\end{array} \qquad (29)$$

Since the Hamilton formula (27) has a value of zero, it is also possible to evaluate $\lambda_x^1$, which is necessary for the next layer.

Figs. 6 and 7 show the variation of the slope and the Mach number in the course of the climb for the case of a flight corresponding to a mass of 120 t at the beginning of the climb.

Notes

1. The solution given by (29) is the optimal solution calculated in each altitude layer. At the altitude h, $V^1$ and $\gamma^1$ are calculated, and from this F and Z, then a new h is calculated by integration of the equation h = V sin $\gamma$. The horizontal distance travelled during this part of the climb is calculated by $\dot{X} = V\cos\gamma$.

   In practice, a procedural constraint must be allowed for. According to the method, the cruising height is attained asymptotically, but this is incompatible with the regulation of the controller. When the rate of climb becomes less than a certain limit (400 feet per minute), the optimisation process is therefore stopped and the slope is kept constant up to $\dot{h}$.

2. At the present time, the climb procedures are type VI procedures: (indicated velocity) = const. = 300 knots, M = const. = 0.78, with a thrust defined by the N1 climb regime. It is therefore interesting to compare the optimal calculated climb profiles ($V^1$ $\gamma^1$) with the actual profile. Overall, the optimal profile is slower, in other words the distance travelled to reach the cruising altitude is greater, but it provides a gain of the order of 5 to 7%.

b) - **Descent**

The descent is calculated in much the same way as the climb (the solution for the first layer is obtained), with the differences that in this case we are concerned with the final layer, the change of variable made is therefore $\sigma = (t_f - t)/\epsilon$, the thrust is no longer greater than but less than the drag, and in equation (29) $\dot{\lambda}_x(o)$ must be replaced by $\dot{\lambda}_x(t_f)$.

In order to link the Mach number with the slope between the end of the cruising phase and the beginning of the descent, the thrust is only gradually reduced (F = X + mg sin $\gamma$). When the thrust calculated in this way becomes equal to the residual thrust (for the current altitude), the constraint F = X + mg sin $\gamma$ is replaced by F = F residual in equation (29).

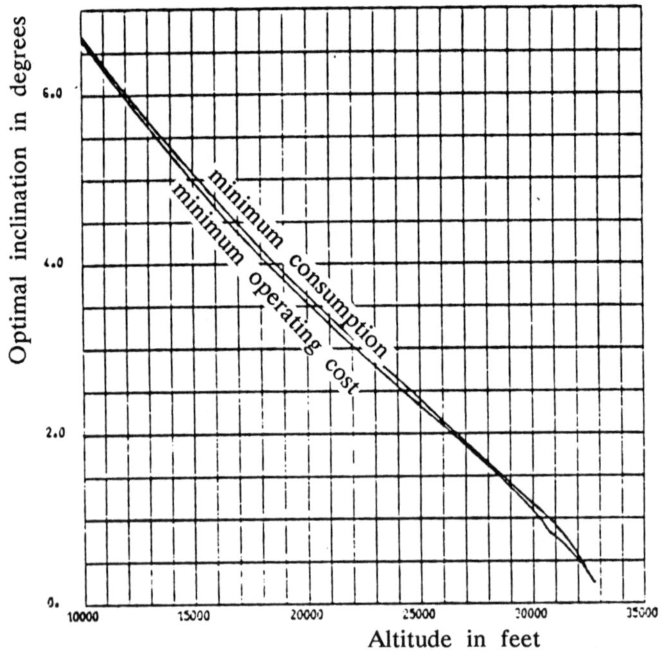
Optimal inclination in degrees
6.0
4.0
2.0
0.
minimum consumption
minimum operating cost
10000
15000
20000
25000
30000
35000
Altitude in feet

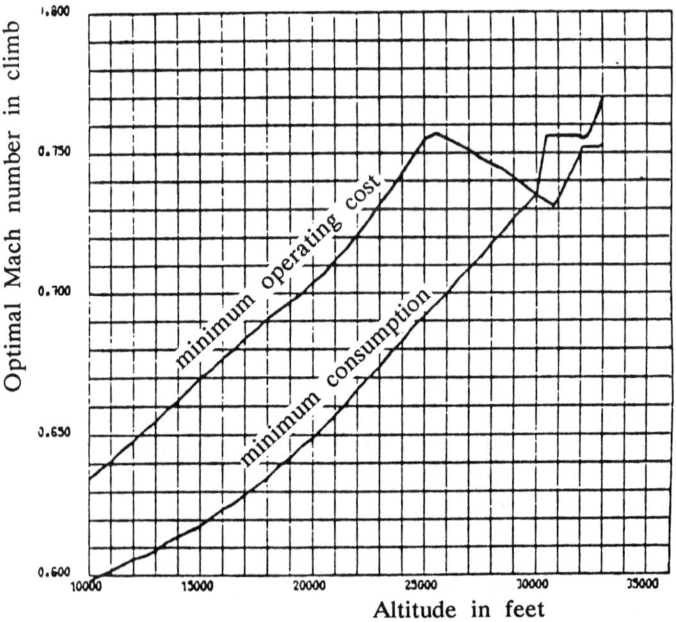
Optimal Mach number in climb
0.750
0.700
0.650
0.600
minimum operating cost
minimum consumption
10000
15000
20000
25000
30000
35000
Altitude in feet

Figs. 6 and 7

Notes

As in the case of climbing, the calculated optimal profile corresponds to a different procedure from that currently used, in which the descent is carried out at M, $V_i$ = const. and with residual thrust. It is also possible to envisage a profile which may be called sub-optimal, where the thrust used is the residual thrust, but the velocity is optimised. Fig. 8 shows the variations of the Mach number in the three cases.

Standard atmosphere, wind zero

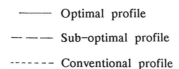

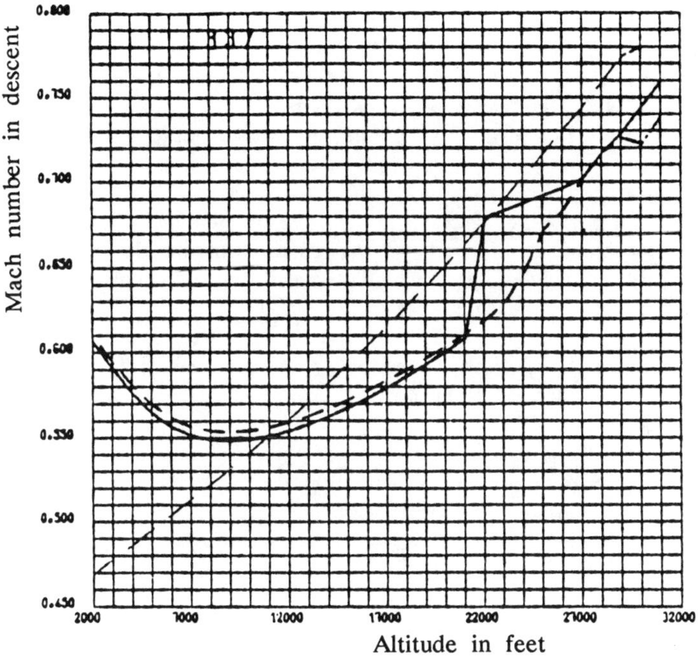


——— Optimal profile

— — — Sub-optimal profile

- - - - - - Conventional profile

Fig. 8

### 14.5.3 – SOLUTION IN THE SECOND LAYER

The solution in the second layer should correspond to the linking of the slope at its initial value $\gamma(o)$ with the value calculated at the beginning of the climb $\gamma^1(h_o)$. It is clear that the calculation of this phase is only of theoretical interest, and has no practical effect on the value of the criterion. Consequently it will not be considered here.

### 14.5.4 – SYNTHESIS OF THE OVERALL TRAJECTORY

In the preceding sections, the optimal profiles for cruising, climb, and descent were determined. Although the method makes it possible to provide links between the different phases with respect to the altitude, the velocity and the slope, it is still necessary to allow for the fact that the total distance travelled is set beforehand. As regards the matters discussed in 14.4, the complete separation of the dynamics has made it possible to avoid the difficulties connected with $t_f$ in equation (15b).

If it is assumed, as here, that the distance to be travelled is sufficiently long for the aircraft to be able to reach its optimal cruising level, the problem is reduced to the calculation of the instant of the beginning of the descent, which causes no difficulties since the characteristics of the descent profile are known.

The results have been found (Figs. 9 and 10) for an aircraft of 140 t over a distance of 1000 n.m.

The aircraft takes five minutes longer to complete the optimal trajectory (in the sense of the direct operating cost characterised by $q = 0.66$) than to complete the conventional trajectory (for a stage of 1000 n.m.) but consumes 835 kg less fuel, corresponding to an overall gain very close to 4%.

## *14.6 – CONCLUSION*

As indicated in the introduction, the objective here was to evaluate a method of sub-optimal optimisation enabling the control to be determined:

1) in a form that is sufficiently simple for real-time implementation to be feasible;

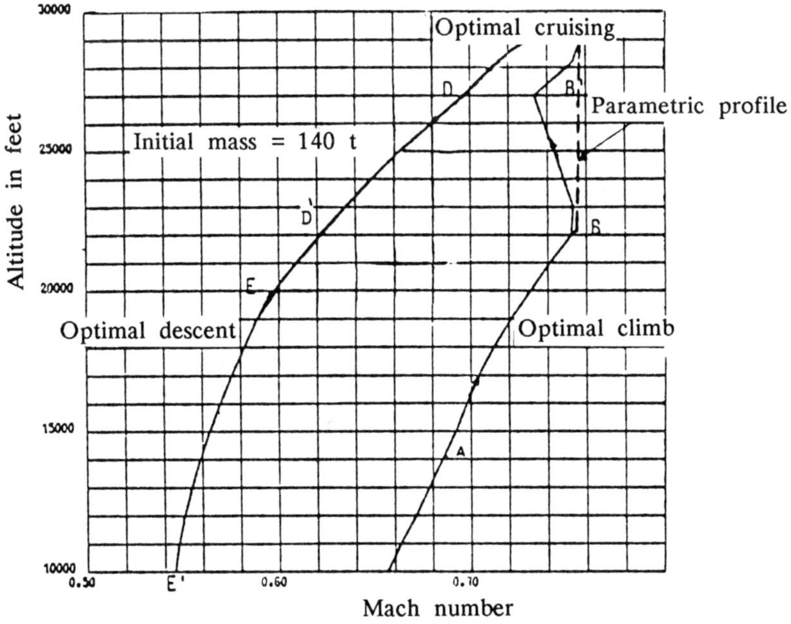
Optimal cruising
Parametric profile
Initial mass = 140 t
Optimal descent
Optimal climb
Altitude in feet
30000
25000
20000
15000
10000
0.50
0.60
0.70
Mach number
D
D'
E
E'
B
B'
A

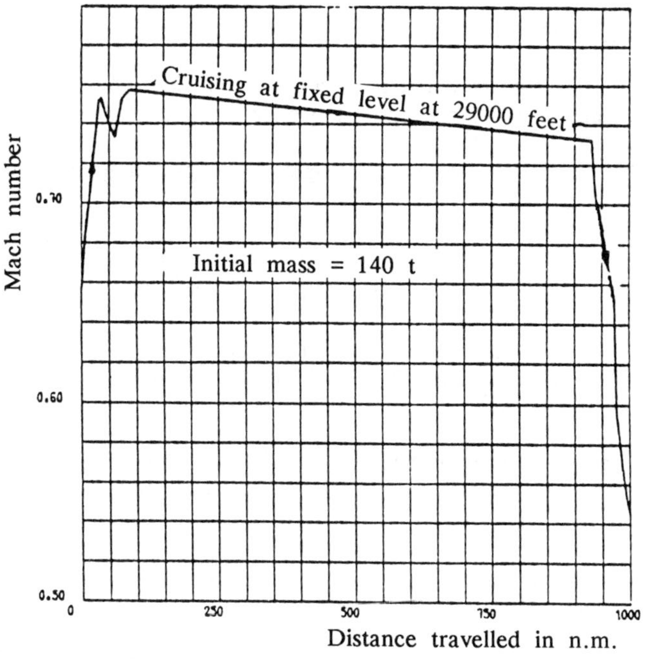
Cruising at fixed level at 29000 feet
Initial mass = 140 t
Mach number
0.70
0.60
0.50
0
250
500
750
1000
Distance travelled in n.m.

Figs. 9 and 10

2) as a function of the state, rather than the time, so that closed loop application is possible. The method of forced singular perturbations meets this requirement. It has been used here in its extreme form for illustrative purposes, and obviously the complete separation of the dynamics adopted may surprise some readers. It is justified a posteriori, to the extent that it enables a solution to be found which provides an appreciable gain, while reducing the overall problem of dynamic optimisation to simple problems of parametric optimisation.

Note:

The present study is an extract from the research carried out at CERT by A.J. Fossard and M. Clique under a contract with Intertechnique, whom we wish to thank here, and by Y.A. Ibrahim in connection with a doctoral thesis for ENSAE.

## *REFERENCES*

M.D. Ardema, Solution of the minimum time to climb problem by matched asymptotic expansions, AIAA Journal, v. 14, No. 7, 1976

D. Blank and J. Shinar, Efficient combinations of numerical techniques applied for aircraft turning performance optimization, Journal of Guidance and Control, v. 5, No. 2, 1982

Calise, Singular perturbation techniques for on line optimal flight path control, Journal of Guidance and Control, v. 4, No. 4, 1981

H. Erzberger and J.D. McLean, Fixed range optimum trajectories for short haul aircraft, NASA TN D 8115, 1975

A.J. Fossard and J.G. Magni, Modélisation, commande et application des systèmes à échelles de temps multiples, RAIRO, v. 16, No. 1, 1982

A.J. Fossard, M. Clique, and G. Lample, Faisabilité d'un calculateur de performances pour avion civil, Rapport de contrat Intertechnique/CERT-DERA, 1981

Y.A. Ibrahim, Optimisation par la méthode des perturbations singulières. Application à la détermination des trajectoires optimales, Thèse de Docteur Ingénieur, ENSAE, 1983

H.G. Jacob, Rechnergestutzte Optimierung statischer und dynamischer Systeme, Springer Verlag, 1982

H.T. Huynh and J.Y. Michel, Optimisation de trajectoires à moyen terme, RT 9/7224 SY ONERA, 1983

P.V. Kokotovic, Singular perturbations and optimal control: outils et modèles mathématiques pour l'automatique, l'analyse de systèmes et le traitement du signal, Editions CNRS, v. 3, 1983

J.F. Magni and A.J. Fossard, Commande en deux étapes des systèmes linéaires à deux dynamiques, RAIRO, v. 16, No. 1, 1982

R.K. Mehra and R.B. Washbrun, A study of the application of singular perturbation theory, NASA TR 15113, Aug. 1979

E.S. Rutkowski, Energy approach to the general aircraft performance problem, Journal of Aeronautical Science, v. 21, No. 3, 1954

K.H. Well, B. Faber, and E. Berger, Optimization of tactical aircraft maneuvers utilizing high angles of attack, Journal of Guidance and Control, v. 5, No. 2, 1982

# CHAPTER 15

# INTERCEPTION IN MINIMUM TIME WITH SPECIFIED FINAL CONDITIONS

This chapter is concerned with the application of the maximum principle for the determination of the guidance algorithm for a missile which is to intercept a target in minimum time with specified final conditions.

## 15.1 - DESCRIPTION OF THE PROBLEM

Let the target have variable velocity and direction. The guidance laws must be determined for the missile flying at constant velocity, in such a way that the interception takes place in minimum time and at a given angle between the velocities of the two moving bodies.

To simplify the representation of the problem, it is assumed that the missile and the target are flying in the same plane and in a homogeneous environment. Let each moving body be characterised by its Cartesian position (x,y), its direction, $\theta$, and its modulus of velocity V, the fixed point (X,Y) being chosen in such a way that at the initial instant the missile is at the origin with its velocity in direction X.

The motion of the bodies is described by the following equations:

Missile (1)

$$\begin{bmatrix} \dot{x}_a = V_A \cos \theta_A \\ \dot{y}_A = V_A \sin \theta_A \\ \dot{\theta}_A = U \quad \text{with} \quad |U| \leqslant U_M \end{bmatrix}$$

Target (2)

$$\begin{bmatrix} \dot{x}_c = V_c \cos \theta_c \\ \dot{y}_c = V_c \sin \theta_c \\ \dot{V}_c = \gamma_c \\ \dot{\theta}_c = r_c \end{bmatrix}$$

where U is the missile control to be determined and $\gamma_c$, $r_c$ are the target controls assumed to be identified by the missile. The initial conditions are as follows:

$$\begin{bmatrix} x_A(0) = 0 \\ y_A(0) = 0 \\ \theta_A(0) = 0 \end{bmatrix} \quad \text{and} \quad \begin{bmatrix} x_c(0) = x_o \\ y_c(0) = y_o \\ V_c(0) = V_o \\ \theta_c(0) = \theta_o \end{bmatrix}$$

Therefore the problem is to determine the control U of the missile which minimises the criterion

$$J = \int_0^{T_f} dt$$

with the final time $T_f$ being such that

$$(3) \qquad \begin{aligned} x_A(T_f) &= x_c(T_f) \\ y_A(T_f) &= y_c(T_f) \\ \theta_A(T_f) - \theta_c(T_f) &= \theta_i \end{aligned}$$

## *15.2 – OPTIMALITY EQUATIONS*

To simplify the introduction of the terminal conditions (3), let the variables be changed as follows:

$$\epsilon_x = x_A - x_c$$

$$\epsilon_y = y_A - y_c$$

$$\epsilon_\theta = \theta_A - \theta_c$$

The equations for the system are written thus:

$$\begin{array}{lll} \dot{x}_A = V_A \cos \theta_A & x_A(0) = 0 & x_A(t_f) \text{ free} \\ \dot{y}_A = V_A \sin \theta_A & y_A(0) = 0 & y_A(t_f) \text{ free} \\ \dot{\theta}_A = U & \theta_A(0) = 0 & \theta_A(t_f) \text{ free} \\ \dot{\epsilon}_x = V_A \cos \theta_A - V_c \cos(\theta_A - \epsilon_\theta) & \epsilon_x(0) = -x_{c0} & \epsilon_x(t_f) = 0 \\ \dot{\epsilon}_y = V_A \sin \theta_A - V_c \sin(\theta_A - \epsilon_\theta) & \epsilon_y(0) = -y_{c0} & \epsilon_y(t_f) = 0 \\ \dot{\epsilon}_\theta = U - r_c & \epsilon_\theta(0) = -\theta_o & \epsilon_\theta(t_f) = \theta_i \\ \dot{V}_c = \gamma_c & V_c(0) = V_o & V_c(t_f) \text{ free} \end{array}$$

If $\psi$ is the adjoint state of the state of this system, the Hamiltonian associated with the optimality problem under consideration is written thus:

$$\begin{aligned} H = {} & \psi_x V_A \cos \theta_A + \psi_y V_A \sin \theta_A + \psi_\theta u + \psi_{\epsilon x}(V_A \cos \theta_A \\ & - V_c \cos (\theta_A - \epsilon_\theta)) + \psi_{\epsilon y}(V_A \sin \theta_A - V_c \sin (\theta_A - \epsilon_\theta)) \\ & + \psi_{\epsilon\theta}(U - r_c) + \psi_V \gamma_c - 1 \end{aligned}$$

and therefore the adjoint system is:

$$\dot{\psi}_x = 0$$
$$\dot{\psi}_y = 0$$
$$\dot{\psi}_\theta = +\psi_x V_A \sin\theta_A - \psi_y V_A \cos\theta_A$$
$$+ \psi_{\epsilon x}(V_A \sin\theta_A - V_c \sin(\theta_A - \epsilon_\theta))$$
$$-\psi_{\epsilon y}(V_A \cos\theta_A - V_c \cos(\theta_A - \epsilon_\theta))$$
$$\dot{\psi}_{\epsilon x} = 0$$
$$\dot{\psi}_{\epsilon y} = 0$$
$$\dot{\psi}_{\epsilon\theta} = +\psi_{\epsilon x} V_c \sin(\theta_A - \epsilon_\theta) - \psi_{\epsilon y} V_c \cos(\theta_A - \epsilon_\theta)$$
$$\dot{\psi}_V = 0$$

The transversality conditions imposed by the terminal conditions imply that

$$\psi_x(T_f) = 0$$
$$\psi_y(T_f) = 0$$
$$\psi_\theta(T_f) = 0$$
$$\psi_V(T_f) = 0$$

These conditions complete the initial and final conditions of the direct system, thus completely defining the system.

The controller must maximise the Hamiltonian which in this case is linear in U, and therefore

$$\hat{U} = U_M \text{ sign } (\psi_\theta + \psi_{\epsilon\theta})$$

Also, since the system is stationary and the final time is free, the Hamiltonian is zero on the optimal trajectory $\hat{H} = 0$:

Theoretically, all these conditions make it possible to find the control $\hat{U}$ by determining the adjoint state $\psi$. However, the integration of the system meeting the conditions at both ends is difficult, especially if it has to be done in real time. Consequently, the problem is divided into two simpler sub-problems as follows:

a) – determination of the control $\hat{U}$ leading the missile to a given point $(x_f, y_f, \theta_f)$ in minimum time.

b) – determination of the end point $(x_f, y_f, \theta_f)$ satisfying the terminal conditions (3).

## *15.3 – DETERMINATION OF THE OPTIMAL CONTROL*

In this section, it is assumed that the point $(x_f, y_f, \theta_f)$ to be reached by the missile is known. In this case, the dynamic of the target no longer affects the solution of the optimality problem, which is written thus:

- System:

$$\dot{x}_A = V_A \cos \theta_A$$
$$\dot{y}_A = V_A \sin \theta_A$$
$$\dot{\theta}_A = U$$
$$\dot{\psi}_x = 0 \rightarrow \psi_x = \text{const.}$$
$$\dot{\psi}_y = 0 \rightarrow \psi_y = \text{const.}$$
$$\dot{\psi}_\theta = +\psi_x V_A \sin \theta_A - \psi_y V_A \cos \theta_A$$

- Limit conditions

$$x_A(0) = 0 \qquad x_A(T_f) = x_f$$
$$y_A(0) = 0 \qquad y_A(T_f) = y_f$$
$$\theta_A(0) = 0 \qquad \theta_A(T_f) = \theta_f$$

- Hamiltonian:

$$H = \psi_x V_A \cos \theta_A + \psi_y V_A \sin \theta_A + \psi_\theta U - 1$$

The optimal control U must maximise the Hamiltonian, and therefore

* if

$$\psi_\theta \neq 0 \qquad \hat{U} = U_M \text{ sign } (\psi_\theta)$$

In this case, the optimal trajectory is a circle of radius $V_A/U_M$ corresponding to the maximum capabilities of the missile.

* if

$\psi_\theta = 0$ with $\dot{\psi}_\theta \neq 0$, $\epsilon U_M$ is switched to $-\epsilon U_M$; with $\dot{\psi}_\theta = 0$, the Hamiltonian becomes independent of U.

In the latter case, the control $\hat{U}$ is such that the trajectory generated complies with $\psi_\theta = \dot{\psi}_\theta = 0$, corresponding to the singular trajectory condition of the optimality problem.

$$\dot{\psi}_\theta = 0 \longrightarrow \tan \theta_A = \frac{\psi_y}{\psi_x} = \text{const}.$$

The singular trajectory is a straight line on which $\hat{U} = 0$.

The optimal trajectory is therefore a series of arcs and straight lines which are completely determined when $\psi_\theta$ is known. Since the final time $T_f$ is free and the system is stationary, the Hamiltonian is zero on the optimal trajectory:

$$\hat{H} = \psi_x V_A \cos \theta_A + \psi_y V_A \sin \theta_A + \psi_\theta \hat{U} - 1 = 0$$

If $\theta_s$ is the direction of the velocity on the singular trajectory, then

$$\psi_x V_A \cos\theta_s + \psi_y V_A \sin\theta_s - 1 = 0$$

with

$$\tan\theta_s = \psi_y/\psi_x$$

and therefore

$$\psi_x = \frac{\cos\theta_s}{V_A} \qquad \psi_y = \frac{\sin\theta_s}{V_A}$$

then

$$\hat{H} = \cos(\theta_A - \theta_s)\,\psi_\theta\hat{U} - 1 = 0 \qquad \text{where } \theta_A = \int_0^t \hat{U}\,dt = \hat{U}t$$

and therefore

$$\psi_\theta = \frac{1-\cos(\hat{U}t-\theta_s)}{\hat{U}}$$

The possible instants of switching are then

$$\tau_i = \frac{|\theta_s + 2k\pi|}{U_M}$$

The condition of cancellation of $\psi_\theta$, $\theta_A = \theta_s$ is identical to that of $\dot{\psi}_\theta$:

$$\dot{\psi}_\theta = \sin(\theta_A - \theta_s)$$

Consequently, if switching to the control takes place, it can only be $(\epsilon U_M, 0)$ or $(0, \epsilon U_M)$, in other words a change from a circular trajectory to a straight one or vice versa. The form of the optimal trajectory to be found is therefore circle–straight line–circle (Fig. 1). The trajectory can be determined from this information either by geometry or by integration of the system.

Given that

$$\epsilon_1 = \text{sign}(\psi_\theta(0))$$
$$\epsilon_2 = \text{sign}(\psi_\theta(t_f))$$

and $\tau_1$, $\tau_2$ are the instants of switching.

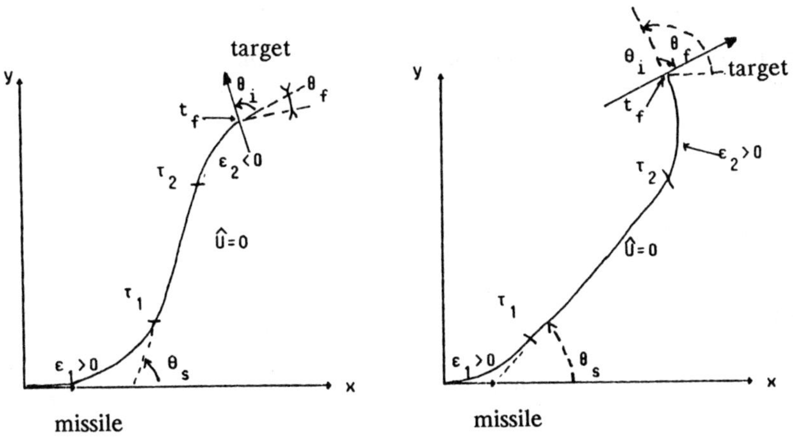


$$\begin{cases} \psi_\theta(0) > 0 \\ \psi_\theta(t_f) < 0 \end{cases} \qquad \begin{cases} \psi_\theta(0) > 0 \\ \psi_\theta(t_f) > 0 \end{cases}$$

| | |
|---|---|
| $\tau_1$, $\tau_2$ | instants of switching |
| $\theta_s$ | course of the missile on rectilinear trajectory |
| $\theta_f$ | angle of impact |
| $\theta_i$ | course of the missile on impact |

Optimal trajectories

Fig. 1

$$x_A(t_f) = V_A\left[ (\tau_2-\tau_1) \cos \theta_s + \epsilon_2 \frac{\sin \theta_f}{U_M} + (\epsilon_1-\epsilon_2) \frac{\sin \theta_s}{U_M} \right]$$

$$y_A(t_f) = V_A[ (\tau_2-\tau_1) \sin \theta_s + \frac{1}{U_M} (\epsilon_1-\epsilon_2 \cos \theta_f)$$

$$- (\epsilon_1 - \epsilon_2) \frac{\cos \theta_s}{U_M} ]$$

$$\theta_A(t_f) = U_M (\epsilon_1 \tau_1 + \epsilon_2 (t_f - \tau_2))$$

Bringing the expression of the final position into the final conditions, we obtain:

$$\theta_s = \text{arc tan} \frac{y_1}{x_1} - \text{arc sin} \frac{\frac{V_A}{U_M} (\epsilon_1 - \epsilon_2)}{\sqrt{x_1^2 + y_1^2}}$$

where

$$x_1 = x_f - \frac{V_A}{U_M} \epsilon_2 \sin \theta_f \qquad y_1 = y_f - \frac{V_A}{U_M} (\epsilon_1 - \epsilon_2 \cos \theta_f)$$

Now, by definition,

$$\theta_s = \theta_A(\tau_1)$$

i.e.

$$\theta_s = \epsilon_1 U_M \tau_1 \rightarrow \epsilon_1 = \text{sign } \theta_s$$

and similarly

$$\theta_f - \theta_s = \epsilon_2 U_M(t_f - \tau_2) \rightarrow \epsilon_2 = \text{sign } (\theta_f - \theta_s)$$

Consequently, to have $\epsilon_1 = \epsilon_2$, it is necessary that

$$\text{sign } \theta_s = \text{sign } (\theta_f - \theta_s)$$

i.e. $0 \leqslant |\theta_s| \leqslant |\theta_f|$

and $\epsilon_1 = \text{sign } \theta_s = \text{sign } \theta_f$

where

$$\theta_s = \text{arc tan} \frac{y_1}{x_1}$$

Similarly, to have $\epsilon_1 = - \epsilon_2$, it is necessary that

$$\text{sign}(\theta_s) = - \text{sign } (\theta_f - \theta_s)$$

where

$$\theta_s = \text{arc tan} \frac{y_1}{x_1} - \text{arc sin} \frac{2 \frac{V_A}{U_M}}{\sqrt{x_1^2 + y_1^2}}$$

By examining the above conditions as a function of the point to be reached, $\epsilon_1$ and $\epsilon_2$ can be defined. The three remaining unknown quantities, $\tau_1$, $\tau_2$, $\tau_f$, are then found by solving the following system:

$$x_A(T_f) = x_f$$
$$y_A(T_f) = y_f$$
$$\theta_A(T_f) = \theta_f$$

## *15.4 – DETERMINATION OF THE END POINT* ($x_f$, $Y_f$, $\theta_f$)

This end point, used for the determination of the control U, must belong to the target trajectory, in other words must be a solution of system (2), and must also comply with the identity of the flight time of the target and the missile.

The first condition is satisfied by calculating the final conditions by integration of the target system on a certain horizon $T_f^*$. These conditions can be used to calculate the trajectory of the missile and, in particular, the final time $T_f$. This solution of the optimality problem may be written symbolically as follows:

$$T_f = F(T_f^*)$$

Consequently, the requisite final time (which determines the end point) is the solution of the formal equation

$$X = F(X)$$

This type of solution is found by an iterative procedure, using variational methods.

Let

$$Y = F(X) - X$$

X must be found such that Y is zero. The gradient $\Delta X/\Delta Y$ will be calculated at an initial point $x_1$ by solving the system for $X_1$ and $X_1 + \Delta X$; it is found that

$$Y_1 = F(X_1) - X_1$$

$$Y_2 = F(X_2) - X_2$$

If the operator F is linear, the required solution is:

$$X_3 = X_2 - Y_2 \left(\frac{X_2 - X_1}{Y_2 - Y_1}\right)$$

Since F is not linear, $Y_3$ is different from zero, and the solution is then calculated by successive iterations, using the formula:

$$X_{n+1} = X_n - Y_n \frac{X_n - X_{n-1}}{Y_n - Y_{n-1}}$$

Although this algorithm for the solution is very simple, it provides the solution in a few iterations. This convergence is explained by the quasi-linearity of the operator F. This operator is related to that found on the assumption of the linearity of the missile and target trajectories, which is written thus:

$$T_f = \frac{V_c}{V_A} T'_f + \frac{x_c}{V_A}$$

## *15.5 - CONCLUSION*

The application of the maximum principle makes it possible to solve the problem of an interception in minimum time with specified final conditions.

However, the solution of the optimality equations requires a large amount of calculation. In the plane, the system to be integrated is of 14th order (reduced to 9th order when the constants are taken into account), with the limit conditions at both ends. The division of the initial problem into two sub-problems enables the dimension of the system to be reduced (to 6th order in the plane, or 4th order if the constants are taken into account), thus simplifying the application of the missile guidance algorithm.

It should be noted that the problem set here requires the anticipation of the movement of the target. Consequently, the guidance algorithm must be linked with an algorithm for the estimation of the target dynamic. Since these estimates are never perfect, it is necessary to recalculate the guidance trajectory in real time, and therefore to store a fast closed-loop guidance algorithm in the missile computer, thus justifying the simplifications made for the solution of the optimality problem.

# CHAPTER 16

# ROBOT CONTROL TECHNIQUES

## *16.1 - GENERAL*

The purpose of this chapter is to define the possibilities of theories and techniques of automation and digital control applied to the analysis of the behaviour, to the programming, and to the real-time feedback control of robot manipulators. These machines may be treated as articulated chains of bodies which are rigid as a first approximation, designed to position, orientate, and move a "terminal device" (such as a gripper or a tool) which acts in a concrete way on the environment by moving objects, welding, assembling, generating forces, etc.

A robot therefore has a number of degrees of freedom, in the terminology of automation; the number is often six, to enable the position and attitude of the terminal device to be completely controlled, with more or less direct actuation by motors. The robot must be capable of "fine" motions, in the case of assembly for example, and also of rapid large-scale motions, as well as the precise following of trajectories for arc welding or shaping.

In general, fine motions require simultaneous monitoring of the position of the terminal device and of the forces exerted by it on its environment.

The transitory motions must generally be rapid, while any necessary trajectory following to prevent collisions of the robot with its environment must be maintained. The control system must generate corresponding position and velocity references, and must match the motions to these references. With modern robots which may have an end-of-arm velocity in excess of three or four metres per second, the dynamic couplings between the component parts of the robots are not negligible, and must be taken into consideration.

The same applies to the case of shaping with a small radius, while in following quasi-rectilinear trajectories with a low linear and angular velocity of the terminal device a kinetic control system is generally sufficient.

The techniques of analysis and synthesis of the control laws required for these various functions of robots are described in the following sections. They are based on the use of mathematical models describing a robot considered as a generator of position, motion, and force. For this reason, the first step is to formalise the corresponding models and the functions which the robot must perform. This introduction also demonstrates the complexity of the problems associated with robotics which must be solved, arising from the presence of multivariable systems with strong coupling, major non-linearities, singularities and structural variations which are often poorly known in advance, and the necessity of calculating the control signals within very short time limits of the order of a few milliseconds.

## 16.1.1 - GENERAL NOTES

### 16.1.1.1 - Definitions and notation

Consider a mechanical articulated structure (Fig. 1) consisting of N rigid bodies connected by simple joints of the rotary type (with one degree of freedom of rotation) or of the prismatic type (with one degree of freedom of translation).

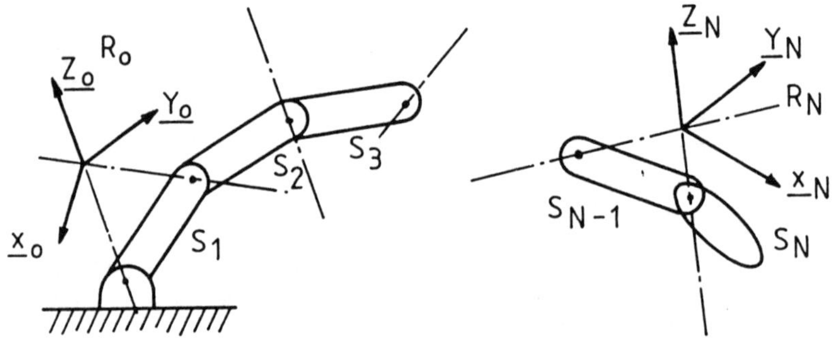


Diagram of a simple articulated chain

Fig. 1

Each body is associated with an orthonormalised frame of reference. The "zero" body, which is generally fixed, is the **base** of the robot. The final body is the **terminal device.** The position and orientation of $R_N$ in $R_o$ form the location of the terminal device. The terminal device may be defined by six independent variables, the **operational coordinates,** for example the cartesian coordinates of $O_N$ in $R_o$ and the Eulerian angles associated with the rotation which causes $R_N$ to be parallel to $R_o$. The situation of the terminal device is a function of the **joint coordinates** $q_1$, $q_2 \ldots q_N$. In the case of a rotary joint, the joint coordinate is the angle of rotation between the two bodies. In the case of a prismatic joint, it is the amplitude of translation. In both cases, the origin of the variations is arbitrary. The situation is defined by the column vector of the operational coordinates, denoted $\underline{X}$, while the vector of the joint coordinates, denoted $\underline{q}$, specifies the **configuration** (or posture) of the robot at each instant.

The joints are moved by motorisation systems consisting of actuators and transmissions (reduction gearing, rods, belts, etc.). A set, $\underline{a}$, of **actuator position coordinates**, $a_1$, $a_2 \ldots a_N$ is associated with these.

The $q_i$ are linked to the $a_i$ by the model of the transmission system:

$$\underline{a} = \underline{\psi}\ (\underline{q}) \tag{1}$$

which is generally one-to-one but not always linear, and can be chosen in such a way that $\underline{\psi}\ (\underline{0}) = \underline{0}$.

The variables $a_i$ and $q_i$ are limited by end-of-travel indicators and mechanical stops respectively, allowing for some exceptions:

$$a_{im} \leqslant a_i \leqslant a_{iM} \tag{2}$$

$$q_{im} \leqslant q_i \leqslant q_{iM} \tag{3}$$

The vectors defined above change with time, and it is therefore also necessary to consider the velocities $\dot{\underline{a}}$, $\dot{\underline{q}}$ and the kinetic torsor $\underline{V}$ which specifies the translation velocity $\dot{\underline{O}}_N$ and the instantaneous rotation velocity $\underline{\Omega}_N$ of the terminal device with respect to $R_o$:

$$\underline{V} = \begin{bmatrix} \dot{\underline{O}}_N \\ \cdots \\ \underline{\Omega}_N \end{bmatrix} \tag{4}$$

Finally, when dealing with the forces and torques acting in the system, $\underline{C}$ denotes the column vector of the motor torques and/or forces, $\underline{\Gamma}$ denotes the vector of the torques and/or forces at the joints, and $\underline{F}$ denotes the torsor of the resultant force and moment of the forces between the terminal device and the environment, assuming that no other part of the robot is in contact with the environment.

### 16.1.1.2 - Geometric, kinematic and dynamic models

The "direct" geometric model describes the situation of $R_N$ in $R_o$ as a function of the joint coordinates. A very widely used formulation is based on the homogeneous representation with four components of the vector

$$\overrightarrow{O_o O_N}: \quad [\underline{O}_N \ \vdots\ 1]^T$$

where the index "T" denotes the matrix transposition. In these conditions, the above vector may generally be represented by:

$$[w\ \underline{O}_N \ \vdots\ w]^T$$

where w is a non-zero scalar.

Thus, if $\underline{X}_N$, $\underline{Y}_N$, and $\underline{Z}_N$ denote the unit vectors of $R_N$ expressed in $R_o$, the matrix defining the transformation from $R_o$ to $R_N$ is written as follows:

$$T_{O,N} = \begin{bmatrix} \underline{X}_N & \underline{Y}_N & \underline{Z}_N & \underline{O}_N \\ 0 & 0 & 0 & 1 \end{bmatrix} \tag{5}$$

The last line makes it possible to have a uniform notation, but does not play an active part in the matrix calculations considered here, which are therefore in practice concerned with matrices of dimensions 3x4. The matrix $T_{O,N}$ is the product of the elementary matrices $T_{i,i+1}$ associated with the transformations between the components of a point in $R_i$ and those in $R_{i+1}$. Each matrix of the type $T_{i,i+1}$ is a function of the joint variable $q_{i+1}$ and of the geometrical parameters of the robot.

$$T_{O,N} = T_{0,1}\ T_{1,2} \cdots T_{N-1,N} \tag{6}$$

This equation, which provides the situation of the terminal device in $R_o$ as a function of the configuration variables, leads to a **direct geometric model** having the following form:

$$\underline{X} = \underline{f}\ (\underline{q}) \tag{7}$$

There are many methods available for obtaining this model automatically in numerical or symbolic form.

It is more difficult to obtain an inverse geometric model, in other words the solution(s) of (7), $\underline{q}\ (\underline{X})$ when it (or they) exist.

The **direct differential model** of the robot is deduced from the geometric model:

$$\underline{dX} = J(\underline{q})\ d\underline{q} \tag{8}$$

where $J(\underline{q})$ is the Jacobian matrix of the vector function $\underline{f}$. this matrix can be expressed in various ways, depending on the choice of the variables defining the situation $\underline{X}$. By contrast, the **kinematic model** of the same type,

$$\underline{V} = J_c(\underline{q})\ \dot{q} \tag{9}$$

is obtained after $R_n$ has been chosen. To take the transmission systems into account, it is sufficient to know the following:

$$\underline{q} = \underline{\psi}^{-1}\ (\underline{a}) \tag{10}$$

and therefore

$$\dot{\underline{q}} = \Lambda(\underline{a})\ \dot{\underline{a}} \tag{11}$$

to obtain

$$\underline{V} = J_A\ (\underline{a})\ \dot{\underline{a}} \tag{12}$$

which forms the direct kinematic model for the actuators.

For robots which are actuated by rotating electric motors and which have simple transmission kinematics, the relationship $\underline{\psi}^{-1}$ is linear and its Jacobian $\Lambda$ is constant. When there are six degrees of freedom, the matrices J, $J_C$, and $J_A$ are square, having dimensions equal to six. When the robot has more degrees of freedom ($N>6$), they are rectangular, of dimension 6xN.

Rectangular matrices are also obtained when a sub-space of the situation space, of dimension $M<N$, rather than the complete situation, is investigated.

It will be understood, therefore, that, although they are locally linear, the differential and kinematic models may involve problems of inversion. A robust method, tolerating changes in rank and dimension, is to use the pseudo-inverse matrix of the matrix under consideration. For example, let J be this matrix and $J^+$ its pseudo-inverse; the general solution of (8) is then written:

$$d\underline{q} = J^+ \, d\underline{X} + (I_N - J^+ J) \, \underline{v} \tag{13}$$

where $\underline{v}$ is the arbitrary vector with N dimensions and $I_N$ is the unit matrix in this space.

The pseudo-inverse matrix is unique, among the set of generalised inverses of J.

The homogeneous term $J^+ \, \underline{dX}$ belongs to the range space of $J^+$, while the second term of (13) belongs to the kernel of J, making it possible to obtain all the configurations leading to (8) when $M<N$. When J is non-singular, $J^+ = J^{-1}$. Finally, when (8) is impossible, the solution $d\tilde{\underline{q}}$ such that

$$d\tilde{\underline{q}} = J^+ \, d\underline{X} \tag{14}$$

is the best in the sense of the least squares:

$$\| J \, d\tilde{\underline{q}} - d\underline{X} \| = \min_{d\underline{q} \in R^n} \| J \, d\underline{q} - d\underline{X} \| \tag{15}$$

Equation (13) for the special case of $\underline{v} = \underline{0}$ is an **inverse differential model**. An **inverse kinematic model** is defined in the same way.

In the stationary state, the Jacobian matrices have an equal effect in the calculations. The principle of the virtual power is written:

$$\underline{F}^T \, \dot{\underline{X}} = \underline{C}^T \, \dot{\underline{a}} = \underline{\Gamma}^T \, \dot{\underline{q}} \tag{16}$$

Therefore, taking (9) into account:

$$\underline{\Gamma} = J_c^T \, \underline{F}$$

This equation can be used to calculate the torques or forces required at the joints to exert the force and moment defining $\underline{F}$ on the environment of the robot. Similarly, at the motor level, it is found that

$\underline{C} = J_A^T \, \underline{F}$. Inversely, therefore, in regular cases, $\underline{F} = J_A^{-T} \, \underline{C}$.

The relationships in these cases between the rigidity matrix $K_A$ of the robot "seen by the motors" and the matrix K "seen by the environment" can be expressed in the same way. The potential energy of deformation can be written thus:

$$K_A = J_A^T \, K \, J_A \tag{18}$$

The above statements presuppose virtual displacements $\underline{dX}$ or virtual velocities $\underline{V}$ which are compatible with the constraints of the mechanism and with the connections with the environment. Within certain limits, these static models may be used when the robot – in an assembly task, for example – is considered simultaneously as a generator of motions in certain directions, and as a generator of forces in other directions. In this case the model is known as a **hybrid model**. It should be noted at this point that, since each degree of freedom is subject to feedback control, the rigidity matrix $K_A$ is a function of the static gains of the feedback controllers, by means of which the rigidity (or its reciprocal, the compliance) of the robot as seen by the environment can be controlled.

The above statements presuppose that the manipulator follows the references in a very precise way. The synthesis of these feedback controllers and the analysis of their performance, and subsequently their actual implementation in the data processing system controlling the robot, can only be facilitated by taking into account the relationships between the state variables of the mechanism and the control variables. The state variables for a motion are the $q_i$ and $\dot{q}_i$, and for the sake of simplicity it will be assumed that the control signals are the torques and forces $\Gamma_i$ exerted by the motors at the corresponding joints. The relationships between the inputs and the state are therefore dependent on the dynamics of the multiple-jointed system. In formal terms, we are dealing with differential equations whose second members are the generalised forces $F_i = \Gamma_i + f_i$, where $f_i$ is the torque (or the force) arising from various dissipative terms such as friction factors. It is assumed that it is possible to model or to disregard these factors (Coulomb friction, viscous friction), and we will only examine the theoretical model representing the change of the state $\{\underline{q} \; \dot{\underline{q}}\}$ of the mechanism under the effect of the torques or forces $\{\Gamma_1 \ldots \Gamma_N\}^T = \underline{\Gamma}$; this is the **dynamic model.**

The equations of motion of robot manipulators consisting of a series of rigid articulated bodies have been determined by many authors. For the most part, they use the Alembert principle of virtual work, the Newton–Euler equations (general theorems of mechanics), or the Lagrange equations.

The most simple motorised mechanisms can be treated as chains of rigid bodies (Fig. 2) whose kinematics has been defined above by the transformation matrices which make it possible to express, within a framework connected to any one of the bodies, including the base, the components of a vector associated with another body. For the dynamics, other data are necessary:

- the mass of each body and the position of its centre of inertia (assumed to be identical to its centre of gravity);

- the inertia matrix of each body, expressed within its frame of reference.

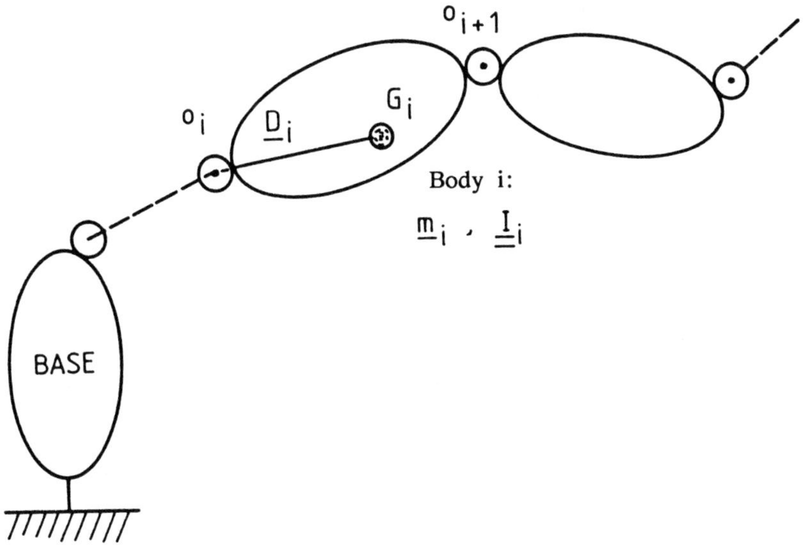


Parameters of the dynamics

Fig. 2

The state vector of the mechanism can be assumed to be formed by the N joint variables $q_i$ and by their N derivatives $\dot{q}_i$ with respect to time. The $q_i$, assumed to be independent, form a set of "generalised coordinates" in mechanics terminology.

The Lagrange function of the system is written thus:

$$L = T - U \tag{19}$$

where $T$ is the total kinetic energy and $U$ is the total potential energy of the mechanism. These scalar functions are the sums of the kinetic and potential energies respectively of all the bodies in the chain:

$$T = \sum_{i=1}^{N} T_i \qquad U = \sum_{i=1}^{N} U_i \tag{20}$$

The potential energy is a function of generalised coordinates $q_i$. In the following text it will be assumed that it is due only to the gravitational field, although it would be possible to take the presence of balancing springs into account.

The kinetic energy is a quadratic form in $\dot{q}_i$ whose coefficients are generally complicated functions of the variables $q_i$.

The Lagrange equations are written thus:

$$\frac{d}{dt}\left(\frac{\partial L}{d\dot{q}_i}\right) - \frac{\partial L}{dq_i} = \Gamma_i \qquad i = 1, 2, \ldots, N \tag{21}$$

where $\Gamma_i$ is the generalised force corresponding to the i-th joint, which, for the sake of simplicity, is often assumed to be identical to the force or to the torque exerted by the motor i on this joint.

The expression of the Lagrange function which is compatible with the formulations of the homogeneous coordinates used above has been established by Uicker and Kahn. If $r_{o,i}$ is the vector defining the position of the elementary particle, of mass $dm_i$, of rigid body number i, expressed in an absolute fixed frame of reference, the elementary kinetic energy is:

$$dT_i = \frac{1}{2}\, \dot{\underline{r}}_{o,i}^{T}\, \dot{\underline{r}}_{o,i}\, dm_i \tag{22}$$

where $\dot{\underline{r}}_{o,i}$ represents the velocity of the particle with respect to the fixed frame. When this frame is the "zero" one linked to the base of the mechanism, and the following transformation matrix is introduced,

$$T_{o,i} = T_{0,1}\ T_{1,2},\ T_{2,3} \ldots\ T_{i-1,i} \tag{23}$$

we obtain

$$\underline{r}_i = T_{o,i}\ \underline{f}_i \tag{24}$$

where the components of $f_i$ are constants in the i-th body since it is assumed to be undeformable. Therefore,

$$dT_i = \frac{1}{2} \text{Tr} \left(\dot{T}_i \underline{f}_i \underline{f}_i^T \dot{T}_i^T\right) \tag{25}$$

and therefore the kinetic energy of the body $C_i$:

$$T_i = \frac{1}{2} \text{Tr} \left(\dot{T}_i \left[ \int_{C_i} f_i f_i^T dm_i \right] \dot{T}_i^T\right) \tag{26}$$

The inertia matrix $J_i$ is a constant. If $\underline{f}_i = [x_i \; y_i \; z_i \; 1]^T$, then

$$\begin{bmatrix} \int x_i^2 \, dm_i & \int x_i y_i \, dm_i & \int x_i z_i \, dm_i & m_i D_{ix} \\ \int x_i y_i \, dm_i & \int y_i^2 \, dm_i & \int y_i z_i \, dm_i & m_i D_{iy} \\ \int x_i z_i \, dm_i & \int y_i z_i \, dm_i & \int z_i^2 \, dm_i & m_i D_{iz} \\ m_i D_{ix} & m_i D_{iy} & m_i D_{iz} & m_i \end{bmatrix}$$

where $m_i$ = mass of the body i and

$$[D_{ix} \quad D_{iy} \quad D_{iz} \quad 1]^T = \underline{D}_i,$$

the vector defining the position of the centre of inertia of the body in its own frame of reference. The potential energy of the body is written thus:

$$U_i = - m_i \, g^T \, T_i \, \underline{D}_i \tag{28}$$

where g is the gravity acceleration vector, expressed in the fixed frame of reference.

Using the above relationships, Uicker has established the form of the Lagrange equations:

$$\sum_{j=1}^{N} \left\{ \sum_{k=1}^{j} \left\{ \left( \text{Tr} . \left( \frac{\partial T_j}{\partial q_i} J_j \frac{\partial T_j^T}{\partial q_k} \right) \right\} \ddot{q}_k + \sum_{l=1}^{j} \left\{ \text{Tr} . \left( \frac{\partial T_j}{\partial q_i} J_j \frac{\partial^2 T_j^T}{\partial q_k q_l} \right\} \dot{q}_k \dot{q}_l \right\} \right.$$

$$- m_j \, g^T \frac{\partial T_j}{\partial q i} \underline{D}_j \} = F_i \tag{29}$$

These equations include the following forces or torques:

- inertial forces (terms in $\ddot{q}_k$)

- centrifugal forces (terms in $\dot{q}_k^2$)

– Coriolis forces (terms in $\dot{q}_k \dot{q}_l$ with $l \neq k$)

– gravitational forces.

The derivatives of the transformation matrices can be calculated from the expression

$$\frac{\partial T_i}{\partial q_i} = L_i \, T_i \qquad (30)$$

where

$$L_i = \begin{bmatrix} 0 & 0 & 0 & 0 \\ 0 & 0 & 0 & 0 \\ 0 & 0 & 0 & 1 \\ 0 & 0 & 0 & 0 \end{bmatrix} \quad \text{if } q_i \text{ is a translation}$$

and

$$L_i = \begin{bmatrix} 0 & 1 & 0 & 0 \\ -1 & 0 & 0 & 0 \\ 0 & 0 & 0 & 0 \\ 0 & 0 & 0 & 0 \end{bmatrix} \quad \text{if } q_i \text{ is a rotation}$$

since the axis of the joint is always in coincidence with $z_i$ when the Denavit–Hartenberg description is used.

The formulation of equations (29) is very general and is compatible with the transformations of references already used to describe the geometry and kinematics. However, it requires a large number of useless calculations, such as multiplications by zeros or ones, the calculation of 4x4 matrices to use only the traces Tr., etc. A method given in the references facilitates the automatic calculation of the coefficients of the Lagrange equations by preliminary calculation of constant and variable terms which will appear as factors in the equations, by automatic regrouping of the variables $q_i$ whose sum is the argument of trigonometric cosine and sine functions, and by using 3x1 vectors and 3x3 matrices.

### 16.1.1.3 – Taking electrical actuators into account

Considering only armature–controlled direct–current motors, the following models are generally used:

* current control:

$$C_i = K_{ci} \, I_i \qquad (31)$$

where $I_i$ is the current in motor number i and $K_{ci}$ is the corresponding torque constant, with the constraints

Maximum intensity) $|I_i| \leqslant I_{imax}$. Maximum voltage $|\dot{C}_i| \leqslant \dot{C}_{imax}$.

* voltage control

$$C_i + \tau_{ei} \dot{C}_i = \frac{K_{ci}}{R_i} (u_i - K_{ei} \omega_i) \tag{32}$$

where $\tau_{ei}$ is the electric time constant, $R_i$ is the resistance of the armature, $u_i$ is the armature voltage, $K_{ei}$ is the counter electromotive force, and $\omega_i$ is the rotation speed of the motor shaft.

The limits are then:

Maximum intensity $|C_i| \leqslant C_{imax}$

Maximum voltage $|u_i| \leqslant u_{imax}$

The relationships between $C_i$, $\Gamma_i$, $q_i$ and their derivatives cause the kinematics and efficiencies of the transmissions between the motors and the joints to be taken into account.

## 16.1.2 - PRINCIPAL FUNCTIONS TO BE PERFORMED

M. Llibre has distinguished two essential functions of the control system of a robot:

- the stabilisation function which is produced by the position feedback, and also the velocity and force feeback systems;
- the guidance and piloting function, which has the purpose of creating the reference inputs and their variations to the feedback units.

### 16.1.2.1 - **Stabilisation of the manipulator**

The purpose of this function is to calculate the control signals for the motors ($u_i$ or $C_i$) to execute a task with the best possible accuracy.
The simplest case is the feedback system for joint variables, which necessitates the measurement of these variables (or of the position coordinates of the actuators), and also, in general, the measurement of their velocities (tachometers). The external forces acting on the mechanism of the robot are then disturbances whose effects must be minimised at the level of the position and/or velocities of the terminal device.

In the most general case, the robot is considered to be simultaneously a generator of positions and a generator of contact forces. Here again, the control signals to be calculated are $u_i$ or $\Gamma_i$, but the latter now include not only terms which are functions of position and velocity, but also terms which are functions of the contact forces to be generated. These terms may be determined either in the tool reference, $R_N$, or in the operational space linked to $R_o$.

16.1.2.2 – **Guidance and fine control of the manipulator**

The most simple case is that of the following of predetermined trajectories and motions, which are recorded in the robot's memory in the form of tables containing the positions and/or the velocities at predetermined, closely spaced instants of time, for example at all the sampling periods. When this information concerns the space of the actuators, the guidance function is reduced to simple interpolation. When this information concerns the operation space, and in particular when it originates from a CAD database, the guidance function must also provide an inverse transformation of coordinates of the $f^{-1}$ or $J^{-1}$ type, preceded if necessary by a geometrical transformation, rotation and translation, to move from the reference of the CAD database to the operations reference linked to $R_0$.

Three further classes of guidance may be identified:

- direct guidance in the configuration space;
- open–loop Cartesian guidance, in other words guidance under the control of the trajectory in Cartesian space;
- guidance and fine control in a closed–loop configuration in Cartesian space, where the modifications of trajectories are calculated on the basis of information from proximity and/or force detectors.

a) – **Direct configuration guidance**

This is concerned with the generation of point–to–point motions, where a sequence of points in the configuration space, between the initial point $\underline{q}^{\circ}$ and the end point $\underline{q}^{f}$ is specified, with, in general, $\dot{\underline{q}}^{\circ}$ = $\dot{\underline{q}}^{f}$ = $\underline{0}$. The limits of the actuators appear in the form of velocity and acceleration constraints relating to the joints, and the following can be distinguished as a function of the other constraints:

- entirely free motion, without points of passage through intermediate configurations and without synchronisation of the motions of all the joints;
- synchronised motion (fixed duration or fixed maximum velocity);
- synchronised motion in minimum time.

The guidance system must then generate the reference signals to be sent to the position and velocity feedback control systems.

b) – **Open–loop Cartesian guidance**

The initial point, the end point, and any intermediate points are specified in the space $R_0$. The control system must then provide a desired motion $\underline{X}^*(t)$ in accordance with the permissible velocities and accelerations, while following as closely as possible the polygon formed by the specified points; in geometrical terms, therefore, we are concerned with linear interpolation, with circular or parabolic transitions in the vicinity of the intermediate points when these points must be passed without stoppage of the motion. Here again, constraints of duration may be specified, generally by setting a percentage of the maximum velocity, or the minimum time may be sought.

On the basis of the law X*(t) thus generated, the position control reference inputs are calculated by inverse coordinate transformation:

$$\underline{q}^*(t) = \underline{f}^{-1} (\underline{X}^*(t)) \qquad (33)$$

which must show compliance with the velocity and acceleration constraints in the joint space; if this is not the case, the constraints in the Cartesian space must be relaxed.

c) - **Closed-loop Cartesian guidance**

This type of guidance must be used when the trajectory is poorly known a priori and must be generated or corrected on the basis of information from exteroceptive sensors, in other words from sensors which measure the relationships between the terminal device and its environment, such as proximity sensors, detectors of joints to be welded, force sensors, etc.

These sensors detect the deviations between the desired position, velocity, and forces and the real state of the relationship between the terminal device and the environment, obtained either directly by measurement in the tool space or by coordinate transformations. The deviations and/or their variations in Cartesian space are then transformed by the inverse Jacobian matrix $J^{-1}$ into variations of the reference signals $\underline{q}^*$ to the feedback control devices.

The control techniques and the theoretical configurations of the various systems of stabilisation and guidance introduced above will be explained in the following paragraphs.

## *16.2 - USE OF KINEMATIC AND GEOMETRICAL MODELS*

The generation of references by a purely kinematic control signal is only used when the motion is defined by velocities $\underline{\dot{X}}$ or $\underline{V}$ in the operation space. The inverse transformation of the the differential or kinematic model then provides, for each instant, the velocities of the joint variables or the actuator variables; for example,

$$\underline{\dot{q}}^* = J^{-1} \underline{V}^* \qquad (34)$$

However, the joints must be subject to position feedback control, implying (numerical) integration to produce the corresponding reference signals, and therefore this type of guidance can be represented by the diagram in Fig. 3:

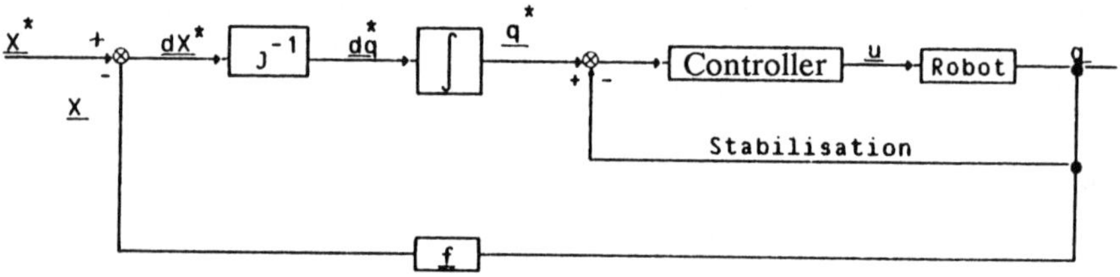


Closed–loop Cartesian guidance (Jacobian controller)

Fig. 3

This is a general diagram; in practice a digitised motion

$$d\underline{X}^*_n = \underline{X}^*_{n+1} - \underline{X}_n \tag{35}$$

where n represents the number of the sampling instant, and where the following relationship is assumed:

$$d\underline{X}^*_n = J(\underline{q}^*_n)\ d\underline{q}^*_n \tag{36}$$

The various methods of generation of trajectories and motions shown above can be used to obtain the sequences $\underline{q}^*_n$ or $\underline{X}^*_n$. In Fig. 3, the stabilisation function has been reduced to a conventional feedback position controller but it will be shown in the next section that the dynamic model can be used to provide better precision at high velocities and when the structural variations of the process are large.

To avoid complicating the notation, the superscript "*" will be omitted from the following text.

## 16.2.1 – GENERATION OF MOVEMENTS IN THE CONFIGURATION SPACE

### 16.2.1.1 – Motion without a fixed trajectory

Consider a robot with N joints for which the initial and end configurations, $\underline{q}^o$ and $\underline{q}^f$, have been recorded during the learning phase. If the trajectory connecting these two states is disregarded, the motion may be generated directly, as follows:

– Let $\delta_q$ be the vector of the joint motions to be made during the sampling period $\Delta T$. For each joint, i, the number of steps, $NP_i$, required to move from $q_i^o$ to $q_i^f$ is then determined, taking the end stop constraints into account. This gives:

$$q_i^f - q_i^o = NP_i \times \delta q_i \tag{37}$$

– In order to synchronise the joints up to the final state, the number of steps of motion of the robot must be the highest value of $NP_i$. Consequently it is specified that

$$NP = \underset{1 \leqslant i \leqslant N}{\text{Max}} \; (NP_i) \tag{38}$$

From this the elementary joint variable for each joint during $\Delta T$ is deduced:

$$\Delta q_i = \frac{NP_i}{NP} \times \delta q_i \tag{39}$$

– Assume that the reference vectors $\underline{X}^{\circ}$ and $\underline{X}^f$ are associated with the configurations $\underline{q}^{\circ}$ and $\underline{q}^f$: the motion from $\underline{q}^{\circ}$ to $\underline{q}^f$ will then be generated for a variation of the reference vector, in each sampling period, of $\underline{\Delta X}$ corresponding to $\underline{\Delta q}$.

| Configuration | | Reference vector |
|---|---|---|
| Initial | $\underline{q}^{\circ}$ | $\underline{X}^{\circ}$ |
| Intermediate No. k ($1 \leqslant k \leqslant NP-1$) | $\underline{q}^{\circ} + k\underline{\Delta q}$ | $\underline{X}^{\circ} + k\underline{\Delta X}$ |
| Final | $\underline{q}^{\circ} + NP\underline{\Delta q} = \underline{q}^f$ | $\underline{X}^{\circ} + NP\underline{\Delta X} = \underline{X}^f$ |

### 16.2.1.2 – Motion with a fixed trajectory

The direct development of the reference vectors implies that the kinematic behaviour of the robot between the initial and final configurations is disregarded. When the motion of the robot is subject to kinematic constraints (such as a known travel time, allowance for points to be passed through, constraints of velocity and acceleration) it becomes necessary to determine the trajectory.

The trajectory will be considered to be unconstrained when no points apart from the initial and final points are specified; in other cases it will be considered to be constrained.

#### a) – Motion with unconstrained trajectory

Minimum time

In this case the aim is to define a trajectory joining the two configurations $\underline{q}^{\circ}$ and $\underline{q}^f$ whose motion is such that

$$
\left[\begin{array}{ll}
\underline{\ddot{q}}^{f} = \underline{0} & \\
\underline{\dot{q}}^{f} = \underline{0} & \\
|\dot{q}_i| \leqslant Kv_i \text{ for } 1<i<N, \text{ where } Kv_i & \text{is the maximum velocity for joint no. i} \\
|\ddot{q}_i| \leqslant Ka_i \text{ for } 1<i<N, \text{ where } Ka_i & \text{is the maximum acceleration for joint no. i}
\end{array}\right. \qquad (40)
$$

If the travel time is given, the trajectory and its motion can be determined globally by interpolation [Binford (19 77)].

For the problem of finding a minimum time trajectory, it is necessary to determine the motion in minimum time of each joint considered in isolation.

For a joint i, it is therefore necessary to determine the control signal $U_i(t)$ of class $C^2$ which causes the state vector

$$\begin{bmatrix} q_i \\ \dot{q}_i \end{bmatrix} \text{ to move from its initial value, } \begin{bmatrix} q_i^o \\ 0 \end{bmatrix} \text{ to its final value } \begin{bmatrix} q_i^f \\ 0 \end{bmatrix}$$

in minimum time. The state equation is as follows:

$$X_i = \begin{bmatrix} q_i \\ \dot{q}_i \end{bmatrix}$$

$$\dot{X}_i = \begin{bmatrix} 0 & 1 \\ 0 & 0 \end{bmatrix} X_i + \begin{bmatrix} 0 \\ 1 \end{bmatrix} U_i \quad \text{under the constraints} \quad \left| \begin{array}{l} |\dot{q}_i| \leqslant Kv_i \\ |\ddot{q}_i| \leqslant Ka_i \end{array} \right. \qquad (41)$$

Optimal control theory [Boudarel (1968)] can be used to demonstrate that the control signal $U_i$ which minimises the time is of the form

$$U_i^* = \pm Ka_i \text{ under the constraint } |\dot{q}_i| \leqslant Kv_i \qquad (42)$$

The resulting motion law is the bang-bang law (Fig. 4) leading to the following expression of the optimal time:

$$T_i^{opt} = 2\sqrt{\frac{q_i^f - q_i^o}{Ka_i}} \quad \text{if } |q_i^f - q_i^o| \leqslant \frac{Kv_i^2}{Ka_i}$$

$$T_i^{opt} = \frac{Kv_i}{Ka_i} + \frac{q_i^f - q_i^o}{Kv_i} \quad \text{if } |q_i^f - q_i^o| \geqslant \frac{Kv_i^2}{Ka_i}$$

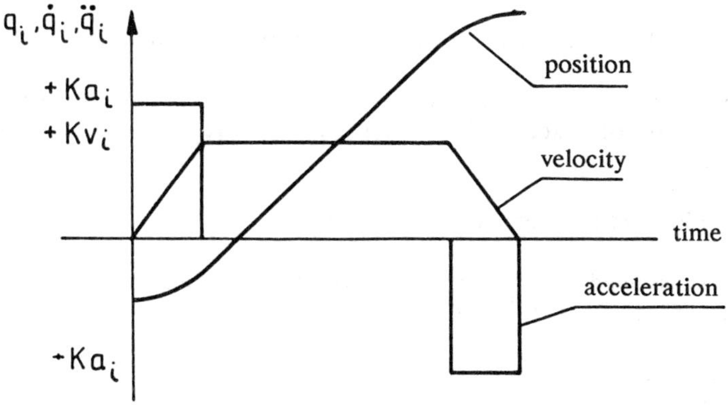


Bang–bang law for the case $> q_i^f - q_i^o > / \, kv_i^2 \, / Ka_i$

Fig. 4

Given the minimum time, $T_i^{opt}$, for each joint, the lower limit of the travel time is given by:

$$\max_{1 \leqslant i \leqslant N} (T_i^{opt}) \tag{44}$$

Coordination of joint movements:

The overall motion is determined from the individual motions of the different joints by synchronisation with the slowest joint (constraining joint). Parent and Laurgeau distinguish a number of methods of coordinating the joint motions, notably the proportional method on the constraining axis. In this method, the velocity laws of the different joints are obtained by transformations of similitude of the velocity law of the constraining axis. This is illustrated in Fig. 5.

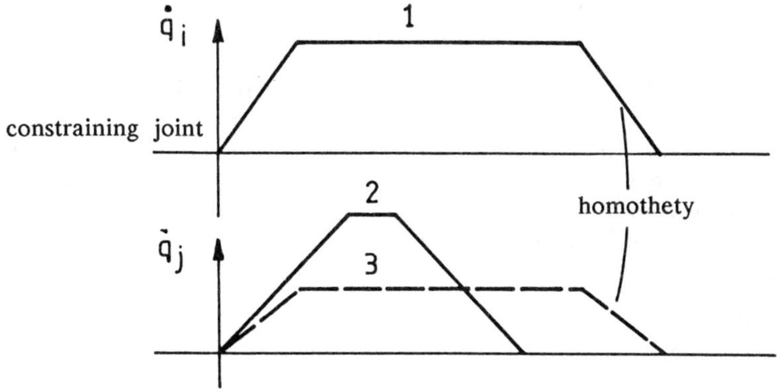


Fig. 5: Coordination by the proportional method on the constraining axis

Law 2 is the velocity law of the joint j, considered in isolation, and is replaced by law 3 obtained by a transformation of similitude of 1.

In certain cases, this method causes the velocity or acceleration constraint to be exceeded. If this is so, the travel time must be increased.

b) - **Motion with constrained trajectory**

It is assumed that there are n points in the manipulator space, and these are denoted by $(\underline{q}^1 \ldots . \underline{q}^n)$.

Methods of interpolation

These methods consist in the adjustment of a number of polynomials at the intermediate points, while providing a continuity of position, velocity, and (if necessary) acceleration at these points.

The method of Khalil and Liegeois generates a trajectory in pseudo-minimum time, providing a passage through the intermediate points at zero acceleration. This is done by introducing the midpoint of each segment: the trajectory interpolated between two successive points consists of two polynomials of the fourth degree.

The pseudo-minimum time is found by determining at each specified point the maximum velocity which can be reached by saturation of the constraint of velocity and/or acceleration.

The procedure is as follows:

1. For the joint i, the sequence of specified points $(q_i^1,\ldots,q_i^n)$ is broken into monotonic increasing or decreasing sequences, corresponding to the various "forward and backward" motions of the joint. In order to avoid exceeding the limits at the "turning points", a zero velocity is specified at these points.

2. Let $(q_i^K,\ldots,q_i^l)$ be a monotonic sequence as defined in 1.

   At each point $q_i^f$ of this sequence, and also at the midpoints $q_i^{jm}$, the limit attainable velocities $V_i^j$ and $V_i^{jm}$ imposed by the motion at zero velocity at the end points of the sequence, and by the acceleration constraint, are determined.

   This gives:

$$\begin{cases} V_i^k = V_i^l = 0 & \text{and, by recursion,} \\ V_i^{j-1} = \text{Min}\left[\sqrt{4\,Ka_i\,|q_i^{j-1}-q_i^j|/3 + (V_i^j)^2},\ Kv_i\right] \\ V_i^{jm} = \text{Min}\left[\sqrt{2\,Ka_i\,|q_i^{j+1}-q_i^j|/3 + (V_i^{j+1})^2},\ Kv_i\right] \end{cases} \quad (45)$$

3. The maximum permissible velocities at each point $q_i^j$ and $q_i^{jm}$, found by saturation of the acceleration and/or of the velocity constrained by $V_i^j$ and $V_i^{jm}$, are then determined. A recursive procedure is used as in 2. The minimum time $T_m{}^j$ for the travel from $q_i^j$ to $q_i^{j+1}$ is deduced from this.

The overall travel time from the configuration $\underline{q}^j$ to the configuration $\underline{q}^{j+1}$ is then

$$T^j = \underset{1 \leqslant i \leqslant N}{\text{Max}} \left(T^j_{m_i}\right) \quad (46)$$

When $T^j$, $\dot{\underline{q}}^j$, $\dot{\underline{q}}^{j+1}$, and $\dot{\underline{q}}^{jm}$ are known, the polynomial interpolation is performed. An example of this method is given in the next section.

General method of synchronisation

Consider the joint i. As in the previous method, the sequence of specified points $(q_i^1, \ldots, q_i^n)$ is decomposed into monotonic sequences; each monotonic sequence can then be described by an optimal trajectory of the bang-bang type (Fig. 6).

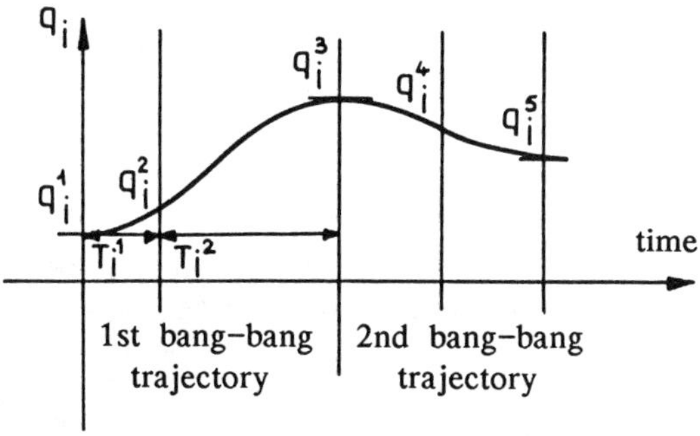


Decomposition into monotonic sequences

Fig. 6

After the optimal motion of each joint has been determined in this way, the time $T_i^j$ for the travel from $q_i^j$ to $q_i^{j+1}$ is easily calculated.

Agreement at the specified points then requires the synchronisation of the different joints with the slowest joint. This is achieved by introducing in each travel from $q_i^j$ to $q_i^{j+1}$ the margin $\Delta T_i^j$ given by:

$$\Delta T_i^j = \max_{1 \leqslant i \leqslant N} (T_i^j) - T_i^j \tag{47}$$

The difficulty of the problem of synchronisation lies in the introduction of these margins, while ensuring that the state (position and velocity) remains the same at the specified points, and that the velocity and acceleration constraints are complied with (Fig. 7).

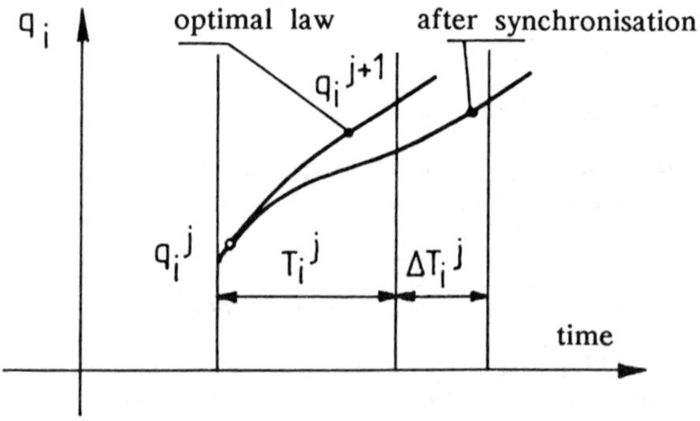


Synchronisation by introduction of the margin

Fig. 7

The solution which is adopted consists in the introduction into the travel from $q_i^j$ to $q_i^{j+1}$ of an acceleration stage supplementary to the optimal acceleration law. The procedure is as follows: after the motion law synchronised to the point $q_i^j$ reached at the instant $t_i^j$ has been determined, a switch, occurring at $t_i^{jc}$ and corresponding to the intermediate point $q_i^{jc}$, is introduced into the following acceleration step leading to the point $q_i^{j+1}$; the motion of second degree from $q_i^j$ to $q_i^{j+}$ is replaced by motions of the second degree from $q_i^j$ to $q_i^{jc}$ and from $q_i^{jc}$ to $q_i^{j+}$ (Fig. 8)

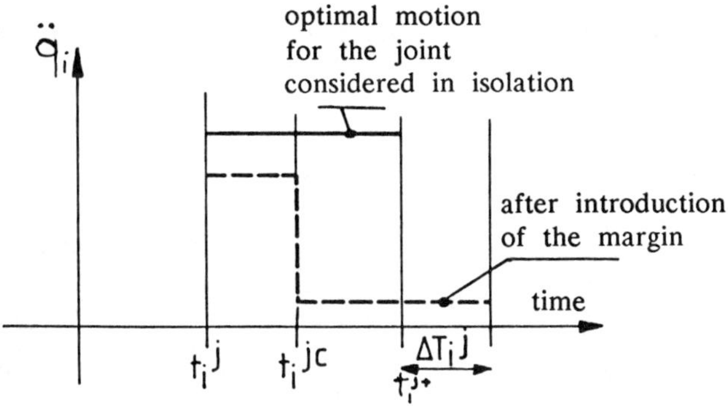


Introduction of the supplementary acceleration step

Fig. 8

Since the instant $t_i^{jc}$ is considered to be a parameter, $q_i^{jc}$ is determined in such a way as to satisfy the interpolation constraints, which require the same initial and final state (position and velocity), and continuity of the position and velocity at the intermediate point. Let there be a total of six constraints which determine the two polynomials of second degree joining $q_i^j$ to $q_i^{j+}$.

The freedom of choice of $t_i^{jc}$ must enable the constraints of velocity and acceleration to be satisfied. However, in cases where the determination of $t_i^{jc}$ is impossible, the margin, and consequently the travel time, must be increased.

c) – **Simulations**

An example of the application of the two methods will now be given. The data are taken from Khalil (1983). The robot under consideration has two joints. The specified points are as follows:

$$\underline{q}^1 = (0.0 \quad 1.0)^T \text{ rad}$$

$$\underline{q}^2 = (-1.0 \quad 0.0)^T \text{ rad}$$

$$\underline{q}^3 = (-1.8 \quad 0.5)^T \text{ rad}$$

$$\underline{q}^4 = (-2.3 \quad 1.0)^T \text{ rad}$$

under the following constraints of velocity and acceleration:

$\underline{K}v = (1.0 \quad 1.0)^T$ rad/s

$$\underline{K}a = (1.0 \quad 1.0)^T \quad \text{rad/s}$$

The first method results in a travel time of 5 s, the second results in a travel time of 4 s. Fig. 9 shows the variation of the joint variables $(q, \dot{q}, \ddot{q})$.

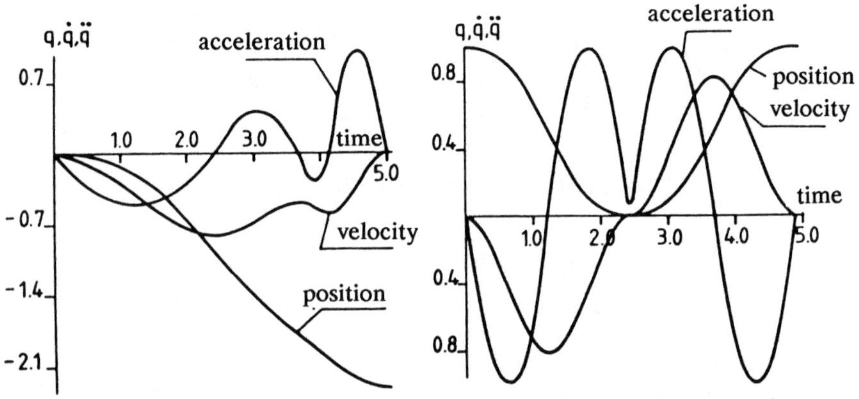


a) Interpolation method

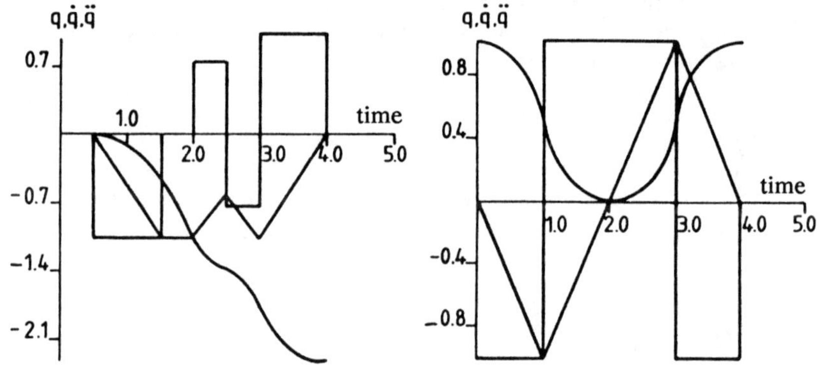


b) Method of synchronisation of bang–bang laws

Change of the joint variables

Fig. 9

The net difference in the travel times arises from the hypotheses concerning the acceleration law:

* in the first method, this is continuous, and provides flexible motion without jerks;
* in the second method, it is discontinuous, resulting in a shorter time, but entailing the risk of imposing major variations of torque on the motors.

## 16.2.2 - GENERATION OF MOVEMENTS IN THE OPERATION SPACE

As in the manipulator space, the trajectory of the effector point may be found by different methods of interpolation. The Khalil–Liégeois method described below is also applicable to the work space.

More attention will be paid here to an automatic approach in which a minimum time trajectory is found, based on the model of Paul, Luh and Lin.

### 16.2.2.1 - Paul model

n points of the Cartesian space, denoted ($\underline{P}^1$,..., $\underline{P}^n$), will be considered. The trajectory consists of segments of straight lines, supported by the polygon joining the specified points, and arcs of parabolas providing a transition between the straight parts (Fig. 10).

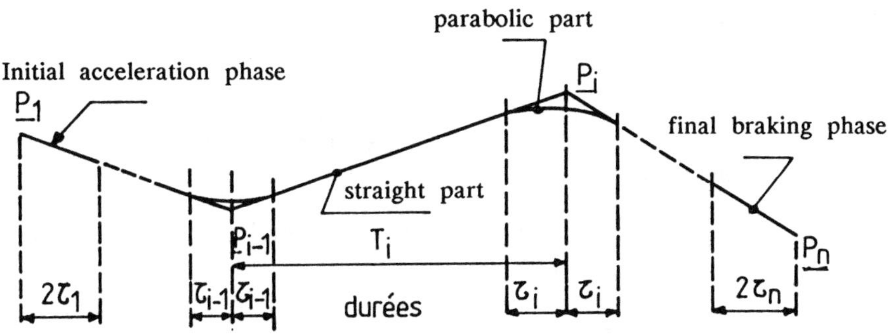


Model of the Paul trajectory

Fig. 10

The first consequence of the model is the existence of deviations at the specified points (except at the initial and final points). The motion along the trajectory is as follows:

– of the first order, of duration $T_i - \tau_i - \tau_{i-1}$ (see figure) on the straight segment of the travel $\underline{P}_{i-1}$, $\underline{P}_i$ ($2 \leqslant i \leqslant n$). The constant velocity along the segment is given by:

$$\underline{V}_i = \frac{\underline{P}_i - \underline{P}_{i-1}}{T_i} \quad \text{for } 2 \leqslant i \leqslant n \tag{48}$$

At the initial and final points, $\underline{V}_1 = \underline{V}_{n+1} = \underline{0}$

– of the second order, with a duration of $2\tau_i$, on the parabolic transition associated with the point $\underline{P}_i$ ($2 \leqslant i \leqslant n-1$). The constant acceleration during the transition is given by:

$$\underline{\alpha}_i = \frac{\underline{V}_i - \underline{V}_{i-1}}{2\tau_i} \tag{49}$$

– of the second order, at the initial and final point of the first and the last segment of straight line. These two phases are respectively the initial acceleration phase and the final braking phase of the motion. In symmetry with the case of the parabolic transition, their durations are $2\tau_i$ and $2\tau_n$ respectively, and their accelerations are:

$$\underline{\alpha}_1 = \frac{\underline{V}_2}{2\tau_1} \qquad \underline{\alpha}_n = \frac{-\underline{V}_n}{2\tau_n} \tag{50}$$

Therefore the variables which entirely define the motion are the $\tau_i$ for $1 \leqslant i \leqslant n$ and the $T_i$ for $2 \leqslant i \leqslant n$. The set of these variables will be denoted henceforth by $(\tau_i, T_i)$. The problem of optimisation, as a function of these variables, is posed in the following way:

* minimisation of the "travel time" criterion:

$$T = \tau_1 + \sum_{2 \leqslant i \leqslant n} T_i + \tau_n \tag{51}$$

under the constraints:

$\|\underline{V}_i\| \leqslant Kv$ for $2 \leqslant i \leqslant n$ where Kv is the maximum velocity of the terminal device

(52)

$\|\underline{\alpha}_i\| \leqslant Ka$ for $1 \leqslant i \leqslant n$ where Ka is the maximum acceleration of the terminal device

### 16.2.2.2 – Method of analytical solution

In the first method, the minimum travel time is assumed to be obtained for a constant acceleration norm equal to Ka during each transition.

The algorithm in Fig. 11, similar to that proposed by Paul, leads to this result.

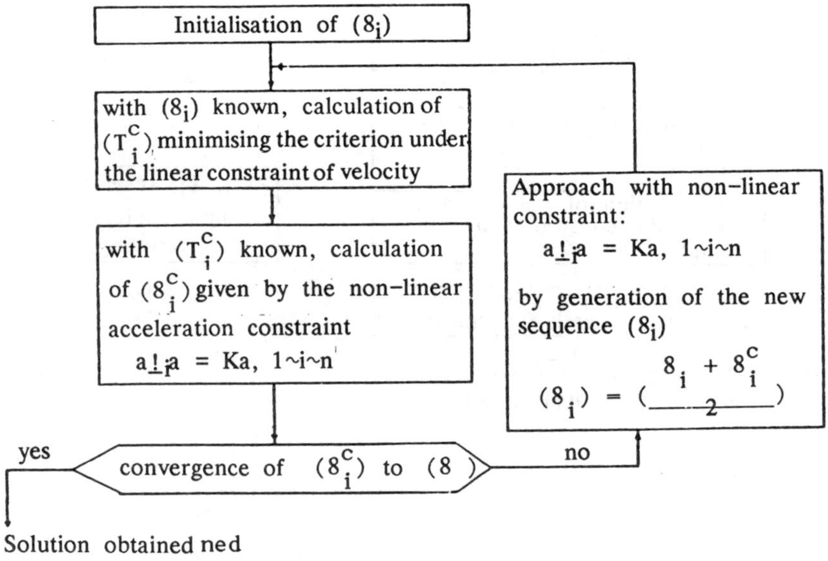


Algorithm for analytical solution

Fig. 11

This first method only allows the variation at the specified points to be overcome by using the velocity constraint, limiting the velocity at entry into the parabolas and consequently their radii of curvature.

### 16.2.2.3 – Method of optimisation

On the basis of this model, Paul, Luh and Lin have proposed a method of optimisation leading to the minimisation of the travel time, under the constraints of velocity and acceleration, with the specification, in the form of a deviation constraint, of a maximum ratio K between the durations of the parabolic transitions and the duration of travel $\underline{P}_{i-1}$, $\underline{P}_i$:

$$\frac{\tau_{i-1}}{T_i} \leqslant K \qquad \text{for } 2 \leqslant i \leqslant n$$
$$\frac{\tau_i}{T_i} \leqslant K \qquad \text{for } 2 \leqslant i \leqslant n \tag{53}$$

The difficulty of the optimisation problem is due to the presence of the non-linear constraint $\|\underline{\alpha}_i\| \leqslant Ka$. The method adopted is that of linearising this constraint and solving the linearised problem by the simplex method. The convergence of the method is demonstrated.

The algorithm in Fig. 12 summarises this procedure.

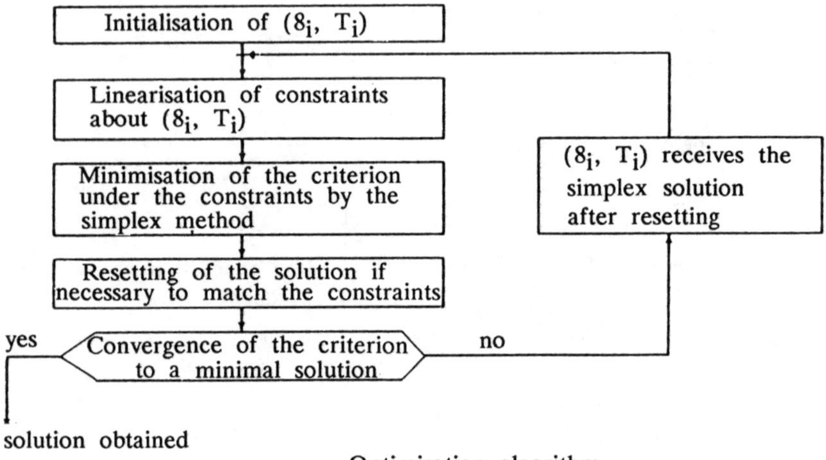


Optimisation algorithm

Fig. 12

16.2.2.4 – **Explicit introduction of a deviation constraint at the points passed through**

The preceding solution does not involve the direct intervention of a deviation constraint. We have adopted the method [Ton 84] of introducing this constraint in the form

$$\|\underline{\epsilon}_i\| \leqslant \epsilon_i^{max} \quad \text{for } 2 \leqslant i \leqslant n-1 \tag{54}$$

where $\epsilon_i^{max}$ is the maximum permissible deviation at the point $\underline{P}_i$ and $\underline{\epsilon}_i$ represents the deviation vector at the point $\underline{P}_i$ calculated at the mid-instant of the parabolic transition. It can be shown that it corresponds geometrically to the point whose tangent is parallel to the base of the parabola (Fig. 13).

If $\underline{P}_e$ and $\underline{P}_s$ are the points of entry to and exit from the parabola, then

$$\underline{P}_e = \underline{P}_i - \underline{V}_{i-1}\,\tau_i \qquad \text{and} \qquad \underline{P}_s = \underline{P}_i - \underline{V}_i\,\tau_i$$

The base of the parabola is given by the vector:

$$\underline{P}_s - \underline{P}_e = (\underline{V}_{i-1} + \underline{V}_i)\,\tau_i$$

It can also be shown that the velocity in the parabola at the relative instant t = 0 is:

$$\frac{\underline{V}_{i-1} + \underline{V}_i}{2}$$

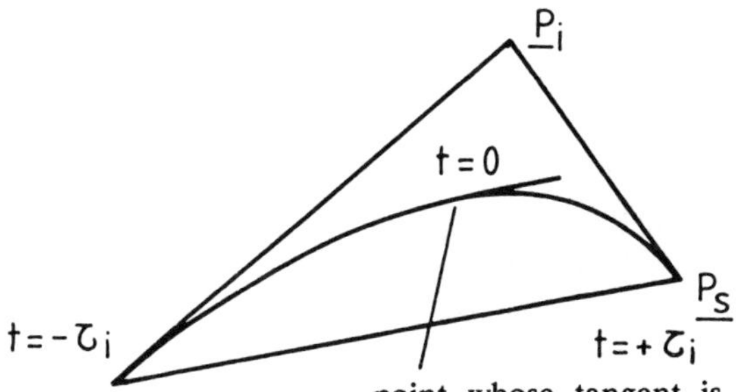


Diagram of the transition parabola

Fig. 13

The problem of optimisation can then be solved by the same algorithm as before, substituting the (n–2) deviation constraints for the constraints in which K has an effect.

It will be noted that the solution obtained by introducing large maximum deviations is that given by the analytical method: the constraints are still satisfied in this case.

Furthermore, there is no risk of the trajectory based on the polygon encountering obstacles if the maximum deviations have been correctly chosen, whereas it is sometimes difficult to predict the result of a trajectory interpolated between the given points (see, for example, the broken line in Fig. 14).

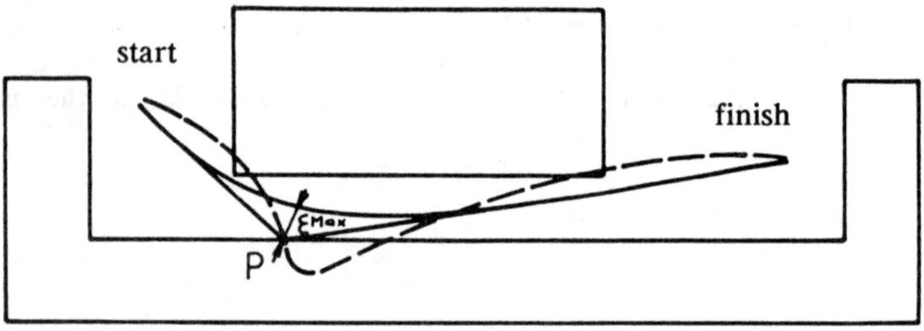


Taking obstacles into account

Fig. 14

16.2.2.5 – **Simulations**

Consider the following example:

– The points specified in Cartesian space are:

$$\underline{P}_1 = (\ 0. \quad 0. \quad 0.)^T \text{ cm}$$

$$\underline{P}_2 = (10. \quad 10. \quad 5.)^T \text{ cm}$$

$$\underline{P}_3 = (10. \quad 10. \quad 10.)^T \text{ cm}$$

$$\underline{P}_4 = (\ 5. \quad -10. \quad 30.)^T \text{ cm}$$

$$\underline{P}_5 = (\ 0. \quad 0. \quad 0.)^T \text{ cm}$$

under the following velocity, acceleration and deviation constraints:

$$Kv = 40 \text{ cm/s}$$

$$Ka = 10 \text{ cm/s}^2$$

$$\epsilon_1^{max} = 2 \text{ cm} \qquad \epsilon_2^{max} = 1 \text{ cm} \qquad \epsilon_3^{max} = 8 \text{ cm}$$

This example was treated successively by the method of Khalil and Liégeois which requires motion strictly through the specified points, the method of Paul, Luh and Lin without control of the deviation at the specified points, and the method derived from the latter by the explicit introduction of deviation constraints.

Table 15 shows the travel times produced by the three methods and the real deviations at the specified points.

Figs. 16 and 17 show the trajectory given by its projections in the three reference planes XY, YZ, and XZ, together with the norm of the vector of the acceleration of the terminal device.

| | Interpolation method | Method without control of deviation | Method with control of deviation |
|---|---|---|---|
| Travel time | 12.2 s | 7.9 s | 8.3 s |
| Real deviations | -<br>-<br>- | 1.6 cm<br>1.7 cm<br>10.0 cm | 1.5 cm<br>1.0 cm<br>8.0 cm |

Fig. 15: Table of results obtained by the three methods

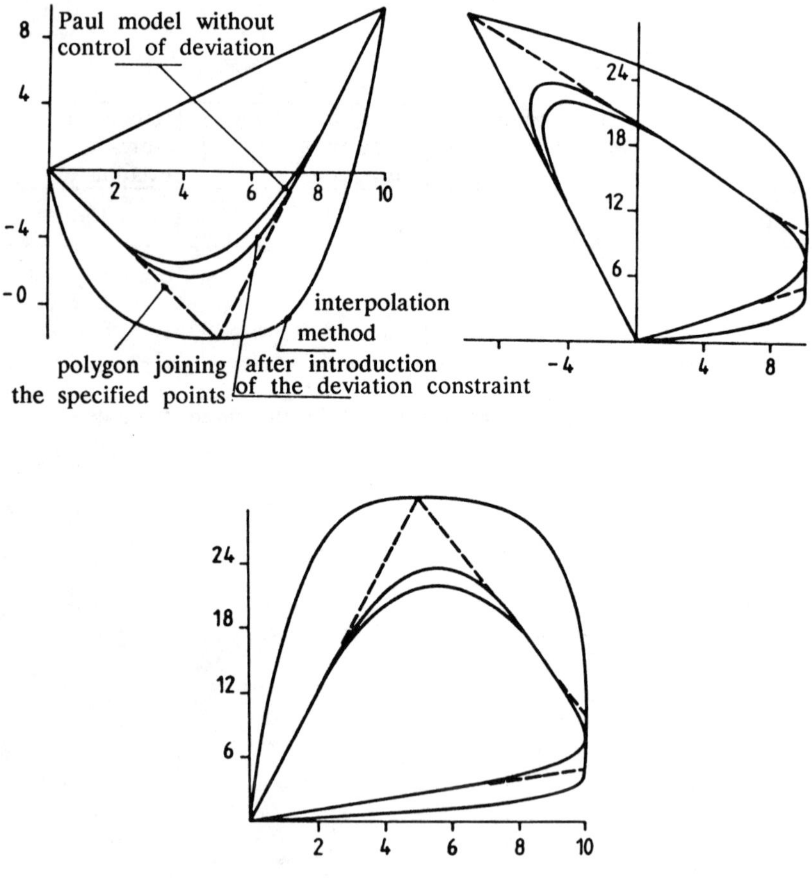


Trajectory of the final element of the robot

Fig. 16

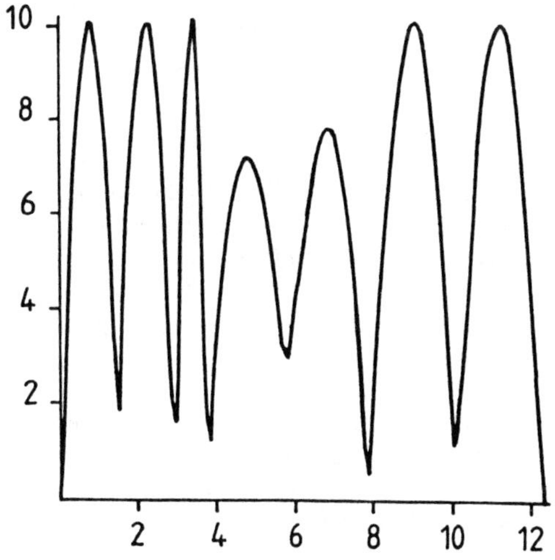


Interpolation method

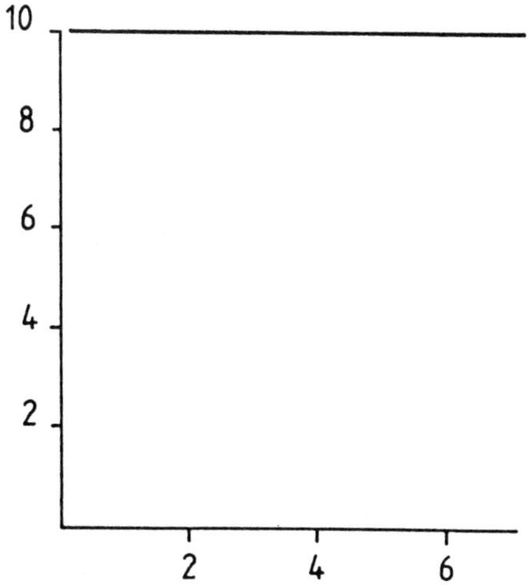


Analytical method without deviation control

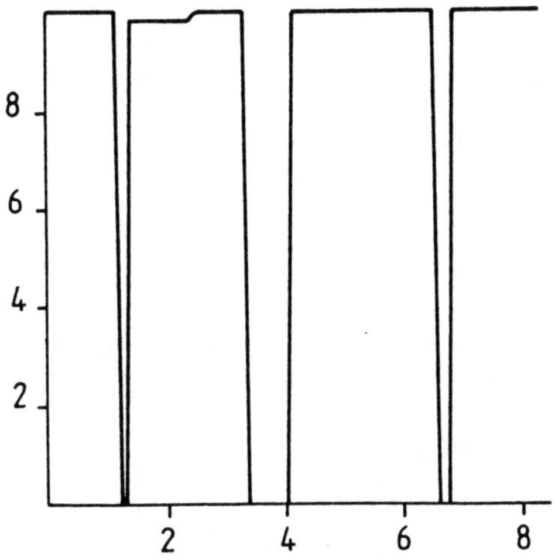


Optimisation method with explicit introduction of a deviation constraint

Variation of the norm of the acceleration of the final element

Fig. 17

## *16.3 - USE OF MODELS OF THE DYNAMICS*

While awaiting the development and application of robotics-orientated computers using specialised processors, such as those with pipe-line and parallel architectures, research and development centres have been forced to reduce the calculation time required by the use of dynamic models for the design of high-performance joint controllers. Two basic approaches are used:

1) The elimination of terms which appear to be negligible by comparison with others as regards most of the robot configurations and for the maximum permitted velocities and accelerations;

2) The tabulation of certain coefficients in the equations for a certain number of robot configurations and payloads.

At the present time, we are not aware of any systematic scientific method which can be used to automate the simplification of the dynamic model. The designer of an automatic control system must proceed in the follow manner, with the aid of digital calculation programs:

* The direct model is used for the calculation of the various terms appearing in the left sides of the Lagrange equations, for a number of "test" motions which represent the normal operating conditions of the robot.

* After the model has been simplified according to the methods described above, the complete inverse model, controlled by a simplified guidance law, is used to analyse, by means of digital simulation, the performance of the controllers and their sensitivity to the variation of different parameters.

### 16.3.1 - THEORETICAL CONTROLLERS

In this section, we shall analyse examples of control laws of increasing complexity, which use models of the dynamics of the process to a greater or lesser extent for the automatic stabilisation of the process at a point or on a reference trajectory. To avoid complicating the description, it will be assumed that the reference $\underline{q}^*(t)$ is expressed in the space of the joint variables $\underline{q}$, while it can be defined in another space (that of the location of the terminal device with respect to the robot base, for example) and that it is then necessary to introduce direct and inverse transformations of coordinates which are the vector functions and Jacobian matrices defined in the first section of this chapter.

**16.3.1.1 – Compensation of gravitational forces and proportional derivative feedback**

It will be shown that, after the forces or torques due to the gravity potential have been compensated, the application of a linear control law of the proportional–derivative type makes any configuration of the mechanism asymptotically stable overall. For a "configuration ramp" reference input, the control is optimal in the sense of the minimisation of a quadratic criterion.

The formulation used here is that of Hamilton, which is more symmetrical than that of Lagrange, but strictly equivalent. It uses the Hamilton formula H = $T + U$, where the kinetic energy is expressed in the following form:

$$T = \frac{1}{2}\,\dot{\underline{q}}^T\,A(\underline{q})\,\dot{\underline{q}} \qquad (55)$$

Introducing the generalised impulses grouped in the vector $\underline{p}$,

$$\underline{p} = \delta T/\delta\dot{\underline{q}} = A(\underline{q})\,\dot{\underline{q}}$$

the Hamilton equations are:

$$\begin{aligned}\dot{\underline{q}} &= \delta H/\delta\underline{p} = A^{-1}(\underline{q})\,\underline{p}\\ \dot{\underline{p}} &= -\delta H/\delta\underline{q} + B\,\underline{u}\end{aligned} \qquad (57)$$

where $\underline{u}$ is the vector of the generalised force control signals and B is the matrix of the application of the control signals.

Let $\underline{q}_f = \underline{q}_d(\infty)$, the final configuration which is required to be asymptotically stable. A new potential function $U^o(\underline{q})$ can be introduced, such that

$$\begin{aligned}U^o(\underline{q}) &\geqslant 0 \qquad \forall\,\underline{q}\\ U^o(\underline{q}_f) &= 0\end{aligned} \qquad (58)$$

and we may specify

$$\underline{u} = -\delta U^o/\delta\underline{q} + \delta U/\delta\underline{q} + \underline{w}$$

where $\underline{w}$ is a new vector of generalised forces.
The function

$$\tilde{H} = T + U^o(\underline{q}) \qquad (59)$$

then acts as a Lyapunov function and it is sufficient to make

$$\underline{w} = -Q\,\dot{\underline{q}} \qquad (60)$$

where Q is a symmetrical matrix defined as positive to obtain complete damping.

Therefore,

$$\dot{\tilde{H}} = -\dot{q}^T Q \dot{q} \leqslant 0$$
$$\tilde{H}(q_f) = 0 \qquad (61)$$

According to the theory of optimal control of continuous processes, the following criterion is minimised:

$$\text{Performance index} = 1/2 \int_0^{\infty} (\dot{q}^T Q \dot{q} + \underline{w}^T Q^{-1} \underline{w}) \; dt \qquad (62)$$

From the various possible potential functions, we may choose

$$U^o(q) = U(q) - U(q_f) - \left[\frac{\delta U}{\delta q}\right]^T_{q_f} (q - q_f) + \ldots$$
$$+ \frac{1}{2} (q - q_f)^T S (q - q_f) \qquad (63)$$

where S is a positive-definite symmetrical matrix, and therefore the optimal controller is:

$$U = \left[\frac{\delta U}{\delta q}\right]_{q_f} - S(q - q_f) - Q(\dot{q} - \dot{q}_f) \qquad (64)$$

The first term compensates the effects of gravity, calculated at the end point, and the others represent a multivariable linear compensating system of the proportional-derivative type. The controller therefore provides asymptotic stability, but does not guarantee the absence or low level of the exceeding of limits in the transitory conditions. Variations in performance according to the configuration of the robot must also be expected, since the matrices S and Q are kept constant while the characteristics of the process "as seen by the controller" vary with the configuration.

However, this control law is easily implemented with a digital computer. The calculation times are short, since the expressions of the gravitational forces are the simplest terms, and at the limit a proportional control signal is sufficient if there are high levels of friction (of the viscous type) within the mechanism itself which can be considered as being derived from a quadratic function analogous to $\dot{\tilde{H}}$.

### 16.3.1.2 – Non-linear decoupling

E. Freund has developed a method which can be used to locally decouple the changes of the variables of a multivariable non-linear system. When applied to the control of a robot, this technique can be compared with other methods based on the idea of using the direct dynamics to compensate the gravitational, inertial, centrifugal, and Coriolis torques or forces, and obtaining a linear decoupled feedback in series with a non-linear compensator to provide decoupling and stabilisation of the deviations of the configuration.

The general vector form of the Lagrange equations when friction and external forces other than gravity are disregarded is as follows:

$$A(\underline{q})\,\ddot{\underline{q}} + B(\underline{q},\,\dot{\underline{q}}) + \underline{Q}(\underline{q}) = \underline{\Gamma} \qquad (65)$$

If it is possible to apply a force (or torque) control signal to the joints, the following will be chosen:

$$\underline{\Gamma} = A(\underline{q})\,\ddot{\underline{q}}_d + B(\underline{q},\,\dot{\underline{q}}) + \underline{Q}(\underline{q}) + A(\underline{q})\,[K_w(\dot{\underline{q}}_d-\dot{\underline{q}}) + K_p(\underline{q}_d-\underline{q})] \qquad (66)$$

In these equations,

$A(\underline{q})$ is the symmetrical matrix of the kinetic energy of the mechanism. It is positive, and therefore regular;

$B(\underline{q},\dot{\underline{q}})$ is the vector of the centrifugal and Coriolis forces;

$\underline{Q}(\underline{q})$ is the vector of the gravitational forces;

$\underline{q}_d = \underline{q}_d(t)$ is the reference vector of the configuration $\underline{q}$;

$K_w$ and $K_p$ are positive matrices which will be determined subsequently.

It will be noted that the control law compensates the gravitational, centrifugal and Coriolis forces by a direct model of these calculated in the instantaneous state $\{\underline{q},\ \dot{\underline{q}}\}$, forming a non-linear feedback loop. The law also "predicts" the inertial forces, and introduces a non-linear compensating network in the controller chains. The theoretical diagram is shown in Fig. 18:

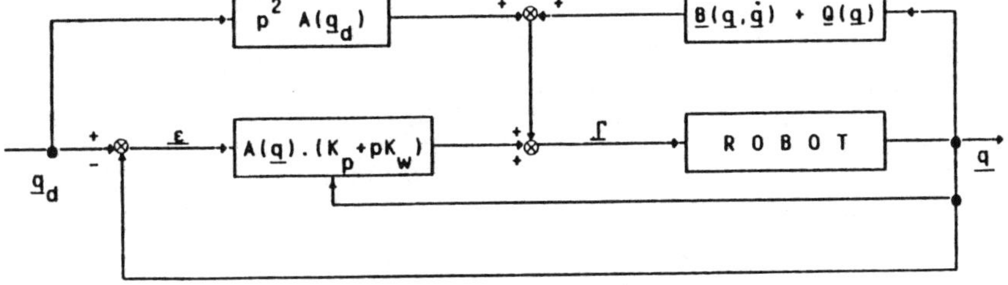


Non-linear decoupling and linearisation

Fig. 18

Given that $\underline{\epsilon} = \underline{q}_d - \underline{q}$, the vector of the configuration deviations, and since $A(\underline{q})$ is a regular matrix, the above equations imply that

$$\ddot{\underline{\epsilon}} + K_w \dot{\underline{\epsilon}} + K_p \underline{\epsilon} = \underline{0} \tag{67}$$

Therefore, by specifying that $K_p$ and $K_w$ are positive diagonals in the form

$$\begin{aligned} K_p &= \text{diag.} [\omega_i^2] \\ K_w &= \text{diag.} [2 \xi_i \omega_i] \end{aligned} \tag{68}$$

it is theoretically possible to stabilise and control the behaviour of each joint error $\epsilon_i$.

## 16.3.2 - DYNAMIC CONTROL APPROACHES IN THE OPERATION SPACE

The approaches described previously demonstrate the separation of the two functions of stabilisation on one hand and guidance and fine control on the other. Since the latter function is generally specified in the operational space, and the errors of the controllers are reflected as functions of the configuration $\underline{q}$ of the robot, some researchers such as Khatib, Samson and Espiau in France have tackled the synthesis of dynamic control laws in the operational space, by attempting to obtain isotropic performance in this space and robustness of the control laws with respect to disturbances and errors in modelling.

In the operational space, the dynamic model (65) may be rewritten to allow for the relationships between $\underline{X}$ and $\underline{q}$ outside the singularities of these relationships. In these conditions, in fact, $\underline{X}$ and $\dot{\underline{X}}$ or $\underline{X}$ and $\underline{V}$ form a set of generalised velocities and coordinates.

By taking into account the relationships of types (7), (8) and (17), established at the beginning of this chapter, the following model is obtained:

$$A_x(\underline{X})\ \ddot{\underline{X}} + \underline{B}_x(\underline{X},\dot{\underline{X}}) + \underline{Q}_x(\underline{X}) = \underline{F} \tag{69}$$

where $A_x(\underline{X})$, $\underline{B}_x(\underline{X},\dot{\underline{X}})$ and $\underline{Q}_x(\underline{X})$ represent the matrix of the kinetic energy, the vector of the centrifugal and Coriolis forces, and the vector of forces due to gravity respectively, $\underline{F}$ being the vector consisting of the operational force and moment.

These matrices are written

$$A_x(\underline{X}) = J^{-T}(\underline{f}^{-1}(\underline{X}))\ A(\underline{f}^{-1}(\underline{X}))\ J^{-1}(\underline{f}^{-1}(\underline{X})) \tag{70}$$

$$\underline{B}_x(\underline{X},\dot{\underline{X}}) = J^{-T}(\underline{q})\ B(\underline{q},\dot{\underline{q}}) - J^{-T}(\underline{q})\ A(\underline{q})\ J^{-1}(\underline{q})\ \dot{J}(\underline{q})\ \dot{\underline{q}} \tag{71}$$

where $\underline{q} = \underline{f}^{-1}(\underline{X})$ and $\dot{\underline{q}} = J^{-1}\ \dot{X}$

and

$$Q_x(\underline{X}) = J^{-T}(\underline{f}^{-1}(\underline{X}))\ \underline{Q}(\underline{f}^{-1}(\underline{X})) \tag{72}$$

The above relationships again show the necessity of the regularity of the transformations of coordinates. The approach of Khatib is similar to that used for control in the configuration space, and in theory permits decoupling. The control law illustrated in Fig. 19 uses the following calculations:

$$\underline{B}_x(\underline{q},\dot{\underline{q}}) = H_1(\underline{q})\ [\dot{\underline{q}}\ \dot{\underline{q}}] + H_2(\underline{q})\ [\dot{\underline{q}}^2] \tag{73}$$

$$\tilde{B}_1(\underline{q}) = B_1(\underline{q}) - J^T(\underline{q})\ A_x(\underline{q})\ H_1(\underline{q}) \tag{74}$$

$$\tilde{B}_2(\underline{q}) = B_2(\underline{q}) - J^{-T}(\underline{q})\ A_x(\underline{q})\ H_2(\underline{q}) \tag{75}$$

where $B_1$ and $B_2$ are the matrices of the Coriolis and centrifugal forces, vectors $[\dot{\underline{q}}\ \dot{\underline{q}}]$ and $[\dot{\underline{q}}^2]$ being formed respectively by the products $q_i\ q_j$, with $j \neq i$, and $\dot{q}_i^2$.

Assuming that there is a force or torque controller of the joints, this controller must be in the following form:

$$\underline{\Gamma} = J^T(\underline{q})\ A_x(\underline{q})\ \underline{F}_m + \tilde{B}_1(\underline{q})\ [\dot{\underline{q}}\ \dot{\underline{q}}] + \tilde{B}_2(\underline{q})\ [\dot{\underline{q}}^2] + \underline{Q}_x\ (\underline{q}) \tag{76}$$

As in the case of the configuration space, the controller is a multivariable one of the proportional derivative type. Its gains can be specified in such a way as to control the dynamics of the error in the operational space:

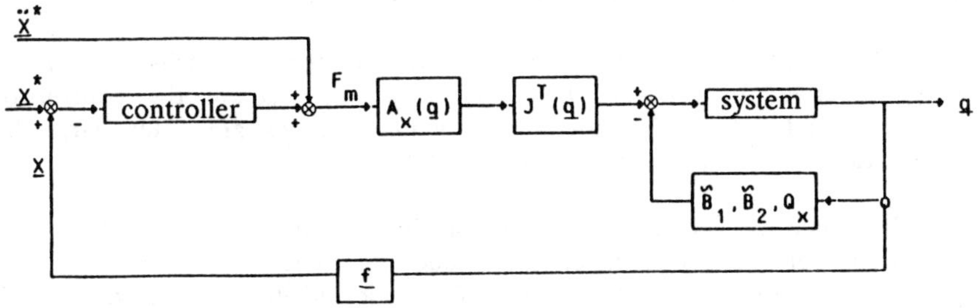


Dynamic control in the operational space

Fig. 19

The approach of Samson and Espiau also uses the formulation of the dynamic model of the process to develop a two-level robust control law.

The first level activates the robot so that it behaves in the operational space as a double integrator under the effect of the external force and moment applied to its terminal device. The second level is that of the proportional derivative controller network.

The dynamic model is approximated by the equation

$$\underline{F} = \hat{A}_x(\underline{X})\ \ddot{\underline{X}}_r\ \hat{B}_x(\underline{X},\ \dot{\underline{X}}) \tag{77}$$

and the reference model is the double integrator

$$\ddot{\underline{X}}_r = \underline{F}_r \tag{78}$$

The aim is therefore to obtain

$$\ddot{\underline{X}} = \underline{F}_m$$

and therefore the control law is

$$\underline{F}_m = -K_P\ \underline{\epsilon} - K_v\ \dot{\epsilon} + \underline{F}_r \tag{79}$$

where $\underline{\epsilon} = (\underline{X} - \underline{X}_r)$ and $\underline{F}_r$ is a function of the deviation $(\underline{X}^* - \underline{X}_r)$ and its derivative with respect to time.

The approach of Samson consists in uses gain matrices $K_p$ and $K_v$ depending in a non-linear way on the states in a form linked to the approximations made on $A_x$ and $\underline{B}_x$. Samson's theorem can then be used to ensure the stability of the system.

### 16.3.3 - CONCLUSION ON DYNAMIC CONTROL

The principle of the dynamic control methods described above is very attractive. In practice, however, it is not used in the control of real industrial robots: in these, the frictional forces are often large by comparison with the "purely mechanical" forces.

In general, however, the study of dynamic control laws, also using procedures of identification of the model parameters if necessary, may prove indispensable in certain fields of robotics likely to be developed in the future, such as high-speed machines and large structures in space. In these cases it will also be necessary to allow for the deformation of the components of the mechanism, which may be too great to disregard.

## *16.4 – CONTROLLERS DRIVEN BY SENSORS*

Controllers driven by sensors are indispensable when the geometrical, kinematic and static relationships between the terminal device of the robot and its environment are poorly known a priori and/or change with time. This is true of the position of certain fixed objects and obstacles and the contact forces. In this case the sensors provide the robot, in real time, with information necessary for the generation of correcting motions or micro-motions. The deviations are most frequently measured in the reference $R_N$ of the terminal device, requiring transformations of the references and the use of geometrical, kinematic, static, or combined models for the calculation of the corrections in the motor space or configuration space (see 16.1).

Depending on the physical connections between the terminal device and the objects in its environment, the following may be observed:

– the case of no contact, or a small amount of contact, with motion of the terminal device;

– the case of a total connection, without motion of the terminal device;

– the case of a connection permitting certain motions of the terminal device.

The first case corresponds to a local geometrical measurement of the form and situation of the immediate environment of the terminal device. It makes use of range sensors which may be mechanical but are more usually ultra-sonic, optical, or other devices arranged in matrix form, in bar form, or singly. The second case is that of a controller of the force and moment which the terminal device of the robot must exert on its environment without significant deformation. The forces are generally measured as near to the connection as possible, by means of strain gauges fitted on the terminal device (on the jaws of a gripper, for example) or on the device coupling this element to the arm (measuring wrist). The measurement of currents in the electric actuators or of pressure in hydraulic motors can only be used when the performance of the transmissions is particularly good and when the measured variables are not subject to excessive noise due to parasitic phenomena.

The third case is the most general one, being that in which the control system of the robot must control both forces and motions simultaneously, for assembling two components, turning a crank, trimming components, transporting an object with two arms, etc. This is the case of combined position and force control, where the current trend is to generate the fine motions and the control of forces at the position of the terminal device itself (active wrist).

### 16.4.1 – PROXIMITY MEASUREMENT FEEDBACK

This may be used, in particular, for:
- the phase of approaching an object before grasping it;
- the following of a joint or a surface.

The diagram in Fig. 20 shows the principle of the control system: the correction vector $d\underline{X}$ is added to the references $d\underline{X}^+$ delivered by the approximate motion generator until the errors $d\bar{\underline{X}}$ determined by the task to be accomplished have been eliminated.

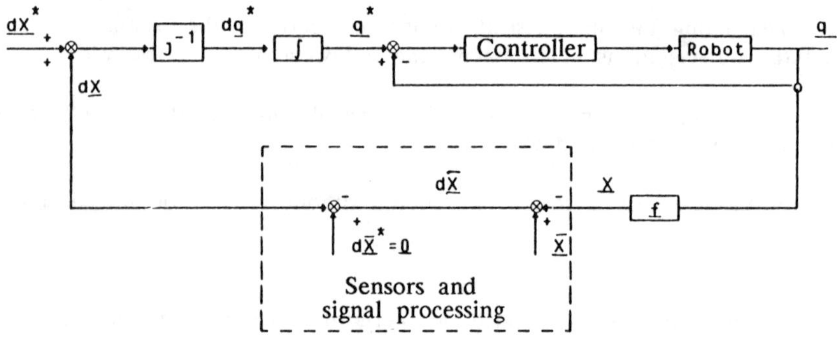


Closed loop guidance

Fig. 20

For example, the sequence of the vectors $d\underline{X}^*$ specifies an approximate nominal trajectory, and the sensors detect the deviation between the real position $\underline{X}$ of the terminal device and the desired position $\bar{\underline{X}}$. It may be noted that in this solution the errors of the sensor reference are merged with those of the reference $R_N$ linked to the terminal device. This type of solution was adopted successfully by the IRISA team, particularly for following a surface with the position and relative orientations of the terminal device controlled precisely despite the presence of high-amplitude oscillations of the base of the manipulator which was fixed to an inaccurately controlled carrier.

In arc welding, however, the control technique may become complicated, since in general the joint detector precedes the torch and may even be movable with respect to it. It is therefore necessary to take into account the advance of the sensor position with respect to the tool and the transformation of coordinates between the sensor reference and the tool reference $R_N$. For the latter, it is necessary to monitor the position $\underline{O}_N$ and the orientation $\varphi_N$ with respect to the joint $R_j$, the measurements being made in the sensor reference $R_c$.

The system shown theoretically in Fig. 21 may therefore be adopted.

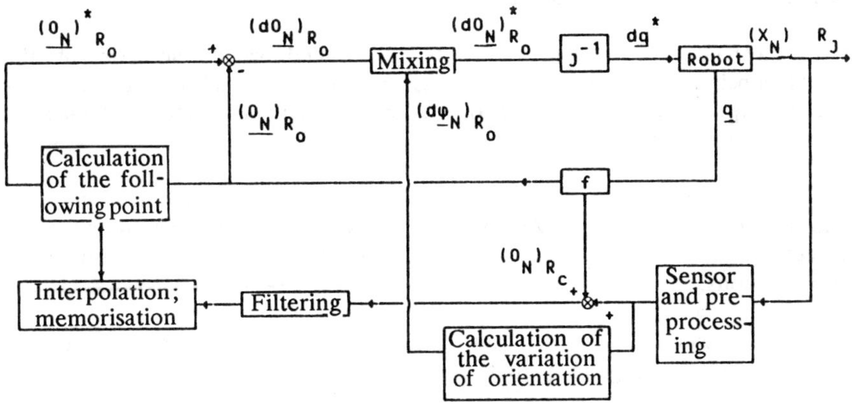


Welding system with joint following

Fig. 21

## 16.4.2 - FORCE FEEDBACK

Very many automatic control methods make more or less direct use of the measurement of contact forces between the robot, or usually its terminal device, and the environment. This is the general case of the hybrid system of simultaneous force and motion control, which combines kinematic and static torsors, and whose principle will be described in the following section. In this section we shall limit the term "force feedback" to the active monitoring of the apparent rigidities of the robot as seen by the environment on which the terminal device acts in a quasi-static way.

In accordance with the notation given in the first section, these rigidities are defined by matrices K, of dimension 6, so that

$$\underline{F} = K\ \underline{dX} \qquad (79)$$

and inversely

$$\underline{dX} = K^{-1}\ \underline{F} \qquad (80)$$

showing the compliance of the mechanical system under the effect of the torsor $\underline{F}$ applied to its terminal device.

This compliance, in the static state, has two origins, as follows:

- the natural mechanical rigidity of the robot and of its transmission systems;

- the "rigidity" of the controllers.

For small deformations, the two phenomena are combined by addition of the effects of each considered in isolation.

The mechanical compliance may be distributed when the arm of the robot consists of deformable segments, or may be localised, at the wrist for example (compliant wrist) or in the transmission systems (cable or belt reduction systems).

To illustrate the general concept of controlled compliance, consider the example of an MA 23 manipulator with six rotations actuated by direct current motor torques through elastic cables and steel band transmissions.

It is assumed that the controllers are linear in the static state and are defined by the gains matrix G:

$$\underline{C} = G(\underline{a}^* - \underline{a}) \tag{81}$$

and that the transmissions are ideal and linear:

$$\underline{\Gamma} = P^T \underline{C} \tag{82}$$

$$\underline{D} = R(d\underline{a} - P^{-1} d\underline{q}) \tag{83}$$

where $\underline{D}$ is the vector of deformations of the transmissions. Therefore $J_c = J_A P$ and $K_A = R^T K R$ and consequently

$$K = J_A^{-T} (K_A^{-1} + G^{-1}) J_A^{-1} \tag{84}$$

By comparing this with relationship (18), it will be noted that the mechanical and electrical compliances are added together.

If K* is the matrix of required rigidity, the synthesis of the motor controllers results in:

$$G = [J_A^T K^* J_A - K_A^{-1}]^{-1} \tag{85}$$

In practice, the controller must also include the damping terms of an active stabiliser whose effects are added to the structural damping of the mechanism. In the case considered here, the optimal controller must include the terms in $\underline{D}$ and $\dot{\underline{D}}$ (measured by strain detectors) to provide state feedback. When the mechanical deformations are negligible, the rigidity is controlled by active means only and damping is provided by the tachometric measurements (or proportional derivative controllers) only.

It should be noted that this control technique requires the identification of the mechanical elasticities of the system, which in practice can only be properly determined when the play and friction between the terminal device and the parts subject to deformation are negligible. Preferably, therefore, the sources of deformations should be located as closely as possible to the point of contact between the terminal device and the environment.

### 16.4.3 - HYBRID CONTROL

This term denotes a controller which provides motions in a sub-space of $\{\underline{X}\}$ defined in an approximate way while "accommodating" the reactive forces appearing in another sub-space between the terminal device and the environment.

In all cases, we are concerned with the control of fine motions and micro-deformations, and therefore, as with pure force feedback, it is desirable that the theoretical assumptions of the linearity of the relationships describing the model should be borne out in practice. This underlines the usefulness of terminal devices or active wrists controlling the forces and motions as closely as possible to the task. Such devices, like that developed by CERT/DERA, have six degrees of freedom, and therefore, for small motions, they reproduce the functions of a hand external to the robot.

In the case of perfect mechanical linkage, the kinematic torsor of the terminal device and the static torsor of its linking forces belong to two complementary sub-spaces of the operations space, which do not vary as long as the type of linkage does not change. When the terminal device is displaced with respect to position and orientation, while maintaining the linkage, the locus of $\underline{q}$ in the configuration space is a "C-surface" (as defined by Mason); conversely, therefore, the task (assembly, trimming, etc.) is carried out by following a suitable trajectory of the vector $\underline{q}$ on the C-surface while complying with the requirements of the static torsor at the contact point.

In hybrid control, the control of each motor represents the contribution of the generalised variable corresponding to the simultaneous satisfaction of the constraints of force and position.

The control law (in the static state) is:

$$C_i = \sum_{j=1}^{N} [(g_{ij})_F \, s_j \, df_j + (g_{ij})_P (1-s_j) \, dx_j] \qquad (86)$$

in which it is possible to identify the errors of force, $dF_j$, and position, $dX_j$, the coefficient $s_j$ (0 or 1) of the desired compliance S according to the direction j of the space $\{\underline{X}\}$ and the gains in force and position $(g_{ij})_F$ and $(g_{ij})_P$.

Fig. 22 shows the structure of the corresponding control system:

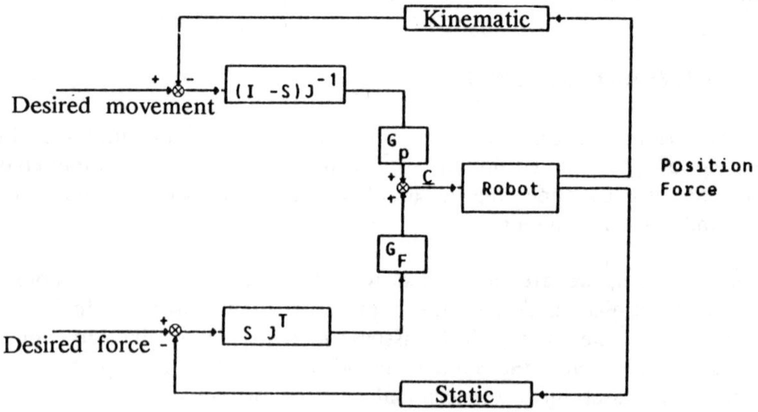


Principle of force/position hybrid control

Fig. 22

This principle, applied to the control of an active wrist with six axes used as a terminal device or as a "left hand", makes it possible to produce the control primitives necessary for the assembly of two parts, the following of surfaces, etc.

## *16.5 - CONCLUSION*

This chapter has described the techniques currently under development for the automatic control of robots, particularly those of the "third generation" fitted with sensors and therefore capable of adaptation to variations in their environment.

Most of these techniques have been tested in the laboratory and have provided excellent results:

- the generation of point-to-point motions in minimum time results in time gains of the order of 10 to 20% by comparison with the conventional control methods;

- the use of a dynamic model improves precision in a ratio of 5 to 10 for very rapid motions, by comparison with the conventional proportional derivative compensation;

- control by reference to sensors can be used to overcome errors in modelling or variations in the kinematic and static relationships between the robot and its environment. The use of an active wrist with hybrid position and force control, for example, has made it possible to assemble prismatic parts with a tolerance of 1/100 mm.

The main difficulties in the practical application of these techniques lie in the complexity of the calculations to be performed in real time by the control computer; these calculations include the solution of problems of mathematical programming for the generation of motions, the calculation of the coefficients of the Lagrange equations for dynamic control, the calculation of Jacobian matrices and their inverses for kinematic or force control, etc.

Faced with this complexity, researchers in this field have attempted to find sub-optimal solutions, robust controllers, feedback structures, etc., which simplify the algorithms while providing good performance in terms of velocity and precision. However, the problem of application in real time remains to be solved, since new electro-mechanical robot structures are being developed, particularly those with direct-drive motors, capable of actuating a joint at speeds of 3 to 10 radians per second. Consequently, the sampling periods, which were of the order of 50 milliseconds ten years ago, and are now from 15 to 20 ms, will have to be decreased to 5, and possibly even 1 millisecond.

For this reason, the search for relatively simple and robust control algorithms must be paralleled by provision for their storage in rapid computer architectures of the parallel processing or pipe-line type, including specialised high-speed processors of the cabled, pre-diffused, pre-characterised, segmented, and other types. This progress in data processing will promote the application of modern methods of information technology and the progressive introduction of artificial intelligence systems.

## *REFERENCES*


M.J. Aldon, Elaboration automatique de modèles dynamiques de robots en vue de leur conception et de leur commande, Thèse d'Etat, USTL Montpellier, 29.10.82

A.K. Bejczy and R.P. Paul, Simplified robot arm dynamics for control, IEEE CDC San Diego, pp. 261-262, Dec. 1981

T.O. Binford et al., Exploratory study of computer integrated assembly systems, Stanford A.I. Memo No. 285.4, June 1977

R. Boudarel, J. Delmas, and P. Guichet, Commande optimale des processus, Dunod, 1968

M. Brady et al., Robot motion, planning and control, MIT Press, 1982

P. Dauchez, Etude de la commande de deux robots manipulateurs lors de tâches coordonnées, Thèse de 3ème cycle, USTL Montpellier, 13.6.83

B. Espiau and M. Leborgne, Modélisation et commande de robots dans un espace opérationnel local, Rapport ISIRA, Rennes, 1984

B. Espiau, G. Andre, and R. Boulic, Utilisation d'informations proximétriques en téléopération, Actes des 3èmes Journées ARA, Pôle Téléopération Avancée, Toulouse, Sept 1984, Ed. CNRS

A. Fournier and W. Khalil, Coordination and reconfiguration of mechanical redundant systems, Int. Conf. on Cybernetics and Society, Washingotn, Sept. 19-25 1977

A. Fournier, Générations des mouvements en robotique, Thèse d'Etat, USTL Montpellier, 2.4.80

E. Freund, Path control for a redundant type of industrial robot, Proc. 7th ISIR, Tokyo, pp. 107-114, 1977

E. Freund, The structure of decoupled nonlinear systems, Int. Journal of Control, v. 21, No. 3, pp. 443-450, 1975

J.M. Hollerbach, A recursive formulation of Lagrangian manipulator dynamics, IEEE Trans. on SMC, v. 10, No. 11, pp. 730-736, 1980

W. Khalil and A. Liégeois, Génération de mouvements optimaux des robots, RAIRO Automatique/systems analysis and control, v. 18, No. 1, pp. 25-31, 1984

W. Khalil, Trajectories calculations in the joint space of robots, Advanced Software in Robotics AIM, Liège, May 1983

A. Liégeois, Automatic supervisory control of the configurations and behaviour of multibody mechanism, IEEE Trans. SMC, v. 12, pp. 868-871, 1977

A. Liégeois and M.J. Aldon, Génération et programmations automatiques des équations de Lagrange des robots et des manipulateurs, Rapport de recherche iNRIA, No. 32, Sept. 1980

A. Liégeois, E. Dombre, and P. Borrel, Learning and control of a compliant computer controlled manipulator, IEEE Trans. Automatic Control, AC 25, No. 6, pp. 1097-1102, Dec. 1980

M. Llibre, R. Mampey, and J.P. Chretien, Simulation de la dynamique des robots manipulateurs motorisés, Actes de congrès AFCET "Productique et robotique intelligente", Besançon, pp. 197-207, Nov. 1983

M. Llibre, Commandes des manipulateurs, Note CERT/DERA, 1984

T. Lozano Perez, Robot programming, Proc. of the IEEE, v. 71, No. 7, July 1983

T. Lozano Perz, M.T. Mason, and R.H. Taylor, automatic synthesis of fine-motion strategies for robots, Int. J. of Robotics Research, v. 1, No. 1, pp. 3-23, 1983

J.Y.S. Luh, M.W. Walker, and R.P.C. Paul, Resolved acceleration control of mechanical manipulators, IEEE Trans. Automatic Control, AC 25, No. 3, June 1980

J.Y.S. Luh and C.S. Lin, Optimum path planning for mechanical manipulators, Trans. of the ASME, June 1981

J.P. Merlet, Commande par retour d'effort en robotique, Rapport de recherche INRIA, No. 351, December 1984

A. Micaelli and J.M. Detriche, Controlling a 6 d.o.f. welding robot along a randomly oriented seam, with a reduced sensor information, Proc. 5th RoManSy, Udine, June 1984, pp. 164-177

R. Paul, Robot manipulators, MIT Press, 1981

M. Parent and C. Laurgeau, Langages et méthodes de programmation, Collection les Robots, v. 5, Hermes Publishing, 1983

C. Samson, Robust non linear control of robot manipulators, IEEE CDC, San Antonio, Texas, December 1983

H. Van Brussel and J. Simons, Active adaptive compliant wrist (AACW) for robot assembly, Proc. 11th ISIR, Tokyo, pp. 377-384, October 1981

# *APPENDIX*

# *PROPERTIES OF POLYNOMIAL MATRICES*

This appendix contains a brief survey of the principal theoretical discoveries concerning polynomial matrices. More detailed expositions can be found in numerous articles and books, some of which are mentioned in the references accompanying Chapter IV.

The study of single-variable linear feedback control systems by the transfer function technique requires the manipulation of polynomials and rational fractions whose mathematical properties are well known. In the general multivariable case, there are several transfer functions, which may be grouped together in matrix form, and it is useful to know which properties of the matrices, the polynomials and the rational fractions are generally conserved.

## *A.1 – POLYNOMIAL MATRICES*

A polynomial matrix is a matrix all of whose elements are polynomials. The polynomial matrix can also be interpreted as a polynomial with matrix coefficients which it is sometimes useful to write in the form $A(x) = A_0 + A_1 x + \ldots.\ A_n\ x^n$. Thus the matrix

$$A(x) = \begin{vmatrix} x + 1 & x + 3 \\ x^2 + 3x + 2 & x^2 + 5x + 4 \end{vmatrix}$$

may be rewritten in the form

$$A(x) = \begin{vmatrix} 1 & 3 \\ 2 & 4 \end{vmatrix} + \begin{vmatrix} 1 & 1 \\ 3 & 5 \end{vmatrix} x + \begin{vmatrix} 0 & 0 \\ 1 & 1 \end{vmatrix} x^2$$

This second form naturally introduces the degree which is the highest integer n such that $A_n \neq 0$.

### A.1.1 – BASIC DEFINITIONS CONCERNING THE MATRICES

a) The **addition** and **multiplication** of two polynomial matrices correspond to the standard definitions. However, by contrast with the case of polynomials with real or complex coefficients, the degree of the product of matrices may be smaller than the sum of the degrees. This is due to the fact that a matrix product may be zero without any of the matrices being equal to zero.

b) The **determinant** of a square polynomial matrix is calculated as normal and represents a polynomial. A matrix is called non-singular and has an **inverse,** generally fractional, if its determinant is not equal to zero. For example, the following matrix

$$A(x) = \begin{vmatrix} x + 1 & x + 3 \\ x^2 + 3x + 2 & x^2 + 5x + 4 \end{vmatrix}$$

has as its determinant det A = − 2(x+1) ≠ 0 and as its inverse:

$$A^{-1}(x) = \begin{vmatrix} -\frac{x+4}{2} & \frac{x+3}{2(x+1)} \\ \frac{x+2}{2} & -\frac{1}{2} \end{vmatrix}$$

c) The **rank** of a matrix is defined as the the highest minor not equivalent to zero. Thus the above matrix is of rank 2 almost throughout. The word "almost" indicates that, for the value x = −1, det A = 0. The rank is thus reduced by a unit. This concept is particularly useful when checking that two polynomial matrices are prime.

d) Three elementary operations may be defined:

* changing the rows (or columns) i and j
* multiplying any row (or column) by a non−zero scalar
* adding to a row (or column) a polynomial multiple of another row (or column)

These operations may be represented by non−singular matrices **whose determinant is scalar**. Matrices of this type are called **unimodular** and have the very important property of possessing polynomial inverses.

## A.1.2 − PRINCIPAL TRANSFORMATIONS

As in the case of scalar matrices, it is interesting from a practical as well as a theoretical viewpoint to discover the simple, triangular or diagonal forms into which any polynomial matrix may be transformed by means of elementary operations.

### A.1.2.1 − **Hermite form**

One method of solving linear equations (A x = y) (the Gauss−Jordan method) consists in putting the matrix into stepped row form by means of elementary operations. This enables the solution(s), if any, to be obtained. In the polynomial case, the analogy of this form is the Hermite form. This is very important not only from a theoretical point of view (for polynomial equations, definition of the highest common factor, etc.) but also from a practical viewpoint.

Theorem

Any polynomial matrix $A_{m,n}(x)$ of rank r may be reduced to a higher quasi-triangular form in which

1) the last m−r rows are zero;
2) in column j $1 \leqslant j \leqslant r$, the element on the diagonal is either zero or a unit polynomial having the highest degree in this column, the elements below being zero.

The representation of A is therefore as follows:

$$\hat{A} = U.A = \begin{vmatrix} p_{11} & \text{-------------} & p_{1n} \\ & p_{rr} \text{--------} & p_{rn} \\ & 0 & \end{vmatrix}$$

where U is a unimodular polynomial matrix.

A dual representation exists and is obtained by reversing the terms, rows and columns.

The proof is relatively simple and is based on the division of polynomials. It consists in replacing the polynomials of a single column with the remainders from their division by the polynomial of lowest degree in this column, until the desired form is obtained.

Example:

The matrix A(x) defined above has the following Hermite form:

$$\hat{A} = \begin{vmatrix} x+1 & 0 \\ 0 & 1 \end{vmatrix}$$

with the following unimodular matrix:

$$U = \frac{1}{2} \begin{vmatrix} -(x+1)(x+4) & x+3 \\ x+2 & -1 \end{vmatrix} \qquad \text{where det } U = -1$$

The uniqueness of this form is only ensured if the matrix A is non-singular or of full column rank.

### A.1.2.2 – Smith form

We shall now consider the form into which a polynomial matrix may be put when the elementary operations on the rows and columns are performed simultaneously.

Theorem

Any polynomial matrix $A_{m,n}(x)$ of rank r may be reduced, by elementary operations on the rows and on the columns, to the following form:

$$
U\,A\,V = \left|\begin{array}{cc:c} i_1 & & \\ & i_r & 0 \\ \hdashline 0 & & 0 \end{array}\right| \begin{array}{l} r \\ \\ m-r \end{array}
$$

$$
\qquad\qquad r \qquad\qquad n-r
$$

where $i_j$ represents a unit polynomial which is a factor of $i_{j+1}$.

If, in addition, $d_j$ is defined as the greatest common divisor of all the minors of order j of A, the following property is found:

$$i_j = d_j/d_{j-1} \quad \text{with} \quad d_o = 1$$

The $i_j$ are called the **invariant polynomials** of A. This form is of basic theoretical interest, comparable to the diagonal forms of scalar matrices (decomposition into singular values or eigenvalues). To demonstrate this, some findings concerned with this form will be listed:

1) If the matrix U is unimodular, its Smith form is the identity matrix (U = U.I.I).

2) The Smith form of ZI−A with A written in the companion form is:

$$
\left|\begin{array}{ccc} 1 & & \\ & 1 & \\ & & |ZI-A| \end{array}\right|
$$

where $|ZI-A|$ denotes the deteminant of the matrix ZI−A.

3) Starting from a transfer function matrix, the Smith form can be used to calculate the Smith−MacMillan form and thus to define the poles and zeros of the corresponding system.

4) It can be used to solve polynomial equations and to test whether two polynomial matrices are prime.

5) The Smith form of a real or complex matrix is

$$
U\,R\,V = \left|\begin{array}{cccc} 1 & & & 0 \\ & & 1 & 0 \\ 0 & \cdots & \cdots & 0 \end{array}\right| ]\,r
$$

where r denotes the rank of the matrix R; this shows that R may be decomposed into the form

$$\underset{(p,m)}{R} = \underset{(p,r)}{B} \;.\; \underset{(r,m)}{C}$$

6) The Smith form of xI−A leads to the Jordan form of A and

provides decomposition into cyclic spaces, in other words decomposition of companion matrices into the diagonal form.

7) ZI−A and ZI−B have the same Smith form if and only if A and B are similar.

Example:

1) The matrix $P \triangleq \begin{vmatrix} 1 & x^2 & -1 \\ 1 & x & 0 \end{vmatrix}$ has the following Smith form:

$$\begin{bmatrix} 0 & 1 \\ -1 & 1 \end{bmatrix} \begin{bmatrix} 1 & x^2 & -1 \\ 1 & x & 0 \end{bmatrix} \begin{bmatrix} 1 & 0 & -x \\ 0 & 0 & 1 \\ 0 & 1 & x^2-x \end{bmatrix} = \begin{bmatrix} 1 & 0 & 0 \\ 0 & 1 & 0 \end{bmatrix}$$

2) The matrix $A = \begin{vmatrix} 0 & 1 & 1 \\ 0 & 1 & 0 \\ -2 & 1 & 3 \end{vmatrix}$ is written in the following forms:

The Smith form of $xI-A$: $\begin{bmatrix} 1 & 0 & 0 \\ 0 & x-2 & 0 \\ 0 & 0 & (x-1)(x-2) \end{bmatrix}$

The Jordan form: $\begin{bmatrix} 1. & 0 & 0 \\ 0 & 2 & 0 \\ 0 & 0 & 2 \end{bmatrix}$

The companion or Froebenius form: $\begin{bmatrix} 0 & 1 & 0 \\ -2 & 3 & 0 \\ 0 & 0 & 2 \end{bmatrix} \begin{matrix} (x-1)(x-2) \\ \\ (x-2) \end{matrix}$

### A.1.2.3 – Reduced column (or row) form

This concept is useful, among other things, for the definition of the property of a transfer function matrix, of the poles and zeros, and of the order of a multivariable system, as well as for the division of two matrices. A polynomial matrix $A_{(m,n)}(x)$ is said to be a reduced column matrix if the matrix consisting of the coefficients of the highest degree in each column is of full rank. A similar definition may be given by replacing the word "column" by "row".

Theorem

Any polynomial matrix of full column rank (particularly a non-singular one) may be put into the reduced column form by means of elementary operations on the columns. The highest degree of all the minors (n,n) is equal to the sum of the degrees in each column.

Example:

The matrix $P = \begin{vmatrix} 1 & 1 \\ x^2 & x \\ 1 & 0 \end{vmatrix}$ is not in reduced column form

since the matrix of the coefficients of highest degree in each column is of rank 1:

$$\text{Rank} \begin{vmatrix} 0 & 0 \\ 1 & 1 \\ 0 & 0 \end{vmatrix} = 1$$

However, the unimodular matrix $\begin{vmatrix} 1 & 0 \\ -x & 1 \end{vmatrix}$ gives it the reduced column form.
Therefore,

$$\begin{bmatrix} 1 & 1 \\ x^2 & x \\ -1 & 0 \end{bmatrix} \begin{bmatrix} 1 & 0 \\ -x & 1 \end{bmatrix} = \begin{bmatrix} 1-x & 1 \\ 0 & x \\ -1 & 0 \end{bmatrix}$$

It may also be noted that the highest degree of the minors of order 2 is 2 and that this corresponds to the sum of the highest degrees in each column.

## A.1.3 – DIVISIBILITY

We shall now show how the principal properties of polynomials – division, g.c.d. – are retained in the case of polynomial matrices.

### A.1.3.1 – **Matrix division**

With the degrees established, the existence of the division of polynomial matrices according to decreasing powers may be investigated. The problem can be set in the following form:

Let A and B be two matrices of dimensions mxr and lxp respectively. Are there unique matrices Q and R such that

$$A = BQ + R \qquad \text{with degree } R < \deg B \text{ ?}$$

As we are dealing with matrices, if the problem is to have any meaning, the dimensions must correspond, in other words A and B must have the same number of rows.

a) In the case where A and B are polynomials of degree n and m:

$$A = a_o + a_1 x + \ldots + a_n x^n$$

$$B = b_o + b_1 x + \ldots + b_m x^m$$

a proof of the existence consists in the performance of the following operations:

– if deg A < deg B it may be specified that Q = 0 and R = A;

– if deg A $\geqslant$ deg B, the polynomial $A_1$ of a lower degree than A is constructed, so that

$$A_1 = A - B\, q_1\, x^{n-m} \quad \text{where} \quad q_1 = \frac{a_n}{b_m}$$

– if deg $A_1$ < deg B, the procedure is stopped; otherwise, it is continued until this condition is present and the required polynomials Q and R are expressed thus:

$$Q = q_1\, x^{n-m} + q_2\, x^{n-m-1} + \ldots + q^j\, x^{n-m-j+1}$$

$$R = A_j$$

b) In the present case concerning matrices, the same method can be used, but the aim is now to find matrices $Q_j$ such that

$$B_m \cdot Q_j = A_{n-j+1}$$

It is known (see the next section) that there is a solution to this problem if and only if it is possible to find a matrix

$B^S$ such that $B_m\, B_m^S\, Y = Y$ regardless of the vector Y. ɹoɟ

The uniqueness is then present if Q is unique, in other words if $B_m^S$ is such that

$$I_p - B_m^S\, B_m = 0 \quad (I_p \text{ is a unit matrix of dimension p})$$

In particular, this is the case where $B_m^{-1}$ exists; some authors describe matrix B as regular.

If matrices A and B have the same number of columns, the division only has meaning in the form

$$A = Q.B + R$$

However, when B is non-singular, another way of presenting the problem is to attempt to find matrices Q and R such that

$$A = B\,Q + R \quad \text{where } B^{-1} R \text{ is strictly proper.}$$

It is then known that these matrices exist and are unique. For this, it is sufficient to decompose each rational fraction forming an element of $B^{-1}$ A into a polynomial plus a rational fraction in which the degree of the numerator is strictly less than that of the denominator. If B is in the reduced row form, uniqueness is guaranteed if the degree of the rows of R is strictly less than the degree of the corresponding rows of B.

### A.1.3.2 – Greatest common divisors (g.c.d.)

As for polynomials, it is possible to define the greatest common divisors of polynomial matrices, with the difference that they exist on the right-hand and left-hand sides.

**Definition**: given two polynomial matrices A and B having the same number of columns, a matrix R is a right-hand divisor of A and B if there are 2 matrices $A_1$ and $B_1$ such that

$$A = A_1\,R$$

$$B = B_1\,R$$

If, additionally, every other divisor $R_1$ on the right of A and B is a divisor of R($R = W\,R_1$), then R is a right-hand g.c.d. of A and B.

It is immediately obvious that this definition does not guarantee the uniqueness of the g.c.d., but it may be deduced that if a g.c.d. is non-singular, and particularly if it is unimodular, the same applies to every g.c.d.

**Construction**

A g.c.d. is obtained simply by transforming the matrix

$$\begin{vmatrix} A \\ B \end{vmatrix}$$

into its Hermite form or into an equivalent quasi-triangular form. There is always a unimodular matrix U such that

$$\begin{matrix} r \\ m+1-r \end{matrix} \begin{vmatrix} U_{11} & U_{12} \\ U_{21} & U_{22} \end{vmatrix}_{\substack{\\ m \quad\; 1}} \begin{vmatrix} A \\ B \end{vmatrix}_{n} \begin{matrix} m \\ 1 \end{matrix} = \begin{vmatrix} r_{11} \cdots & \\ & r_{rr} \\ 0 & \end{vmatrix} \triangleq \begin{vmatrix} R \\ 0 \end{vmatrix}$$

with the rank of $\begin{vmatrix} A \\ B \end{vmatrix}$ equal to r. The partitioning of the matrix U shows that R is in fact a g.c.d. of $\begin{vmatrix} A \\ B \end{vmatrix}$. It may be seen that R is non-singular if r is equal to n.

**Example**

Let the matrices be as follows:

$$A = |1 \quad x^2 \quad -1| \quad \text{and} \quad B = |1 \quad x \quad 0|$$

Then a right-hand g.c.d. R is obtained by finding the

Hermite form of $\begin{vmatrix} A \\ B \end{vmatrix}$.

$$\begin{vmatrix} 0 & 1 \\ -1 & 1 \end{vmatrix} \begin{vmatrix} 1 & x^2 & -1 \\ 1 & x & 0 \end{vmatrix} = \begin{vmatrix} 1 & x & 0 \\ 0 & -x^2+x & 1 \end{vmatrix} \triangleq R$$

We obtain

$$A = |1 \quad -1|R. \quad \text{and} \quad B = |1 \quad 0|.R$$

and

$$\begin{vmatrix} 0 \\ -1 \end{vmatrix} A + \begin{vmatrix} 1 \\ 1 \end{vmatrix} B = R$$

which show that R has the two properties of the definition.

Let the matrices be as follows:

$$A = \begin{vmatrix} x & -1 \\ -x & x^2 \end{vmatrix} \quad \text{and} \quad B = \begin{vmatrix} x & -x \\ 0 & 1 \end{vmatrix}$$

We then obtain

$$\begin{vmatrix} 0 & 0 & 1 & 0 \\ 0 & 0 & 0 & 1 \\ -1 & 0 & x & -(1-x^2) \\ 0 & 1 & 1 & -(x^2-x) \end{vmatrix} \begin{vmatrix} x^2 & -1 \\ -x & x^2 \\ x & -x \\ 0 & 1 \end{vmatrix} = \begin{vmatrix} x & -x \\ 0 & 1 \\ 0 & 0 \\ 0 & 0 \end{vmatrix} = \begin{vmatrix} R \\ \\ 0 \end{vmatrix}$$

which shows that B is a divisor of A. It may also be seen that R is invertible, corresponding to the fact that $\begin{vmatrix} A \\ B \end{vmatrix}$ has a rank of 2.

A.1.3.3 - **Prime matrices**

Two polynomial matrices are said to be right–hand (or left–hand) prime if all their right–hand (or left–hand) g.c.ds. are unimodular. The properties of these prime matrices will now be listed.

1) There are polynomial matrices X and Y such that
$$X A + Y B = I_n$$

2) There is a unimodular matrix U such that

$$U \begin{vmatrix} A \\ B \end{vmatrix} = \begin{vmatrix} I_n \\ 0 \end{vmatrix}$$

3) The invariant polynomials of $\begin{vmatrix} A \\ B \end{vmatrix}$ are all equal to 1.

4) The rank of $\begin{vmatrix} A \\ B \end{vmatrix}$ is equal to n regardless of the value of the indeterminate x.

The generalised Bezout identities can be simply deduced from the second property identities by developing U $U^{-1} = I = U^{-1}$ U, so that after correct partitioning of U and $U^{-1}$

$$\begin{vmatrix} X & Y \\ -B_1 & A_1 \end{vmatrix} \cdot \begin{vmatrix} A & -Y_1 \\ B & X_1 \end{vmatrix} = \begin{vmatrix} I_n & 0 \\ 0 & I_r \end{vmatrix}$$
$$\begin{vmatrix} A & -Y_1 \\ B & X_1 \end{vmatrix} \cdot \begin{vmatrix} X & Y \\ -B_1 & A_1 \end{vmatrix} = \begin{vmatrix} I_m & 0 \\ 0 & I_1 \end{vmatrix}$$

with $r = m+l-n$.

These identities are fundamental to the definition of the optimal controllers, as was shown in Chapter 5.

## A.1.4 – APPLICATIONS

Some direct applications of the above results and properties will now be described.

A.1.4.1 - **Polynomial equations X A + Y B = C (A X + B Y = C)**

Given the matrices $A_{(m,n)}$, $B_{(l,n)}$, $C_{(p,n)}$, the problem consists in finding the matrices X and Y which comply with the equation X A + Y B = C.

From the previous results, it can be stated that there is a solution if A and B are right–hand prime. If $X_p$ and $Y_p$ denote a particular solution, the general solution is given by

$$X = X_p + S\ B_1$$

$$Y = Y_p - S\ A_1$$

where S is any polynomial matrix and $A_1$, $B_1$ are two matrices defined by the condition

$$[-B_1, A_1] \begin{vmatrix} A \\ B \end{vmatrix} = 0$$

The proof of this theorem consists simply in translating the Bezout identity for this particular problem. If A and B are right-hand prime, there is a unimodular matrix U such that

$$U \cdot \begin{vmatrix} A \\ B \end{vmatrix} = \begin{vmatrix} U_1 & U_2 \\ -B_1 & A_1 \end{vmatrix} \begin{vmatrix} A \\ B \end{vmatrix} = \begin{vmatrix} I_n \\ 0 \end{vmatrix}$$

Also, a particular solution is given by

$$X_p = C \cdot U_1 \quad \text{and} \quad Y_p = C \cdot U_2$$

If A and B are not right-hand prime, the equation has a solution if and only if the right-hand g.c.d. of A and B is a right-hand divisor of C.

The equation A X + B Y = C is solved in an identical way.

A.1.4.2 – **Bilateral equation A X + Y B = C**

This bilateral equation is much more difficult to deal with. One method is to transform the matrices A and B into the Smith form which makes it possible to state that the equation has a solution if and only if the matrices

$$\begin{vmatrix} A & 0 \\ 0 & B \end{vmatrix} \text{ and } \begin{vmatrix} A & C \\ 0 & B \end{vmatrix} \text{ are equivalent.}$$

A.1.4.3 – **Primary description of a transfer function**

The input-output description of a stationary linear system passes through a transfer function matrix whose elements are rational fractions. For a system with p outputs and m inputs, the transfer function matrix of dimension p,m may be written in the form

$$H = \frac{\theta}{d} \quad \text{where } \theta \text{ is a polynomial matrix and d is a polynomial.}$$

Equally, we may write

$$H = B.A^{-1} = A_1^{-1}\ B_1$$

where A,B are right-hand prime matrices and $A_1$, $B_1$ are left-hand prime matrices. The transfer function matrix is then said to be in irreducible form, since the realisation in state form is minimal and of an order equal to the degree of the determinant of $A_1$ or A. In other words, this form covers the totally observable and controllable part of the system.

By using the properties of polynomial matrices, it is easy to move from one representation to another. Writing the matrix

$\begin{vmatrix} A \\ B \end{vmatrix}$ in the Hermite form gives the result:

$$\begin{matrix} m \\ l \end{matrix} \begin{vmatrix} U_1 & U_2 \\ -B_1 & A_1 \end{vmatrix} \begin{vmatrix} A \\ B \end{vmatrix} \begin{matrix} m \\ l \end{matrix} = \begin{vmatrix} R \\ 0 \end{vmatrix} \begin{matrix} m \\ l \end{matrix}$$
$$\quad m \qquad l \qquad m$$

If A is non-singular, this leads to the following properties:

1) $A_1$ is non-singular

2) $B\ A^{-1} = A_1^{-1}\ B_1$

3) $A_1$ and $B_1$ are left-hand prime, regardless of the choice of A and B

4) If A and B are right-hand prime, then det A = det $A_1$.

These primary representations of a transfer function matrix also lead to a definition of the poles and zeros of the system represented:

1) The poles of H are the roots of the determinant of A.

2) The zeros of H are the roots of the invariant polynomials of B, which is equivalent to saying that they are the values Z (in the discrete case) or p (in the continuous case) for which the rank of B decreases.

### A.1.4.4 – Property of a transfer function

A rational matrix H(x) is called proper (or strictly proper) if

$$\lim_{x \to \infty} H(x) = C^{te} \quad (= 0)$$

It can be demonstrated that, if A(x) is in the reduced column form, $H(x) = B.A^{-1}$ is strictly proper (or proper) if and only if each column of B is of a lower degree (or a lower or equal degree) than the degree of the corresponding column of A.

## *A.2 - PROBLEMS OF INVERSES*

The search for an optimal controller includes the stage of the inversion of matrices which are sometimes not exactly inversible. To solve this problem, generalised inverse matrices are defined, in particular the Moore-Penrose inverse, which in certain conditions is unique.

### A.2.1 - GENERALISED INVERSE

A generalised inverse matrix $A^s{}_{(n,m)}$ of a matrix $A_{m,n}$ of rank r is defined as a matrix for which

$$A\ A^s\ A = A$$

If it is also required that

. $A^s\ A\ A^s = A^s$

. $(A^s\ A)^T = A^s\ A$

. $(A\ A^s)^T = A\ A^s$

then $A^s$ is unique and is equal to

$$A^s = D^T\ (D\ D^T)^{-1}\ (C^T\ C)^{-1}\ C^T$$

where

$$\underset{(m,n)}{A} = \underset{(m,r)}{C} \cdot \underset{(r,n)}{D}$$

### A.2.2 - INVERSES OF POLYNOMIAL MATRICES

In certain conditions, polynomial matrices have polynomial inverses.

Theorem

Let $A_{m,n}(x)$ be a polynomial matrix of rank m throughout (there is a minor of constant non-zero order m). This matrix has at least one polynomial right-hand inverse.

A can be written in the form

$$A = \left| A_{1_{m,m}} \; , \; A_2 \right|$$

with det $A_1$ = constant and an inverse in the form

$$A^d = \begin{vmatrix} A_1^{-1} ( I_m - A_2 \; X) \\ X \end{vmatrix}$$

where X is any polynomial matrix.

<u>Theorem</u>

Let $A_{m,n}$ be a polynomial matrix of rank m almost throughout; then there is a right-hand fractional inverse.

Matrix A can be written in its Hermite form:

$$A \; . \; U = \begin{vmatrix} R & 0 \end{vmatrix}$$

and a right-hand inverse is then given by

$$U \; \begin{vmatrix} R^{-1} \\ X \end{vmatrix}$$ where X is any matrix.

Equivalent theorems exist for the left-hand inverses. In the case where A has a rank of r, it is necessary to return to the generalised inverses.

### A.2.3 – THE EQUATION M.D = Y

A number of problems require an answer to the question of whether the equation MD = Y, where $D_{n,r}$ and $Y_{m,r}$ are given polynomial or scalar matrices, has a solution M. The answer is contained in the following theorem:

<u>Theorem</u>

A solution M exists if and only if there is a $D^s$ such that $YD^sD$ = Y. The set of solutions M is then given by

$$M = Y \; D^s + X - X \; D \; D^s$$

where X is any matrix.

## *A.3 – SPECTRAL FACTORISATION*

This spectral operation is fundamental to quadratic optimisation. It consists in finding, on the basis of a parahermitian matrix $A.\bar{A}$, a factorisation such that factorisation such that

$$A.\bar{A} \triangleq \underset{(1,m)}{A(Z^{-1})} \; . \; \underset{(m,1)}{A^T(Z)} = \underset{(1,r)}{D(Z^{-1})} \; . \; D(Z)$$

where $D_{(1,r)}$ is a polynomial matrix having a stable left-hand polynomial or fractional inverse.

If A has a rank of $1 \leqslant m$, D exists and is square and invertible.

If A has a rank of r, it is possible to perform transformations to bring $A\bar{A}$ into the equivalent form

$$A\bar{A} \sim \begin{vmatrix} K\bar{K} & 0 \\ 0 & 0 \end{vmatrix}$$

where K has a rank of r. It is then possible to factorise $K\bar{K}$ and subsequently $A\bar{A}$.

# INDEX